# Painting & Decorating

# Painting & Decorating

## Sixth Edition

Derek Butterfield
Alf Fulcher
Brian Rhodes
Bill Stewart
Derick Tickle
John Windsor



WILEY-BLACKWELL
A John Wiley & Sons, Ltd., Publication

This edition first published 2011 by Blackwell Publishing Ltd.


Blackwell Publishing was acquired by John Wiley & Sons in February 2007. Blackwell's publishing program has been merged with Wiley's global Scientific, Technical and Medical business to form Wiley-Blackwell.

*Registered office:* John Wiley & Sons Ltd, The Atrium, Southern Gate, Chichester, West Sussex, PO19 8SQ, UK
*Editorial offices:* 9600 Garsington Road, Oxford, OX4 2DQ
The Atrium, Southern Gate, Chichester, West Sussex, PO19 8SQ, UK
2121 State Avenue, Ames, Iowa 50014-8300, USA

For details of our global editorial offices, for customer services and for information about how to apply for permission to reuse the copyright material in this book please see our website at www.wiley.com/wiley-blackwell.



*Library of Congress Cataloging-in-Publication Data*
Butterfield, Derek.
Painting and decorating / Derek Butterfield, Alf Fulcher, Brian Rhodes, Bill Stewart, Derick Tickle and John Windsor. – Sixth Edition.
pages cm
Includes bibliographical references and index.
ISBN 978-1-4443-3501-9 (pbk. : alk. paper)
1. House painting–Equipment and supplies. 2. National Vocational Qualifications (Great Britain)–Handbooks, manuals, etc. I. Title.
TT320.B88 2011
698′.14–dc22
2010042185

A catalogue record for this book is available from the British Library.

Set in 9 on 13 pt Helvetica by Toppan Best-set Premedia Limited
Printed and bound in Malaysia by Vivar Printing Sdn Bhd

1 2011

# Contents

# Editor's note

I would like to dedicate this, the sixth edition of *Painting & Decorating* to Alf Fulcher, who recently passed away. Alf was one of the original team of five authors who put this book together over thirty-five years ago. In addition to teaching in Further Education for many years he played a major role with the Construction Industry Training Board, revamping the syllabus for Painting & Decorating courses and linking them to competency-based training. Since the last edition, it has been my pleasure to meet up, after many years, with Derick Tickle, again, one of the original team. Having worked with Derick when I first started teaching, I have a great respect for his multiple talents and look forward to working with him again, on future editions.

Derek Butterfield

# Preface to the sixth edition

This book is designed on the principle of a series of concise and simply written technical information sheets, providing essential job knowledge and understanding related to practical training schemes for the painter and decorator. It does not give instruction in practical skills but sets out to augment training and further education courses in the craft. It has again been my privilege to coordinate the revision for the Sixth Edition.

The content aims to cover the knowledge evidence requirements of both National Vocational Qualifications (NVQs) and Scottish Vocational Qualifications (SVQs). As well as NVQ/SVQ students, the book should also be of value to BTEC students studying construction, and to anyone requiring up-to-date knowledge of painting materials and processes. The Sixth Edition of *Painting & Decorating* has been updated by removing materials and practices that have become obsolete, and introducing new materials, tools and plant, which are current in the industry.

Since the last edition, the new 'Working at Height Regulations' have come into force, and the formulation of paint is drastically changing to facilitate legislation regarding Volatile Organic Compounds (VOC) emissions into the atmosphere. These changes are likely to continue as new legislation is introduced and updated in the future.

I would like to thank the many specialists who have generously given their time and knowledge to assist in this revision: Akzo Nobel/ICI paints, Purdy International Corporation, The Lead Paint Safety Association, Lincrusta, Hamilton Acorn, The Health and Safety Executive, Wrights of Lymm, members of both The Association of Painting Craft Teachers and Scottish Association of Painting Craft Teachers and special thanks to Mike Warren of Spraytrain.com for his assistance with the spray section, and specifically his input on HVLP spraying.

Derek Butterfield
Reading

# PART 1

# Tools and Equipment

# 1.1 Small tools and equipment

## 1.1.1 Painters' hand tools

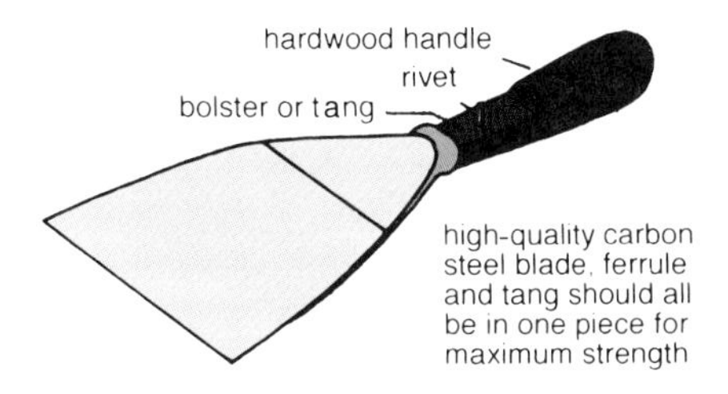


**Fig. 1.1** Construction of a good-quality painter's knife

**Fig. 1.2** Stripping knife

**Fig. 1.3** Filling knife

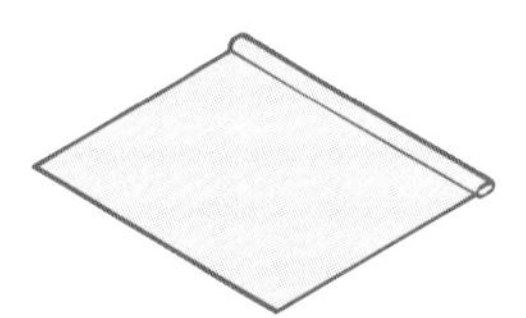

**Fig. 1.4** Continental filling blade

**Fig. 1.5** Caulking tool

### Stripping knife or scraper (Fig. 1.2)

Blades must be capable of taking and retaining a good edge.

***Size*** 25 mm, 38 mm, 50 mm, 68 mm, 75 mm and 100 mm wide.

***Use*** To remove old wallpaper, old paint films or loosely attached deposits.

### Filling knife (Fig. 1.3)

Similar to a stripping knife in appearance but having a thinner gauge, specially treated metal blade which is more flexible. The blade need not be sharp, but must be thin and true, without any nicks.

***Size*** Up to 150 mm wide.

***Use*** To apply filler to open-grain timbers and small holes or shallow indentations in uneven surfaces.

***Care*** When not in use, protect blade edge with a timber, soft aluminium or plastic cover.

### Continental filling blade (Fig. 1.4)

Rectangular, flexible blade set in plastic handle, which gives rigidity.

***Size*** 25 mm, 50 mm, 75 mm and 100 mm wide.

***Use*** To apply filler to open-grained timbers, small holes and shallow indentations in uneven surfaces.

### Caulking tool (Fig. 1.5)

Rigid plastic blade set in a wooden or plastic handle.

***Size*** 200 mm, 250 mm and 300 mm wide.

***Use*** For general filling and bedding-in scrim tape along plasterboard or fibreboard joints before texturing, or other forms of direct decorating on to dry-lined surfaces.

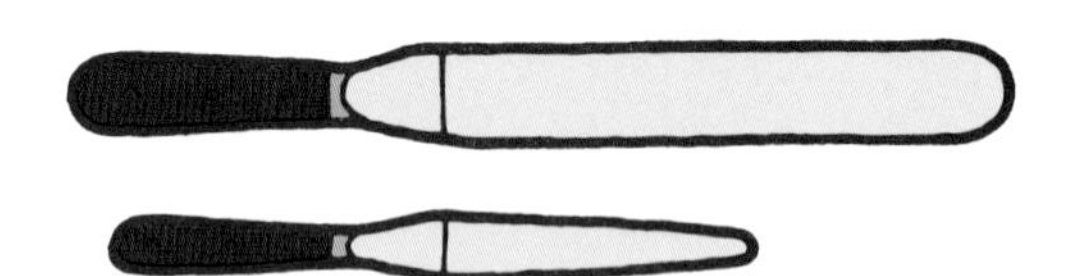

**Fig. 1.6** Palette knives

## Palette knife (Fig. 1.6)

Long, narrow flexible blade with rounded end. Should be so balanced that when laid on the bench the blade will not touch the bench top.

***Size*** Blades from 75 to 300 mm long.

***Use*** For mixing paint both in tins and on palette boards.

***Care*** Ensure that the end does not become bent or burred over.

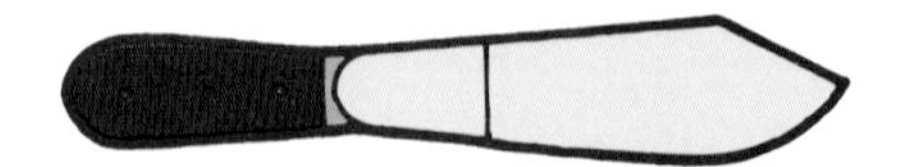

**Fig. 1.7** Stopping knife

## Stopping knife, putty or glazing knife (Fig. 1.7)

Usually has one side of the blade straight and the other curved (clipped), but can be obtained with a double curved blade.

***Size*** Blade length of either 112 mm or 125 mm.

***Use*** (i) To force putty and hard stopper into small holes and cracks.
(ii) To bevel facing putties when glazing.

***Care*** If the point becomes worn or burred over, the blade should be reground.

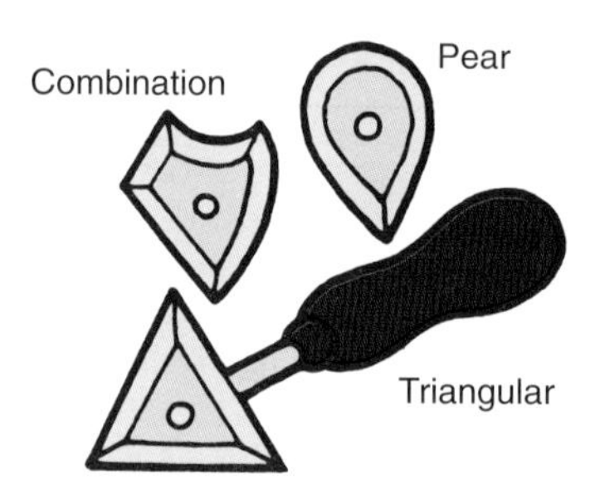


**Fig. 1.8** Shave hooks

## Shave hook (Fig. 1.8)

The head has a bevelled cutting edge round the perimeter and is available in the three shapes illustrated.

***Use*** (i) To scrape paint debris from mouldings, cornices and ornamental beadings, usually in conjunction with paint removers or burning-off equipment.
(ii) To cut out cracks in plaster surfaces prior to stopping.

***Care*** Maintain sharp edge with a file.

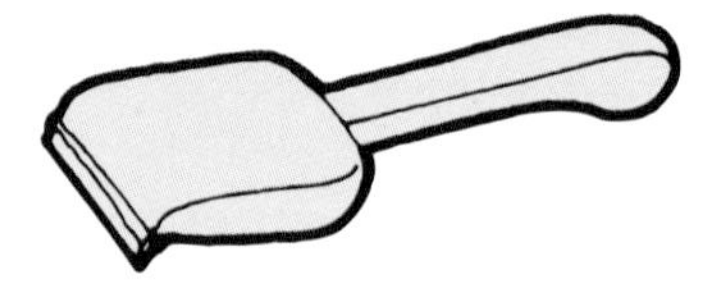

**Fig. 1.9** Universal scraper

## Universal scraper (Fig. 1.9)

Short sharp replaceable blade attached to a large handle.

***Size*** Blades between 45 and 80 mm wide.

***Use*** To remove old paint, varnish or discolouration from timber surfaces.

***Care*** Edge must always be kept sharp.

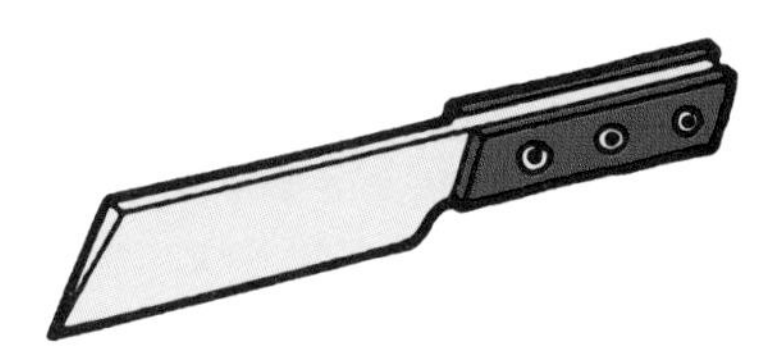

**Fig. 1.10** Hacking knife

## Hacking knife (Fig. 1.10)

A heavy, rigid metal blade with a leather handle. The back of the blade is flat to take hammer blows.

***Size*** Blades either 100 mm or 125 mm long.

***Use*** To remove old hard putties before reglazing.

## Paint stirrer (Fig. 1.11)

A stiff blade with holes cut along its length. The flat end fits the bottom of the tin, and paint flows through

Fig. 1.11 Paint stirrer

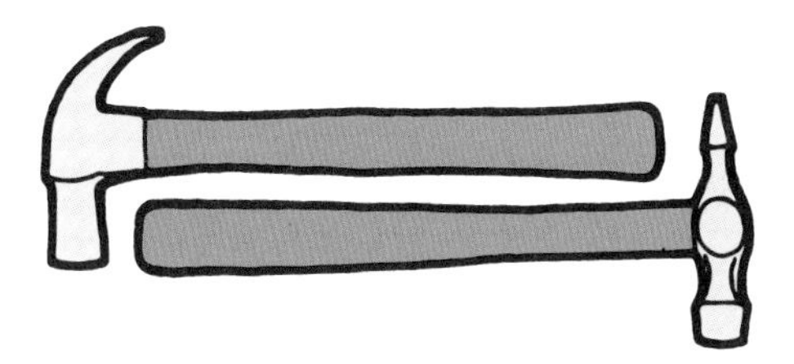

Fig. 1.12 Hammers

Fig. 1.13 Pincers

Fig. 1.14 Nail punch

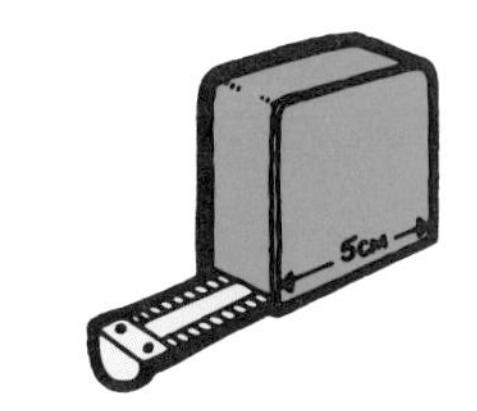


Fig. 1.15 Measuring tape

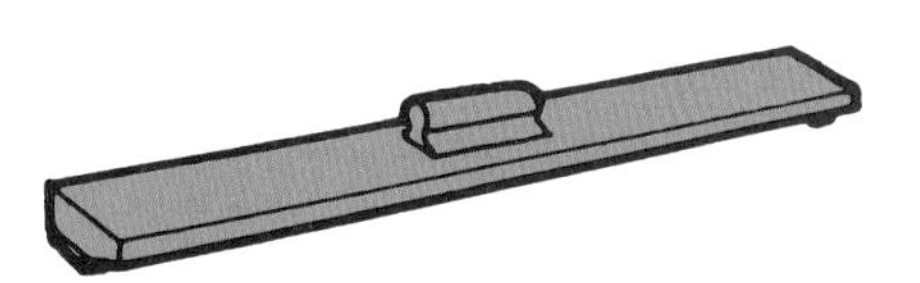

Fig. 1.16 Straight-edge

Fig. 1.17 Screwdriver

the holes, mixing the paint more quickly. Usually made of metal, but wooden stirrers are available from manufacturers.

***Size*** Various sizes available up to 600 mm long.

***Use*** To stir paint.

### Hammer (Fig. 1.12)

***Size*** 6 oz (170 g) to 8 oz (227 g).

***Use*** (i) With nail punch, chisel and hacking knife.
(ii) To drive in nails, sprigs and pins.
(iii) To remove heavy rust scale.

### Pincers (Fig. 1.13)

***Size*** 150 mm, 175 mm or 200 mm.

***Use*** To remove nails and glazing sprigs.

### Nail punch (Fig. 1.14)

***Size*** 2 mm, 3 mm, 5 mm point sizes.

***Use*** To punch nail heads below the surface in woodwork before stopping.

### Measuring tape (Fig. 1.15)

Coated linen, PVC-coated fibreglass or steel tape.

***Size*** From 2 to 30 m long.

***Use*** To measure for estimating.

### Straight-edge (Fig. 1.16)

Straight wooden batten with bevelled edge and two blocks to lift from surface.

***Size*** 300 mm to 1 m long.

***Use*** To guide a lining fitch when running lines (see also Fig. 1.90).

### Screwdriver (Fig. 1.17)

***Size*** Blade length about 125 to 150 mm. (Slotted, Pozi-drive and Phillips varieties available.)

***Use*** To remove fitments, window and door furniture before painting and paperhanging.

### Wire brush (Fig. 1.18)

(i) Hardened and tempered steel wires in a wooden handle.

**Fig. 1.18** Wire brushes

**Fig. 1.19** Pointing trowel

**Fig. 1.20** Wood chisel

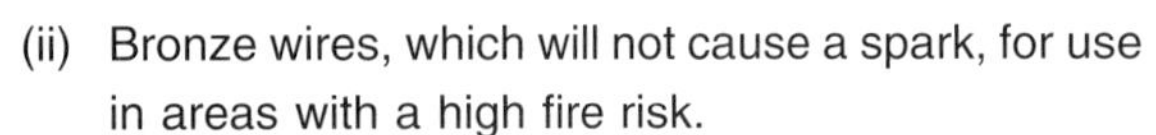

(ii) Bronze wires, which will not cause a spark, for use in areas with a high fire risk.

***Size*** Available in a variety of shapes and sizes from 65 to 285 mm long.

***Use*** (i) To remove corrosion from iron and steel.
(ii) To remove loose deposits from surfaces before painting.

**Pointing trowel (Fig. 1.19)**

***Size*** 125 mm or 150 mm blade.

***Use*** To make good large cracks and holes.

**Wood chisel (Fig. 1.20)**

***Size*** Short blade approximately 20 mm wide.

***Use*** (i) To remove old putties, in place of a hacking knife.
(ii) For general purposes during surface preparation.

**Paint strainer (Fig. 1.21)**

(i) Metal: tinplate, galvanised iron or zinc. The replaceable copper-gauze straining discs are available in three grades, 30-mesh coarse, 40-mesh medium and 60-mesh fine. Must be cleaned thoroughly, immediately after use, to prevent the mesh becoming clogged.
(ii) Cardboard frame and muslin mesh disposable strainer.
(iii) Nylon cloth or fine muslin stretched over the kettle.

***Use*** To remove dirt and skins from paint and varnish.

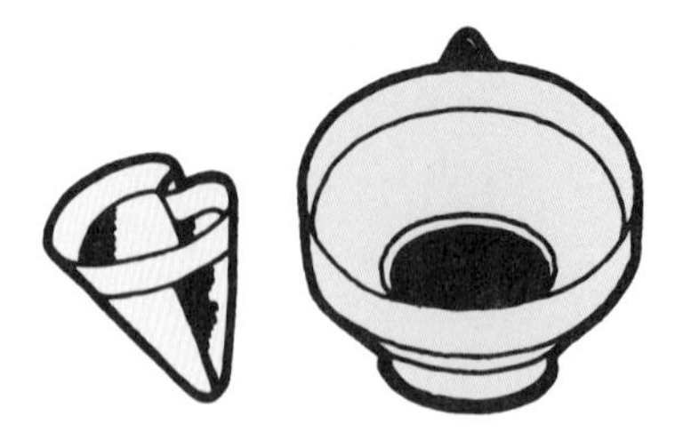

**Fig. 1.21** Paint strainers

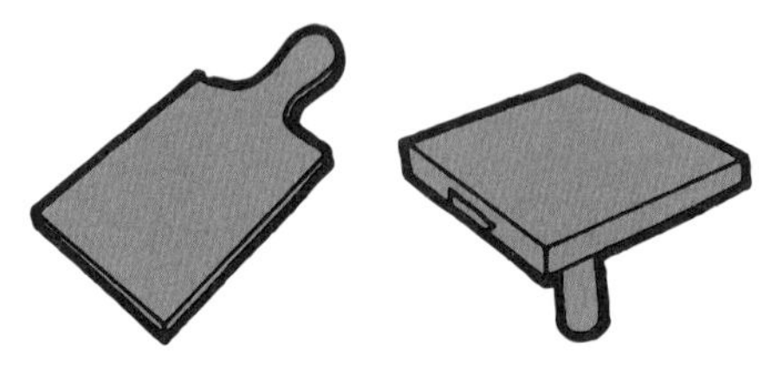

**Fig. 1.22** Filling board and hawk

**Fig. 1.23** Rubbing block

**Filling board and hawk (Fig. 1.22)**

Made of oiled plywood, laminated plywood, metal-covered plywood or thick-gauge plastic.

***Size*** For stopping, approximately 100 × 130 mm; for fillers, approximately 180 × 230 mm – plus the length of the handle.

***Use*** To mix and hold stoppers and fillers before and during application. The hawk is used to hold plaster or sand and cement when filling in large holes and cracks.

**Rubbing block (Fig. 1.23)**

Made of wood, plastic, cork or rubber (the last-named has the longest life).

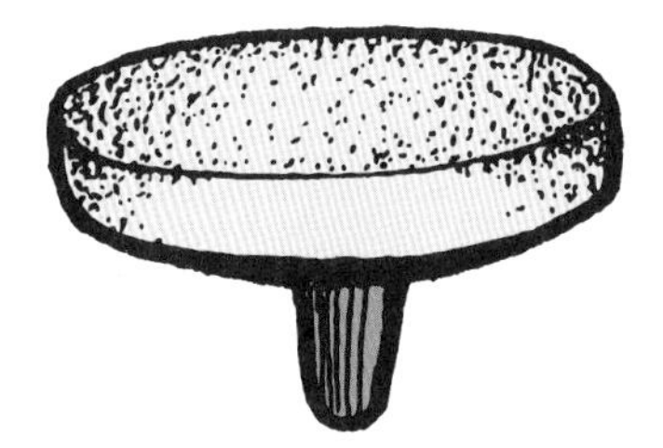

Fig. 1.24 Sponge filling tool

Fig. 1.25 Glasscutter

*Size* A rubbing face of approximately 70 mm wide × 100 mm long.

*Use* To hold abrasive papers, ensuring a flat, smooth surface and providing a more comfortable grip. Also available on a pole for large wall or ceiling areas. Clean-up is minimised with vacuum attachment.

**Sponge filling tool (Fig. 1.24)**

Circular-shaped synthetic sponge set on a flat stock at right angles to a plastic handle.

*Size* 200 mm diameter.

*Use* For smoothing the filling compound used in the caulking of plasterboard joints.

**Glasscutter (Fig. 1.25)**

Single wheel, six wheel and diamond cutters are available. The hardened steel wheel types are most commonly used and are lubricated during use by dipping in white spirit. The six wheel type is in a revolving head to enable blunted wheels to be changed easily.

*Use* For scribing or scoring the glass at the point where it is to be cut.

**Straight-edge (Fig. 1.26)**

In use the glasscutter is guided along a straight-edge of varying lengths and made of wood or plastic.

*Use* Long straight-edges require steadying hands at both ends, and glasscutting becomes a two-person operation, although some have rubber suckers, which grip the glass, holding the straight-edge rigid when in use. When positioning the straight-edge, an allowance of 3 mm is made for the width of the cutting head of the glasscutter.

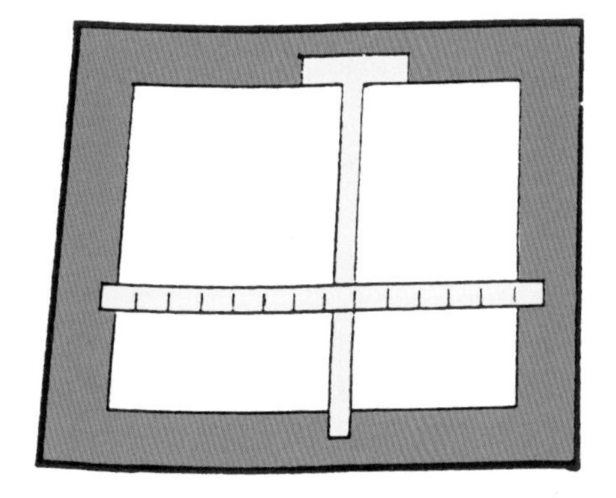

Fig. 1.26 Straight-edge and T-square

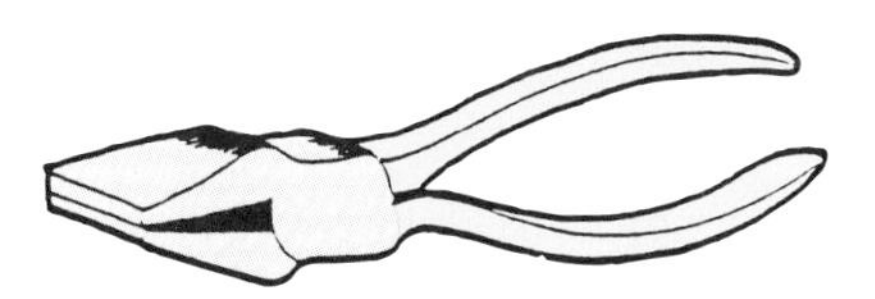

Fig. 1.27 Glazing pliers

**Tee-square (Fig. 1.26)**

A straight-edge of wood or plastic with a tee-piece guide at one end set at right angles to the straight-edge.

*Use* The guide fits the edge of the glass to ensure a cut at right angles to the edge.

**Glazing pliers (Fig. 1.27)**

Metal pliers with flattened ends to grip the glass

*Use* To grip the glass when cutting thin strips. The pliers apply gentle leverage to break the glass along the scored cut made by the glasscutter.

## 1.1.2 Paint containers

**Paint kettle or can (Fig. 1.28)**

Made of galvanised sheet iron, black sheet steel, zinc, aluminium alloy or plastic.

***Size*** Available with 125 mm, 150 mm, 180 mm, 200 mm diameter, or holding approximately $\frac{3}{4}, 1, 1\frac{1}{2}, 2\frac{1}{2}$ litres of paint.

***Use*** To hold a convenient quantity of paint decanted from manufacturer's container.

***Care*** Clean out thoroughly with appropriate solvent immediately after each job. Aluminium kettles will dissolve if cleaned with caustic soda. Burning out weakens welds and joints. Heat and strong solvents may destroy plastic kettles.

**Fig. 1.28** Paint kettle and kettle hook

**Fig. 1.29** Buckets

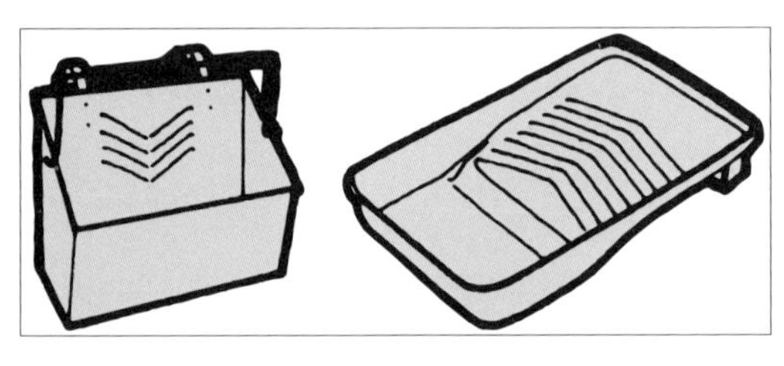

**Fig. 1.30** Paint trays

**Kettle hook or pot hook (Fig. 1.28)**

Wire hook with a double prong, one end used to suspend the kettle from the ladder rung, leaving the operative with both hands free.

**Bucket or pail (Fig. 1.29)**

Made of galvanised iron or plastic.

***Size*** 7 litres, 9 litres, 14 litres.

***Use*** To hold water, washing solution, paste or water-thinned paints.

***Care*** Plastic buckets will melt if subjected to a naked flame.

**Paint tray (Fig. 1.30)**

(i) Rectangular metal or plastic tray; various sizes to take rollers from 100 to 350 mm wide.

(ii) Tank, bucket or scuttle with either a raised side or a wire grid. Several sizes available containing up to 10 litres of paint.

***Use*** To hold paint for roller application, designed to ensure an even take-up of paint.

## 1.1.3 Mechanical hand tools

**Rotary disc sander (Fig. 1.31)**

A disc sander has a flexible rubber or composition sanding head, attached to a motor which can be driven either by electricity or by compressed air. On to the sanding head can be fitted any one of a range of abrasive papers.

***Use*** (i) Fitted with abrasive disc, for general preparation of surfaces including joinery, floor sur-

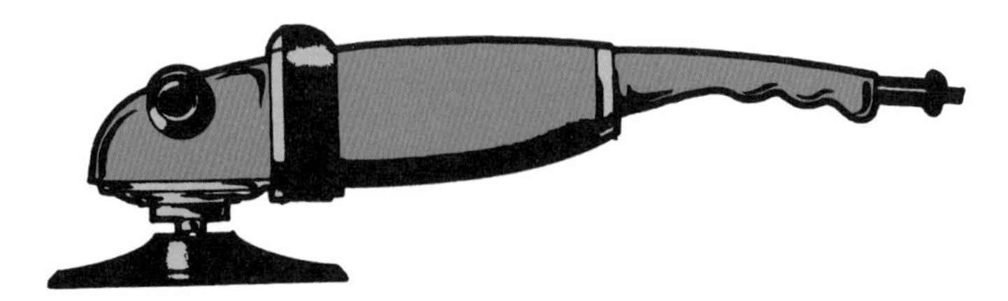

**Fig. 1.31** Rotary disc sander

Fig. 1.32 Rotary wire brush

facing, previously painted surfaces and for removal of rust. Suitable for work on curved surfaces.

(ii) A lambswool mop can be fitted over the sanding head and used for polishing.

(iii) A carborundum grinding wheel can be fitted to remove rough surfaces from welds.

*Note* Difficult to control and liable to damage the surface or leave it uneven.

***Safety precautions*** See 'Rotary wire brush'.

### Rotary wire brush (Fig. 1.32)

Radial cup- or disc-shaped wire brush which can be fitted to air-driven and electrically operated motors.

***Use*** To remove loose and flaking paint and rust from metal surfaces.

### Safety precautions

(i) Goggles must always be worn.

(ii) Hand-held power tools with exposed rotating heads must be switched off and stopped before they are laid down; they may otherwise spin, causing damage and injury.

(iii) Portable abrasive wheels larger than 55 mm diameter must be marked with maximum speed specified by the manufacturer.

(iv) In explosive atmospheres, phosphor-bronze brushes must always be used.

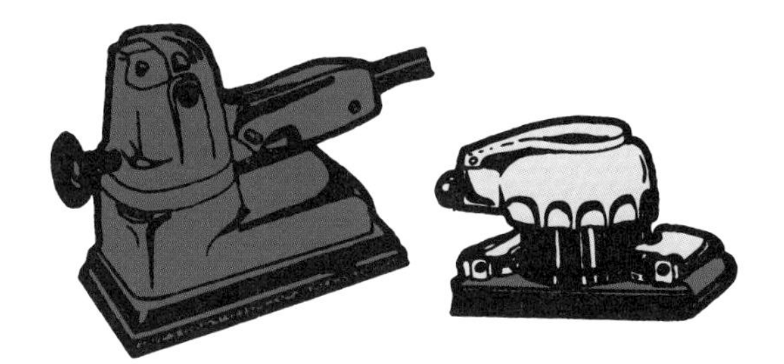
Fig. 1.33 Orbital sander

Fig. 1.34 Belt sander

### Orbital sander (Fig. 1.33)

Electric or air-driven sander consisting of a rectangular flexible platform pad on to which various types of abrasive paper are fixed. The platform moves in a small circular or orbital motion which abrades the surface. The rate of orbit varies according to make and type but can range from approximately 6000 to 20 000 orbits per minute: usually, the faster the motion, the better the finish. Orbital sanders are comparatively light in weight and can be used for long periods without undue fatigue.

***Use*** To prepare and smooth timber, metal, plastic and previously painted surfaces.

***Safety*** Electrically operated types can constitute a safety hazard if used with water. Air-driven types are perfectly safe.

*Note* Although slower in use, they are easier to control and produce finer surfaces than rotary sanders.

### Belt sander (Fig. 1.34)

A sanding machine with a continuous belt of abrasive paper. It maintains a flat sanding action but abrades at a faster rate than orbital sanders.

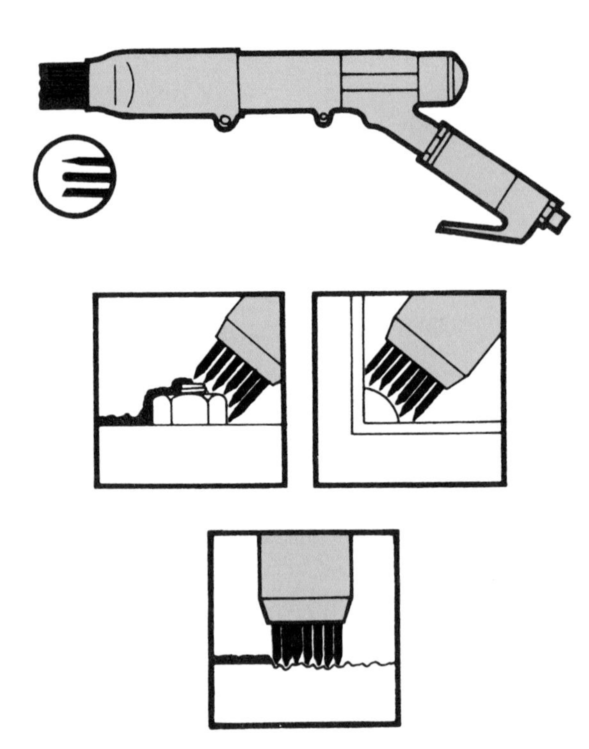

Fig. 1.35 Needle gun

*Size* Belt sizes either 75 or 100 mm wide × 610 mm long. Other large heavy-duty machines are available for sanding floors.

*Use* (i) To sand large areas of timber.
(ii) To sand metal surfaces to remove light rust.

## Needle gun or descaling pistol (Fig. 1.35)

An outer body or casing containing a number of hardened steel needles which are propelled forward by an air-driven spring-loaded piston. On hitting the surface, the needles rebound and are again forced forward by the piston. This action is continuous, working at the rate of approximately 2400 blows per minute. The individual needles are self-adjusting, thus ideal for use on uneven surfaces.

*Use* For the removal of rust, particularly around awkward areas such as nuts and bolts. The method is too slow for economic use on large areas. Also used for cleaning stonework and ornamental ironwork.

*Power* Driven by compressed air with an air consumption of approximately 5 to 10 cfm (2.36 to 4.72 litre/s) according to the size of the gun, at 90 psi (6.1 bar).

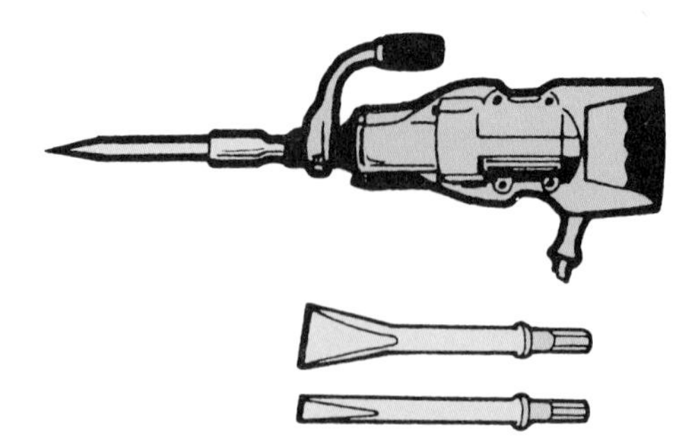

Fig. 1.36 Descaling chisels

*Needles* Three types are available:
(i) Pointed tip for breaking through heavy rust and millscale. These leave a rough surface.
(ii) Chisel-ended. Similar use to pointed tip but less damaging to the surface, leaving only a slight surface mark.
(iii) Flat-ended. These will not mark the metal surface and can be used on thin-gauge metals. Also used on surfaces which require a light treatment, such as concrete aggregate and stonework.

*Safety precautions* Goggles must always be worn when descaling. *Never* use in an explosive atmosphere – sparks cause explosions (special spark-resisting needles can be obtained for this work).

## Descaling hammer and chisel (Fig. 1.36)

This can be a conversion of the needle gun or a specially designed pneumatic or electrically operated descaling hammer. It works by a piston moving backwards and forwards at high speed inside a cylinder. The piston produces hammer blows on the chisel.

*Use* The constant hammering action of the chisel is used to remove heavy rust and scale. The process is relatively slow and not economic on large areas. The surface of the metal may be damaged as the chisel leaves pits and burrs which are difficult to coat with a full paint system.

*Safety precautions* Goggles must always be worn when descaling.

## 1.1.4 Burning-off equipment

Heat is used to soften paint films sufficiently to allow the paint to be scraped off with stripping knives and shave hooks. Must not be used to remove lead paint systems – see 6.4.2.

### LPG gas torch

- *Bottle-type torch* (Fig. 1.37)
  Operated by liquefied petroleum gases (LPG), either butane or propane (see 1.4.3). Portable and lightweight, with a refill gas bottle which screws under the torch. Various nozzles can be fitted with different jets to produce a range of flame shapes and varying degrees of heat. A small bottle of gas can last from 2 to 4 hours depending on the type of nozzle used.
  If this type of lamp is operated at an acute angle, it may flare up or stop.
- *Independent torch* (Fig. 1.38)
  Burners attached to hoses which are fixed to large gas cylinders containing from 10 lb (4.5 kg) to 32 lb (14.5 kg) of propane or butane gas. Two torches can be run from one cylinder. They are lighter to use and provide easier movement than the bottle type, particularly in awkward spaces.
- *Disposable cartridge type* (Fig. 1.39)
  Burner or torch head screwed into a disposable gas cartridge. Very light to use but more expensive than the returnable 'refill' bottles.
  These torches have a short burning time, and the flame produces less heat than the larger types.

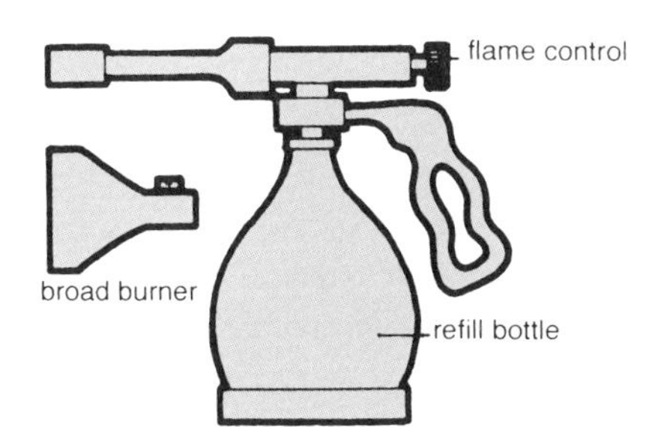


**Fig. 1.37** Bottle-type gas torch

**Mains gas torch or burner (Fig. 1.40)**
A hand torch operated by connection to natural or coke gas supply, convenient in the workshop if gas supply is laid on but with limited use on sites.

**Hot air paint stripper (Figs. 1.41 and 1.42)**
Similar in many ways to an electric hair drier. Hot air is produced by an electric filament which can be easily adjusted to any temperature between 20°C and 600°C.

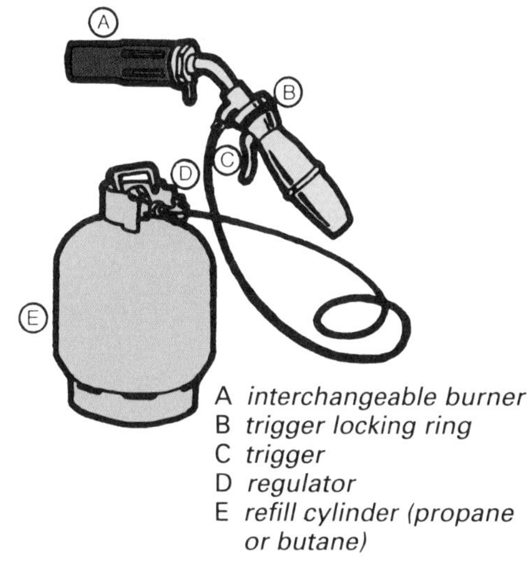


**Fig. 1.38** Independent gas torch

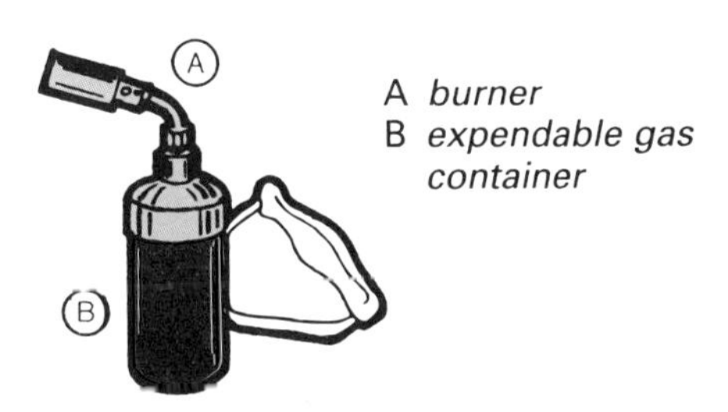


**Fig. 1.39** Disposable cartridge gas torch

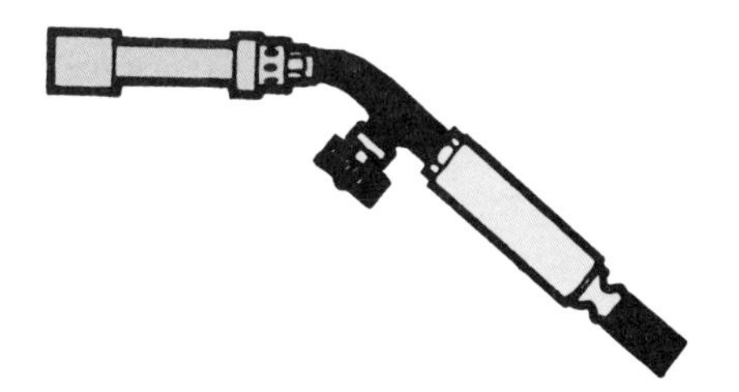

**Fig. 1.40** Mains gas torch

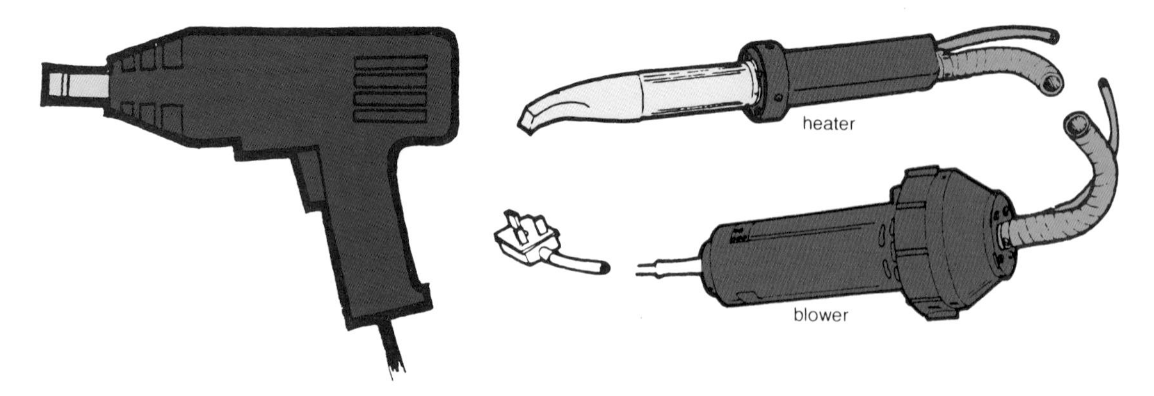


**Figs. 1.41 and 1.42** Hot air paint strippers

With no flame, the risk of fire is much less than with blow lamps and gas torches. There is less chance of cracking glass, and it is very difficult to scorch timber.

Ideal for use on old or delicate surfaces, and in areas of high fire risk such as eaves of old buildings.

### Safety precautions

(i) *Always* use recommended reinforced hose to avoid kinking and damage on site and to withstand effect of LPG.
(ii) *Always* ensure that curtains and soft furnishings are removed from the area before burning off.
(iii) *Always* make sure that no woodwork is left smouldering before leaving the job.
(iv) *Always* keep a bucket of sand or a fire extinguisher at hand to smother any fire immediately.
(v) *Always* make sure all charred and burnt paint is extinguished before placing it in a metal bin (safer to place all debris in a bucket of water).
(vi) *Do not use* to remove old lead paints (see 6.4.2).

## 1.1.5 Brushes: components and types

### Characteristics of fillings

***Pure bristle* (Fig. 1.44)**

This term applies only to hairs from the pig, hog or boar. As domesticated pigs have only a small amount of short, soft hair, the bristles used in paint brushes

**Table 1.1** The construction of a painter's brush (Fig. 1.43)

| | |
|---|---|
| A Handle | Usually made of beech, birch or alder wood. Imported brushes from the Far East and China often use local hardwood. Cheaper varieties will incorporate plastic handles. Some heavily grained timbers are sealed or varnished to reduce splitting and water penetration, whilst alder wood, for example, is suitable untreated. |
| B Stock | The means by which the handle is fixed to the filling: the ferrule may be made from metal or more recently of plastic. Metal ferrules are of nickel-plated steel, copper or copper-covered steel. Nickel-plated steel will rust when in contact with water for prolonged periods. Stainless steel will not. Copper is a flexible metal and is prone to swelling of the stock if brushes are soaked in water. |
| C Setting | An adhesive, which cements the filaments together at the root. Traditionally vulcanised rubber but nowadays two-part epoxy resin is used. |
| D Filling | Four main types: (a) bristle; (b) man-made fibres such as Orel® (polyester), Chinex® and Tynex® (brands of nylon); (c) natural fibres; (d) mixtures of bristle, hair and fibres, may also be referred to as *union*. 'Length out' is the amount of filling visible. |

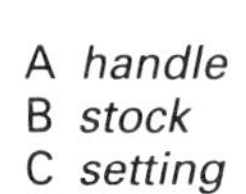


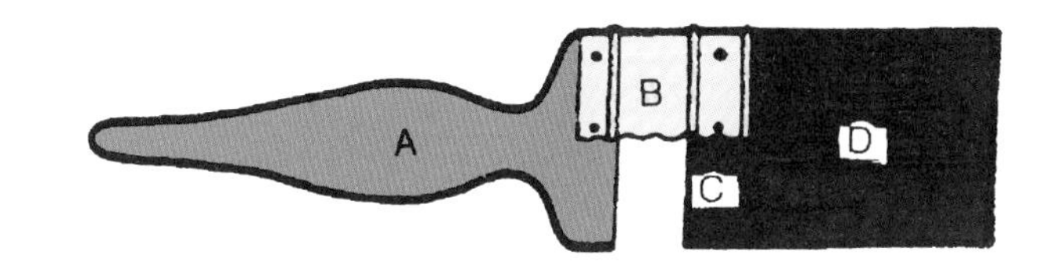


**Fig. 1.43** The construction of a painter's brush

**Fig. 1.44** Pure bristle

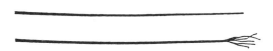

**Fig. 1.45** Synthetic filament

come from wild pigs, hogs or boars found mainly in China and India. Chinese bristles are of high quality, very resilient and black in colour. Indian bristles are rather coarse, longer in length, less resilient than Chinese bristles, and they vary in colour. The main properties of bristle are:

(i) Strength and resilience: they can have a long 'length out' yet still work well and hold the weight of paint.
(ii) A split tip or *flag* which acts as a barb when releasing the paint and provides a soft tip for laying off.
(iii) Serrations or scales like minute teeth along the length of the bristle: these prevent the hairs lying too close together, allowing the brush to hold more paint.
(iv) A natural taper from root to tip which allows a bundle of bristles to lie together in a brush form.
(v) A natural curl or lean: this enables the bristles to be curved inwards and lie together. Dusting brushes are reversed and curve outward.

***Synthetic filament*** **(Fig. 1.45)**

Nylon and polyester are two synthetic filaments (man-made filaments) commonly used in the manufacture of synthetic filament brushes, often blended together.

Orel® is a polyester, Chinex® and Tynex® are brands of nylon. They are extruded into a number of shapes and styles.

Hollow filaments, just like a drinking straw, have limited strength and often become misshapen during use. Solid Round Taper filaments (SRT) possess all the benefits of natural bristle and more besides. They are more hard-wearing, lasting up to five times longer, not affected by ordinary solvents, resistant to many chemicals and not damaged by insects and fungi, require no breaking in prior to use in gloss and are very easy to clean.

With the increased use of water-borne paints, stains and varnishes, synthetic filament brushes reduce tramlines associated with the application of quick-drying products.

Good quality SRT brushes may be used equally well in both water-borne and solvent-based coatings.

### Furs and soft hairs

Very fine hairs used in specialist brushes for signwriting, graining and gilding.

***Sable*** This term applies to hairs cut from the tail of various animals of the weasel family. The most commonly used is from the kolinsky. The best hairs are obtained from animals which live in the wild Arctic regions of Asia, principally China.

Sable is a red-brown hair. Although fine, the hair has great strength and will spring back into shape once pressure upon it has been released. It is tapered and produces a very fine tip that makes it ideal for fine, controlled painting. Used principally for signwriting brushes.

***Ox*** Hair cut from the ear of an ox. They are a little darker in colour than sable and although equally fine, they do not have the strength or springiness of sable. Long lengths are used in coach liners. When mixed

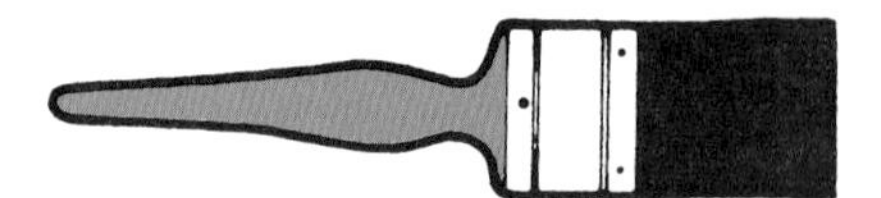

Fig. 1.46 Flat paint brush

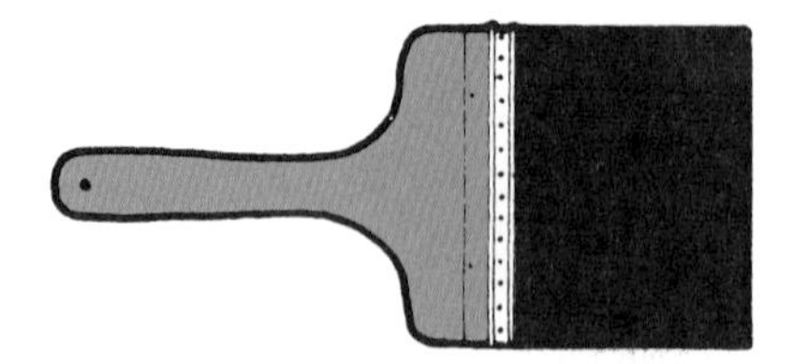

Fig. 1.47 Flat wall brush

with sable, they may be called 'sabox' fillings. Used as a substitute for sable to save costs.

***Squirrel*** Still incorrectly called camel hair and obtained from the blue squirrel, generally those native to the colder regions of Asia and America. It is a dark brown hair and much softer than sable. Squirrel is used as the filling in gilders' tips, sword stripers, mops and graining cutters.

***Badger*** The most commonly available hair is obtained from the Chinese badger. Its long hair, generally off-white in colour with a brown stripe in the centre, has the peculiar characteristic of being slightly thicker in the middle than at its fine tip. This property produces fillings which swell out towards the end, quite opposite to most brush fillings which tend to taper. Badger hair is used in softeners.

***Synthetic filaments*** Most commonly made from nylon. They are very fine hairs with considerable strength and springiness. They can be produced in any colour although bright red-brown is a common colour used. Principally manufactured in Japan. Used in sign-writing brushes and one-strokes.

### Flat paint or varnish brush (Fig. 1.46)

Usually pure bristle but also available in man-made filaments. Quality varies according to the amount and quality of the filling and the 'length out'.

***Size*** 25 mm, 37 mm, 50 mm, 62 mm, 75 mm, 100 mm wide. Also varying in thickness and length of filling.

***Use*** To apply most types of paint and varnish to a wide variety of surfaces including doors, sashes, frames and wall areas.

Fig. 1.48 Dusting brush

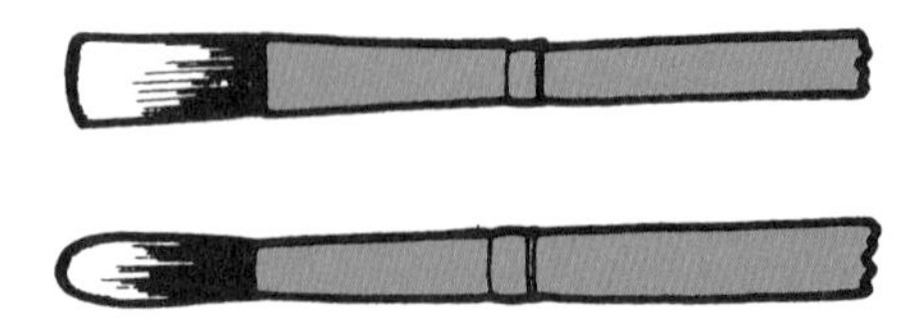

Fig. 1.49 Fitches

### Flat wall brush (Fig. 1.47)

Available in a wide range of varying quality including pure bristle, man-made filament, and a mixture of bristle, hair and fibre (union). Quality is also dependent on weight of filling used and the 'length out'.

***Size*** 100 mm, 125 mm, 150 mm, 175 mm wide.

***Use*** (i) To apply emulsion paints to large areas.
(ii) To apply adhesive to surface coverings.

### Dusting brush (Fig. 1.48)

Black or white filling of either pure bristle or man-made filaments, usually nylon.

***Size*** Available as standard type of brush or as a three- or four-ring type, both 90 or 110 mm wide.

***Use*** To remove dust and dirt from surfaces before painting.

### Fitch (Fig. 1.49)

Bristle filling, usually white, set in either a round or a flat nickel ferrule.

***Size*** Flat type from 5 to 28 mm wide. Round type from 3 to 20 mm in diameter. Available in eight sizes.

***Use*** To apply paint to detailed work and areas difficult to reach with paint brush.

### Radiator or flag brush (Fig. 1.50)

Bristle filling attached to a long wire handle which can be bent to fit into awkward areas.

***Size*** 25 mm, 37 mm, 50 mm wide.

***Use*** To apply paint to restricted spaces such as behind pipes and column radiators.

**Fig. 1.50** Radiator brush

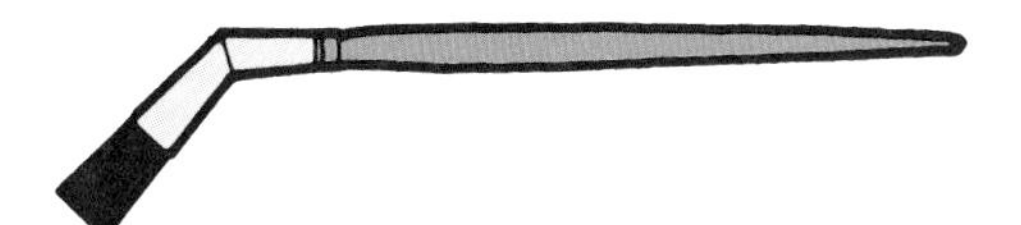

**Fig. 1.51** Crevice brush

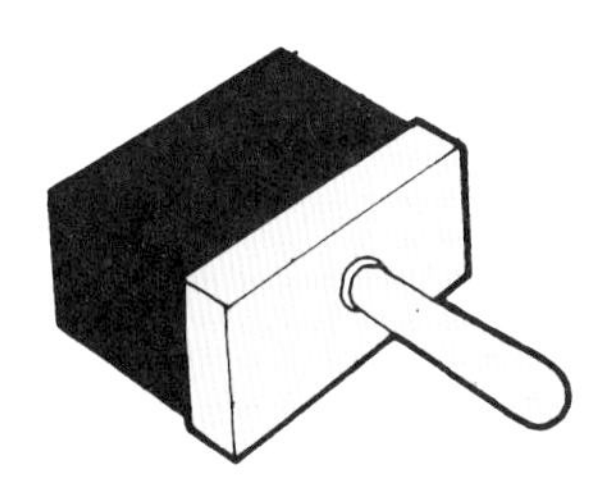

**Fig. 1.52** Cement or block brush

### Crevice brush or bent fitch (Fig. 1.51)

Similar to fitch, but angled to assist working in awkward spaces.

***Size*** Flat or round, in sizes from 9 mm, 12 mm, 15 mm wide and 18 mm, 31 mm in diameter.

***Use*** Similar to radiator brush.

### Cement paint brush (or block brush) (Fig. 1.52)

Man-made filling (synthetic), set in polished wooden handle.

***Size*** 150 mm wide.

***Use*** An inexpensive brush for the application of cement paint and other masonry finishes. The tough filling resists the abrasive action of application to rough surfaces like cement rendering and brickwork. The filling is also unaffected by the lime content of cement paints.

### Care and maintenance of brushes

Before using a new brush, flirt the bristles if applicable to remove any loose hairs.

***Overnight storage*** Brushes used in oil or alkyd paints can be stored overnight by suspending in water or placing in patent 'brush keep'.

***Cleaning brushes after use*** Clean thoroughly to remove all traces of paint, especially from the root where paint tends to accumulate and harden, causing the brush to fan out. A comb is an invaluable aid in this process. *Method:* Remove all traces of paint with a suitable solvent. Wash out the solvent in warm detergent solution, then rinse thoroughly in clean water. Dry the brush by hanging in a well-ventilated room. Bristle brushes stored damp will be ruined by mildew.

***Storage*** Wrap the fillings in waterproof paper and lay flat in boxes. Sprinkle with moth repellent (not required for synthetic brushes). Store should be neither too warm nor too dry (about 18°C).

*Never* stand a brush on the filling as this can damage it; either lay brushes flat or suspend them. *Never* wash or use a pure bristle brush in alkaline materials as they destroy the bristle.

## 1.1.6 Paint rollers

### Types of roller

Paint rollers consist of a plastic core and a fabric cover referred to as a sleeve. It is the fabric type that gives the application characteristics of the roller. The roller sleeve is loaded onto a frame for use; this can be a cage or stick system. Cages are for 175 mm and 230 mm sleeves whereas a double-arm frame is required for rollers 300 mm and above. Double-arm rollers are the preferred option for ceilings and flooring to ensure even coverage when used on an extension

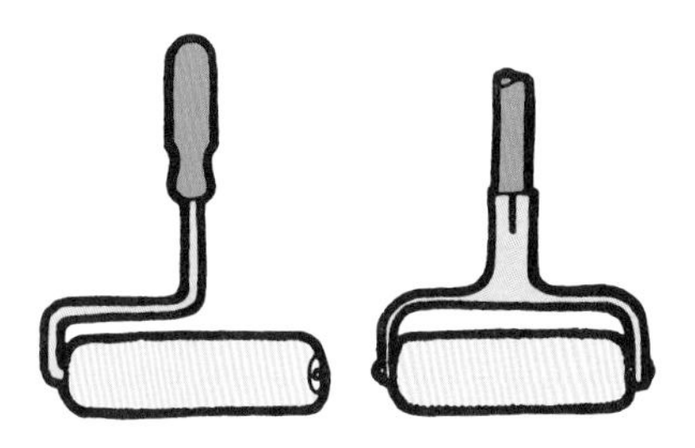

**Fig. 1.53** Cylinder rollers

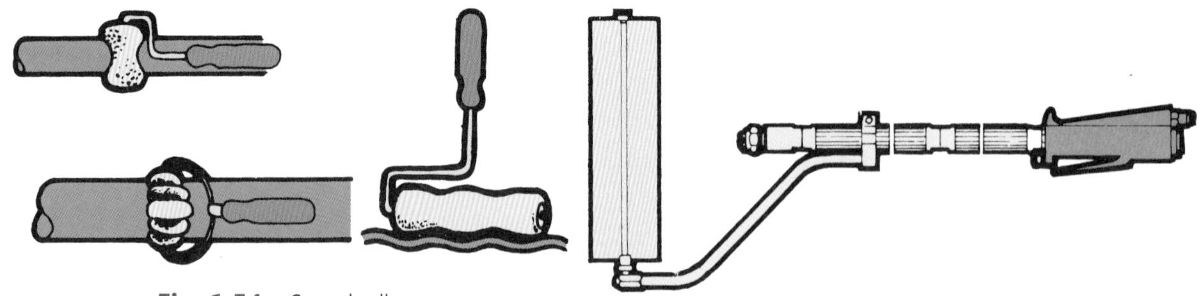

Fig. 1.54 Curved rollers

Fig. 1.55 Airless-fed roller

pole. Stick systems require the sleeve to have an internal mechanism. These are very popular on the continent but ranges are limited in the UK. They are also supplied in a 'padded' version, which increases the fabric area providing greater coverage.

Mini rollers, traditionally known as radiator rollers, now come in a wide range of fabrics and sizes – 100 mm, 125 mm and 150 mm – and fit a standard stick system handle that comes in short and long sizes.

***Curved roller*** **(Fig. 1.54)**

Produced with a hollow section, cut to allow a wrap-round on pipes and poles.

Another type of roller designed for painting pipes and poles has a number of separate roller sections fitted to a spring axle. The axle bends to almost any diameter, and the roller is loaded in an ordinary tray.

Rollers are also produced with several hollow sections designed for painting corrugated sheeting.

These are all now quite rare with people customising standard rollers for these jobs. There are corner rollers available, making it possible to give the same textured effect into corners, but these are not that popular, hence poor availability in stores.

***Airless-fed roller*** **(Fig. 1.55)**

A cylinder roller fed by an airless spray tip. Paint is pumped through the handle from an airless spray pump. The roller is instantly charged and the paint flow controlled by a trigger on the handle.

Application is fast, but this type of roller has a limited use because it requires an airless spray pump to feed paint to the roller.

***Advantages***

(i) Can be used on contracts where application by spray is not allowed.
(ii) No overspray.
(iii) Reduces need for masking.
(iv) Extension arms can be fitted to provide a reach of nearly 2 m. This enables reasonably high ceilings and walls to be coated without using a scaffold.

***Extension/telescopic arms***

These can be used with most rollers to enable reasonably high ceilings and wall areas to be coated without using a scaffold.

## Types of covering

***Woven fabrics*** Generally made from polyamide, a hard-wearing material resistant to tearing on rough surfaces. Other characteristics are enhanced paint pick-up, easy clean, flatter finish (reduced 'orange peel effect'), less tiring to use as pressure is not required for paint discharge.

***Knitted fabrics*** Generally made from polyester, a higher-density fabric will produce a more even finish due to good shape retention. High resilience means good penetration into porous surfaces, but requires pressure to discharge paint, making these more tiring to use especially on rough surfaces. These rollers are hard to clean compared to their woven counterparts.

***Sheepskin*** Made from sheep's pelt, these have a natural high density that cannot be matched by man-made methods. The greatest paint pick-up of all rollers, making these ideal for large areas by reducing loading frequency.

***Lambswool*** Available in a range of pile heights, these are wool woven into a synthetic backing, i.e. no skin. Due to the inconsistent orientation of the fibres and lower density, these do not have as high a paint

**Table 1.2** Coverings and their applications

| *Fabric* | *Pile height* | *Surface* | *Coating* |
|---|---|---|---|
| Knitted polyester | 4–6 mm | Smooth | Primers, sealers and undercoats |
| Knitted polyester | 10–12 mm | Smooth to semi-smooth | Emulsions, floor paints |
| Knitted polyester | 16–18 mm | Rough | Emulsion, masonry paints |
| Knitted polyester | 28–32 mm | Extra rough | Emulsion, masonry paints |
| Woven polyamide | 4–6 mm | Smooth | Primers, sealers and undercoats |
| Woven polyamide | 10–12 mm | Smooth to semi-smooth | Emulsions, floor paints |
| Woven polyamide | 16–18 mm | Rough | Emulsion, masonry paints |
| Sheepskin | 18 mm | Rough | Very high paint pick-up of emulsion and masonry paints |
| Mohair | 4–6 mm | Smooth | Solvent-based paints, stains and varnishes |
| Foam | 2–4 mm | Smooth | Solvent-based coatings without risk of fibre loss |
| Microfibre | 6–8 mm | Smooth to semi-smooth | Wood stains and varnishes |
| Microfibre blend | 10–12 mm | Smooth | Emulsions no spatter, low coating thickness (not good for covering darker colours), fine finish |

pick-up as genuine sheepskin. The short pile is often referred to as simulated mohair.

***Mohair*** Hair of the goat is ideal for a fine finish on smooth surfaces with most solvent-borne coatings and can be used with water-borne, unlike the simulated versions.

***Foam*** Polyurethane foam is ideal for applying solvent-borne coatings without the risk of fibre loss. There may be a need in some coatings to lay off with a brush to remove aeration caused by the foam, and for this reason this is not to be used for water-borne, as their quick-drying nature will leave voids in the applied coating.

***Microfibre*** Fibres of 1 denier provide high pick-up of low-viscosity coatings such as wood stains, and their absorbent nature reduces spatter and runs. These are ideal for water-borne and solvent-borne floor coatings ensuring quick and even application.

## Selection of paint rollers

The choice of whether to use a knitted roller or a woven roller depends on a number of factors, and Table 1.3 aims to help you decide which would be the best for any given situation.

There are two types of fabric used in paint roller manufacture: knitted fabrics and woven fabrics. But there is one specific advantage of woven rollers, which is that they are much easier to clean!

**Fig. 1.56** Woven fabric

***Knitted***

Knitted fabrics rely on density of individual fibres for paint pick-up and resilience but tend to clog at the base. This is particularly noticeable when cleaning.

***Woven* (Fig. 1.56)**

Woven fabrics have twisted yarns split at the tip into a mass of fibres. This means that the density at the base

**Table 1.3** Choosing between knitted and woven rollers

| Property | Knitted | Woven | Comments |
|---|---|---|---|
| Paint pick-up | Good | Very good | Difference is greater for longer pile. The long pile woven is closest to sheepskin, which has the greatest paint pick-up. The sheep's wool (woven wool/polyester) has similar properties to the sheepskin but with greater durability. |
| Coverage rates | Good | Very good | Woven has a greater level of paint release and therefore a greater coverage/loading. |
| Finish | Stipple | Smooth | The knitted will give more even coverage, which is most noticeable on the first coat. The woven will give a flatter finish. |
| Penetration of substrate | Very good | Good | Woven lays the paint on the surface whereas the knitted forces the paint into low spots. This is important for porous/coarse surfaces such as fine textured masonry or breezeblock. |
| Shape retention | Very good | Good | The knitted will maintain a smooth outer diameter and therefore produce an even finish. This is most evident on the medium pile on smooth surfaces. The woven will open up releasing more paint but this will not produce as even a finish as the knitted. |
| Durability | Good | Very good | Woven are made from polyamide yarns, which are harder-wearing than the knitted fibres of polyester. Therefore, the woven sleeves are longer-lasting. |
| Ease of cleaning | Average | Very good | Woven are more open fabrics and therefore the paint is easier to wash out. |

is lower, hence releasing the paint more easily. This also makes cleaning much quicker.

Woven fabrics contain fibres (mainly polyamide), which are spun into pile yarn and woven in a 'W' form to a fabric backing. The void at the base of the woven fabric allows greater paint storage and, as mentioned above, will release it more readily, ensuring more coverage per loading is achieved. The woven fabric also 'opens up' when rolled, which means it will conform to the surface without too much pressure. This makes the job less tiring especially on rough surfaces.

Splits at the tip of the yarns in the woven fabric also regulate the paint release and mean the woven fabric will cover an area uniformly. The split tips also deliver the paint to the surface in a way that minimises the stretching of the paint and therefore leaves a flatter finish, which makes the woven rollers ideal for the recently launched flat paints.

The majority of woven fabrics are polyamide, which is particularly hard-wearing and will withstand roughcast surfaces without breaking up. Woven rollers tend to be more expensive than knitted, but the longer life helps to compensate for this.

### Use

Rollers can be chosen from the wide range available to suit most types of work, including large areas of flat surface, anaglypta and ingrain papers, roughcast, brickwork and concrete. They are also very useful on perforated hardboard (pegboard), acoustic boards, wire netting, corrugated sheeting and pipework.

### Advantages

(i) Paint is applied several times faster than with a brush on flat and textured surfaces.
(ii) Chances of 'flashing' and 'sheariness' are reduced.
(iii) Extension poles can reduce the need for scaffolding.
(iv) No brush marks, although a slight 'orange peel' texture is unavoidable.

### Disadvantages

(i) Rollers cannot paint right into angles and corners (corner rollers are available, but difficult to find).

(ii) They are not suitable for fine, intricate work.

(iii) Some people object to the texture.

### Care of rollers

(i) Rollers must be clean and free from solvents before storing. They can be stored in a 'patent roller keep' if they are to be used in the same material again.

(ii) After cleaning and washing, rollers should always be hung up to dry, never left resting on the pile, as this will crush and distort the covering.

(iii) If stored damp, mildew can ruin lambswool or mohair coverings and the metal parts will corrode. Always store in clean, dry, well-ventilated conditions.

## 1.1.7 Paint pads and mittens

### Paint pad (Fig. 1.57)

A short pile face, usually mohair, attached to a foam pad or cushion which is fitted to a handle. Modified versions may have

*(a)* detachable head to allow for replacements, thus reducing costs;

**Fig. 1.57** Paint pads

*(b)* wheels inserted in the side of the plastic body to make cutting-in easier.

***Size*** Various sizes are available depending on the manufacturer. They range from large (150 × 100 mm) down to a small sash painter sometimes called a 'toothbrush' pad. Some firms also make a crevice pad.

***Use*** Can be used to apply gloss, eggshell, flat and emulsion paints, wood stains and varnishes. Used on flat surfaces such as ceilings, walls, woodwork, building boards and over firmly attached wallpaper. Useful for painting bands or lines.

***Care*** Clean only in water or white spirit and soap and water, as appropriate. Strong solvents and some types of brush cleaner may damage the foam pad. When clean, hang up to dry.

### Paint mitten (Fig. 1.58)

A mitten, made of sheepskin, designed to fit the hand like a normal glove-mitten.

***Use*** Used to apply emulsion paint, primer, undercoats and finishes to surfaces such as iron railings, behind radiators and pipes, which are difficult to coat by normal methods.

***Care*** Clean in similar way to rollers.

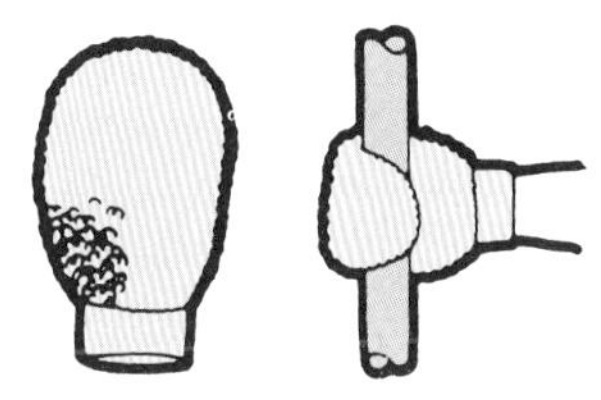

**Fig. 1.58** Paint mitten

## 1.1.8 Paperhangers' tools and equipment

### Paperhanging brush (Fig. 1.59)

Made of white or grey bristle or synthetic filaments set in rubber.

***Size*** 200 mm, 225 mm and 250 mm wide.

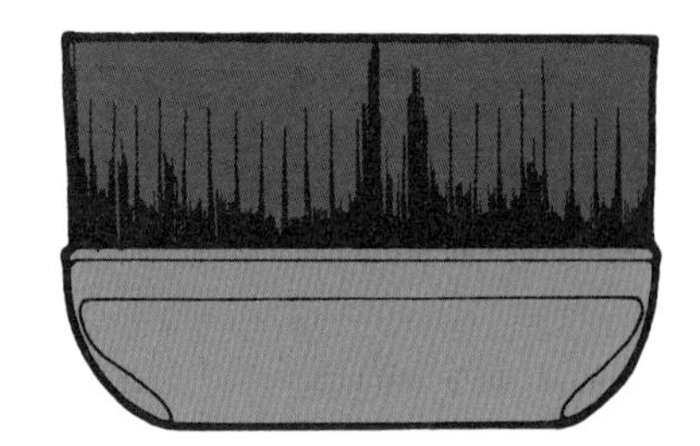

Fig. 1.59 Paperhanging brush

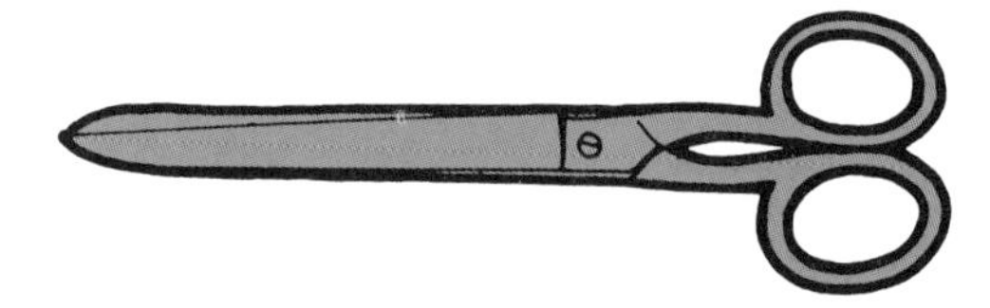

Fig. 1.60 Scissors or shears

***Use*** To apply papers and fabrics to walls and ceilings.

***Care*** Keep clean: wash in soap and tepid water and hang to dry.

**Scissors or shears (Fig. 1.60)**

Made of polished steel with hollow-ground blades.

***Size*** 250 mm, 275 mm and 300 mm long.

***Use*** To cut lengths and trim at angles and around obstacles.

***Care*** Keep clean and sharp. Never clean with abrasive paper as this rounds off the cutting edge. Store dry and clean to prevent rusting.

**Seam and angle roller (Fig. 1.61)**

Made of hardwood, foam rubber or plastic. Side arm or double arm fitting.

***Size*** 25 mm and 37 mm wide.

***Use*** To roll down edges of paper at seams and angles.

***Care*** Keep clean and lubricate sparingly.

**Felt roller (Fig. 1.62)**

Made of a number of felt discs or a felt-covered cylinder.

***Size*** 90 mm and 175 mm wide.

***Use*** In place of paperhanger's brush, it applies a firm, even pressure to the paper without distorting or polishing it.

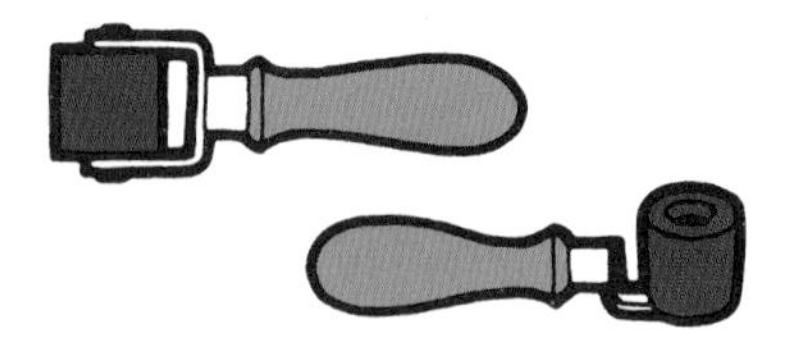

Fig. 1.61 Seam rollers

Fig. 1.62 Felt roller

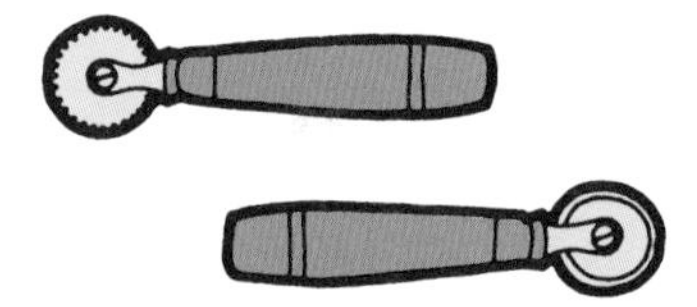

Fig. 1.63 Casing wheels

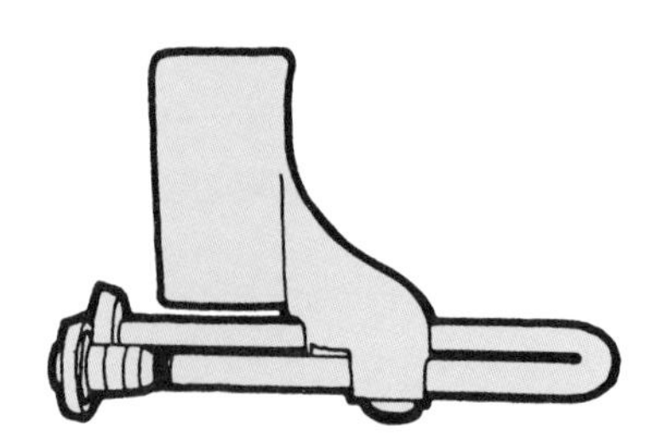

Fig. 1.64 Hand trimmer

***Care*** Wash regularly in warm water.

**Casing wheel (Fig. 1.63)**

Plain or serrated blades.

***Size*** 37 mm wheel.

***Use*** An alternative method to scissors and trimming knives for trimming surplus paper at angles or around obstacles.

***Care*** Keep clean and sharp.

**Hand trimmer (Fig. 1.64)**

Two small cutting wheels attached to a guide which is held against the edge of the paper.

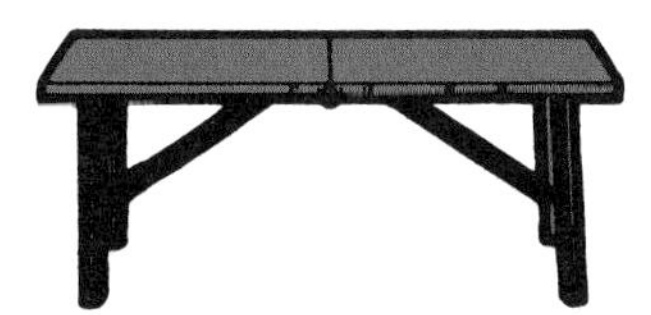

**Fig. 1.65** Paste board

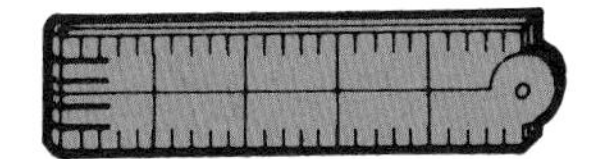

**Fig. 1.66** Rule

**Fig. 1.67** Trimming knife

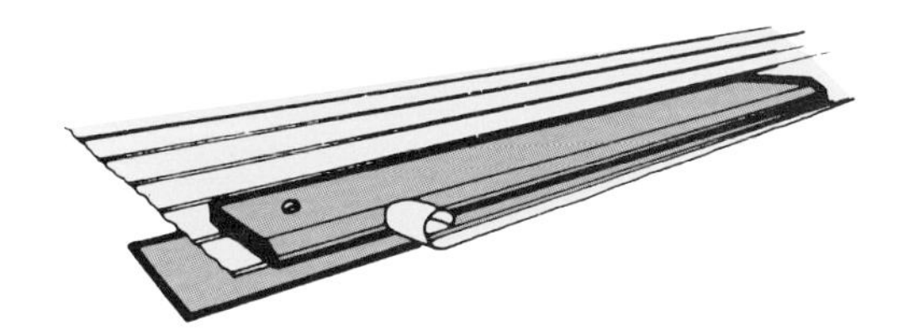

**Fig. 1.68** Steel straight-edge

*Use* To remove the selvedge from untrimmed paper, or to cut narrow strips of paper. It is adjustable and can cut strips up to approximately 75 mm from the edge of the paper.

*Care* Replace blades when blunt.

### Paste board (Fig. 1.65)

Made from wood, usually collapsible for easy transportation. Should be firm.

*Size* 1.830 m long × 560 mm wide.

*Use* For measuring, cutting, matching and pasting surface coverings.

*Care* Keep face and edges clean and free from paste.

### Rule (Fig. 1.66)

One metre long, made of boxwood or plastic and folded into four sections. Working is more accurate if the rule is used on its edge.

*Use* For measuring areas, lengths and widths.

### Trimming knife (Fig. 1.67)

Available with retractable blade. Replacement blades either angled or rounded.

*Use* For trimming and cutting at angles and around obstacles.

*Care* Keep edge sharp or replace blade regularly.

### Steel straight-edge (Fig. 1.68)

Perfectly straight with a bevelled edge.

*Size* 600 mm, 915 mm, 1.400 m and 1.830 m long.

*Use* To guide a knife when trimming papers and fabrics. Used over a zinc strip to protect the paste board and ensure a clean cut.

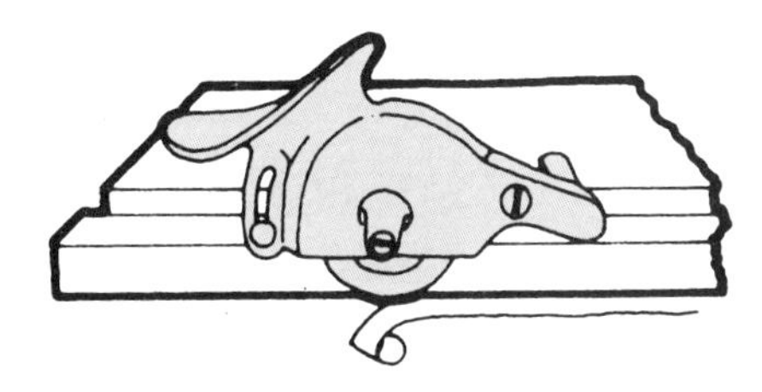

**Fig. 1.69** Ridgely track-trimmer

**Fig. 1.70** Spirit level

### Ridgely track-trimmer (Fig. 1.69)

A combined 1.830 m metal purpose-made straight-edge and cutting wheel used over a zinc strip 75 mm wide × 1.830 m long.

*Use* To trim or cut widths of paper or fabrics before or after pasting.

### Spirit level (Fig. 1.70)

Available in metal or wood to check horizontal and vertical lines.

*Size* Many sizes, but 1 m long is the most useful when paperhanging.

### Plumb bob and line (Fig. 1.71)

A small weight, usually brass or stainless steel, suspended on a length of fine cord.

*Use* To check that the first length of surface covering on every wall is upright.

Fig. 1.71 Plumb bob

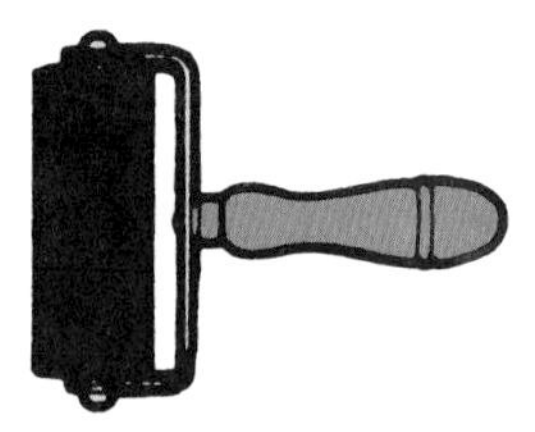

Fig. 1.72 Rubber roller

**Rubber roller (Fig. 1.72)**

Solid rubber roller.

***Size*** 90 mm and 175 mm wide.

***Use*** To hang materials such as vinyl papers and photo murals where a felt roller or brush would not be heavy enough.

***Care*** Keep clean.

**Spatula (Fig. 1.73)**

Straight-edged flexible metal or plastic blade set in a plastic stock and handle.

Fig. 1.73 Metal spatula

***Size*** 250 mm and 450 mm wide.

***Use*** (i) For smoothing vinyl papers and fabric-backed vinyl wall coverings.

(ii) For pressing vinyls and papers into angles and acting as a guide when cutting with a trimming knife.

## 1.1.9 Graining, marbling and broken-colour tools and brushes

**Mottler (Fig. 1.74)**

Pure white bristle filling, set in a metal ferrule. Sometimes has a short wooden handle.

***Size*** 50 mm and 75 mm wide.

*Use* (i) For simulating the soft lights and darks seen beneath the surface grain of many woods (mottling).

(ii) As a general graining brush to simulate the straight grain and heartwood patterns of mahogany, walnut and similar woods.

**Overgrainer (Fig. 1.75)**

Pure white bristle filling, set in a metal ferrule. Similar to a mottler but with a longer filling and much thinner in section.

***Size*** 50 mm and 75 mm wide.

***Use*** To add fine parallel grain patterns to completed work (overgraining). Often used with water-graining colour and passed through a coarse comb to separate the filling.

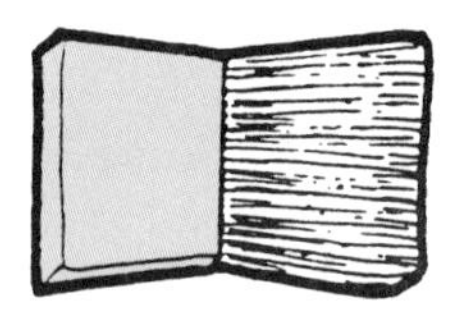

Fig. 1.74 Mottler

Fig. 1.75 Overgrainer

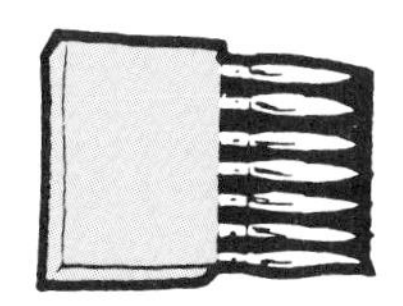

**Fig. 1.76** Pencil overgrainer

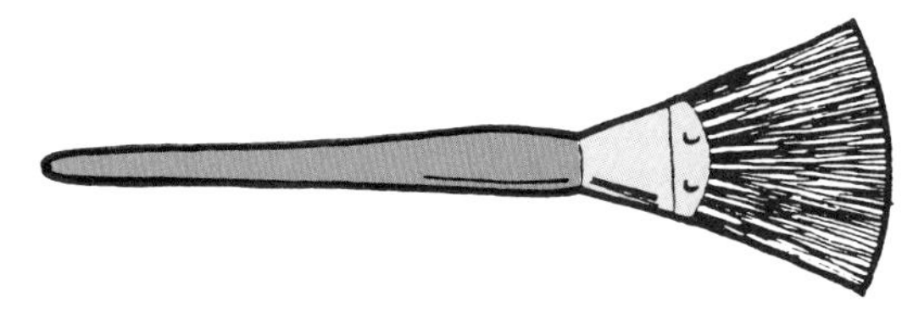

**Fig. 1.77** Fan fitch overgrainer

**Fig. 1.78** Flogger

### Pencil overgrainer (Fig. 1.76)

Separate tufts of pure white bristle filling, set in a metal ferrule. Also available with a pure sable hair filling but expensive to buy.

***Size*** 50 mm and 75 mm wide.

***Use*** Similar to those for the overgrainer.

### Fan fitch overgrainer (Fig. 1.77)

Separate tufts of pure white bristle filling, set in a flattened metal ferrule to produce a fan shape.

***Size*** 25 mm and 38 mm at the widest end.

***Use*** (i) To simulate the curly grain patterns of burr walnut, pollard oak and similar decorative woods.
(ii) Used as an overgrainer.

### Flogger (Fig. 1.78)

Long, pure bristle filling, set in a wooden stock or a metal ferrule.

***Size*** From 75 mm to 100 mm wide.

***Use*** (i) To simulate the pore markings of wood by beating wet graining colour.
(ii) To imitate the decorative texture 'dragging'.

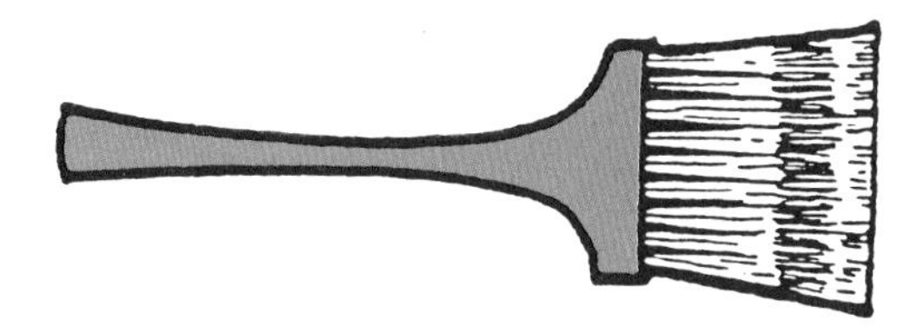

**Fig. 1.79** Badger softener

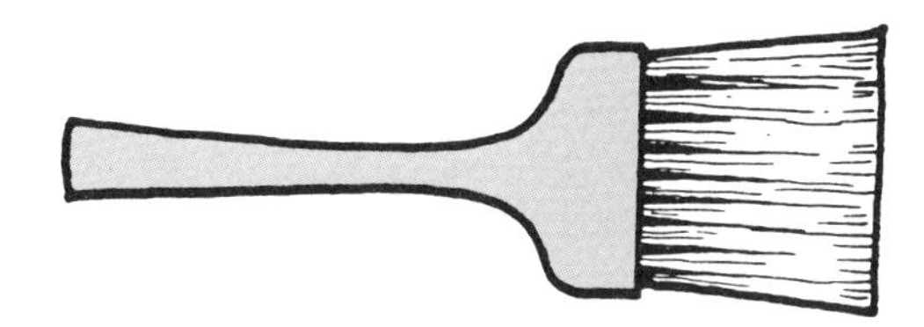

**Fig. 1.80** Hog hair softener

### Badger softener (Fig. 1.79)

Separate tufts of pure badger hair filling, set in a wooden stock. Some smaller varieties are set in a metal ferrule.

***Size*** 75 mm and 100 mm wide. Smaller flat varieties: 25 mm, 38 mm, 50 mm and 75 mm wide. Smaller round varieties: small size No. 2, to large No. 12.

***Use*** To gently fade out or soften the sharp edges of patterns produced in graining and marbling. Although intended only for use with water-based materials, they are often used with oil-graining colour and marbling gilp.

### Hog hair softener (Fig. 1.80)

Similar to the badger softener, but containing a pure white bristle (hog hair) filling.

***Size*** 75 mm and 100 mm wide.

***Use*** As for the badger softener but not as soft and intended for oil-based materials only.

### Drag brush or brush grainer (Fig. 1.81)

A filling of pure bristle on one side and coarse nylon on the other, set in separate tufts in a wooden stock.

***Size*** 100 mm wide.

***Use*** To simulate a variety of straight grain patterns by dragging through wet graining colour.

### Cutter (Fig. 1.82)

Short squirrel hair filling, set in a metal ferrule.

***Size*** 50 mm and 75 mm wide.

**Fig. 1.81** Drag brush or brush grainer

**Fig. 1.82** Cutter

**Fig. 1.83** Filbert fitch

***Use*** To simulate the fine clean highlights common in feather mahogany and similar decorative woods. Used in conjunction with a wet chamois leather to remove water-graining colour in the area of the highlights.

### Filbert fitch (Fig. 1.83)

Short pure white bristle or synthetic filling, set in a flattened metal ferrule.

***Size*** A range from small No. 0 to large No. 16.

***Use*** (i) For simulating heartwood grain patterns of oak, pine and similar woods.
(ii) Painting marble veins.

### Steel comb (Fig. 1.84)

Flexible steel comb with long, parallel, square cut teeth.

***Size*** 25 mm, 50 mm, 75 mm and 100 mm wide, with fine, medium and coarse graded teeth. Also made from cork tiles, with nicks cut out as required.

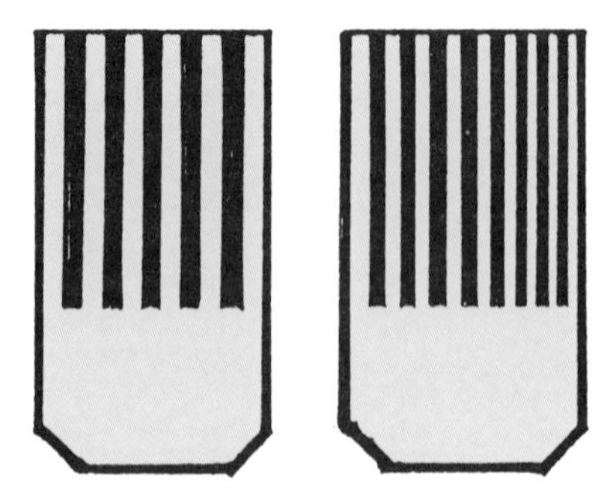

**Fig. 1.84** Steel combs

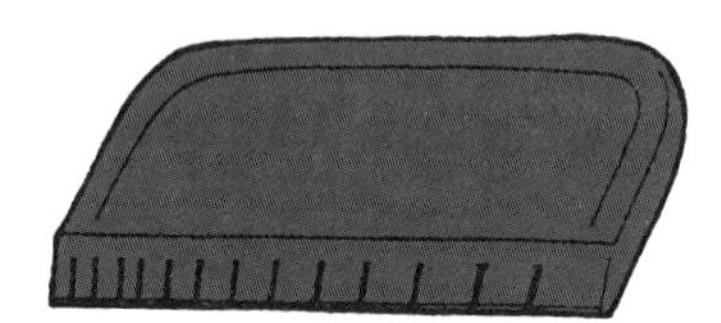

**Fig. 1.85** Rubber combs

***Use*** (i) To simulate the coarse straight grain patterns of oak, pine and similar woods. Used covered in clean linen rag, and dragged through wet graining colour. Finer grades used for 'cross combing' to break up the coarse straight grain.
(ii) To produce the decorative texture 'combing' by dragging through wet scumble applied to walls or panels.

### Rubber comb (Fig. 1.85)

Flexible rubber comb with short, vee-shaped teeth. Also available with square cut teeth.

***Size*** Approximately 75 mm wide. Available with fine, medium and coarse graded teeth. Triangular

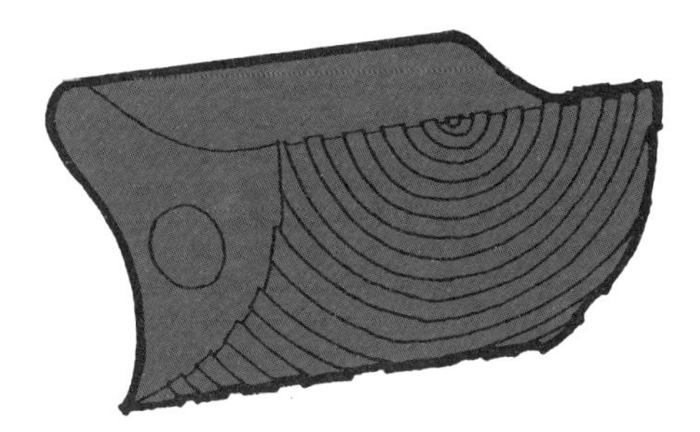

**Fig. 1.86** Heart grain simulator

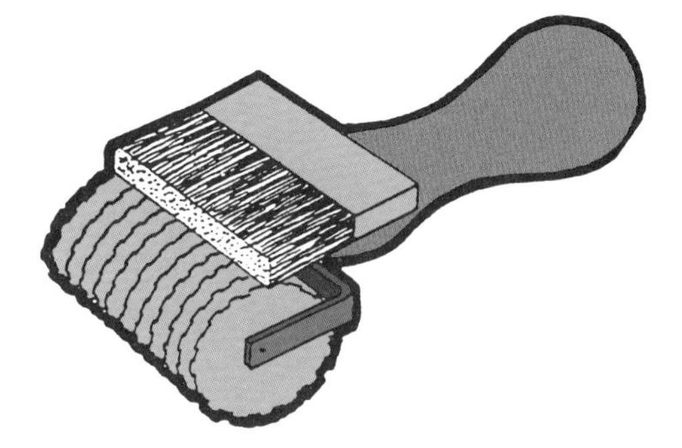

**Fig. 1.87** Check roller

varieties contain all three grades of teeth arranged on the three edges. Also available with teeth graduated from fine to coarse on one edge.

***Use*** (i) To simulate the coarse straight grain patterns of oak, pine and similar woods.
(ii) To imitate combing similar to steel combs.

### Heart grain simulator (Fig. 1.86)

A moulded rubber appliance with a heartwood pattern engraved on the curved working face, and a flat toothed comb on the edges.

***Size*** 75 mm and 125 mm wide, with fine, medium and coarse grain patterns.

***Use*** For the fast and repetitive simulation of heartwood patterns of oak, pine and similar woods. The curved face is dragged with a rocking motion through wet graining colour. The comb is used to simulate the straight grain patterns at each side.

### Check roller (Fig. 1.87)

A roller made from a series of loose-fitting serrated metal discs. Sometimes equipped with a special clip-on mottler to feed graining colour to the discs, and to separate them in use.

***Size*** 50 mm and 75 mm wide.

***Use*** As a printing roller to simulate the dark, broken pore markings of oak and similar woods.

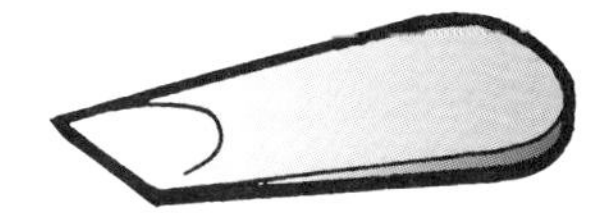

**Fig. 1.88** Veining horn or thumb piece

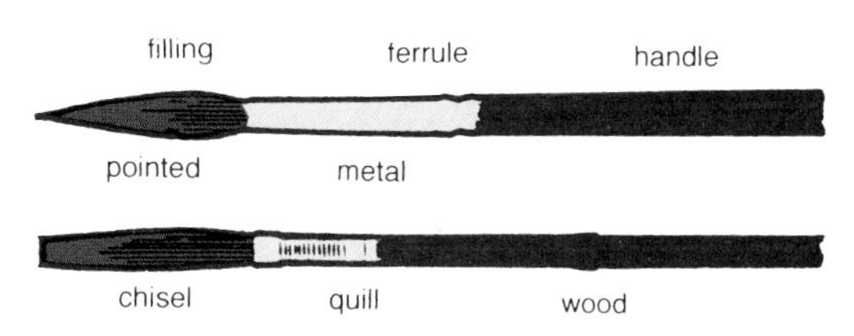


**Fig. 1.89** Signwriting brushes

### Veining horn or thumb piece (Fig. 1.88)

A flexible strip of plastic, or any similar material shaped with one square end, and the other end rounded.

***Size*** Approximately 100 mm long.

***Use*** (i) To simulate the grain patterns of quartered oak and heartwood of oak. Used with clean linen rag to 'wipe out' the patterns from wet graining colour.
(ii) For simulating the veins of some marbles.

### Feathers

Swan, goose or other large, fairly rigid feathers.

***Use*** (i) Applying marbling colour when producing either background textures or veins.
(ii) Producing wavy or curly grain markings of decorative timbers such as walnut.

## 1.1.10 Signwriting and decorative painting tools and brushes

### Signwriting brush (Fig. 1.89)

Also called writers, pencil brushes, or sables. Long-handled brushes with sable, 'sabox' or nylon fillings. They can be obtained through

**Table 1.4** Signwriting brush sizes

| *Ferrule* | *1* | *2* | *3* | *4* | *6* | *8* | *12* |
|---|---|---|---|---|---|---|---|
| Quill | Crow | Small duck | Duck | Large duck | Goose | Ex small swan | Condor |
| Diameter in mm | 1.6 | 1.9 | 2.2 | 2.6 | 3.6 | 4.9 | 7.8 |

**Fig. 1.90** Lining fitch

**Fig. 1.91** Coach liner

**Fig. 1.92** Sword striper

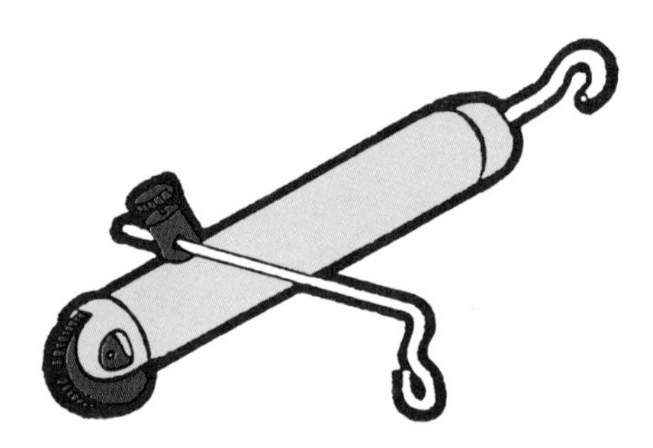

**Fig. 1.93** Lining wheel

***Metal ferrule*** The most common form. The filling is enclosed within and attached to the handle by a cylindrical, seamless, non-rusting metal casing.

***Quill*** Filling enclosed in the base of a bird's feather which is shrunk upon the handle.

***Chisel end*** The tips of the filling finish to a flat end. Suitable for all types of signwriting.

***Pointed end*** The tips of the filling finish to a point. Suitable for fine painting and lettering.

***Size*** Metal ferrule brushes are available in sizes dependent on diameter and 'length out'. Quills are available as the bird's name which relates to the feather size from which they are made (see Table 1.4)

***Use*** (i) Signwriting and decorative painting.
(ii) Veining when marbling.
(iii) Heartwood painting when graining.

## Lining fitch (Fig. 1.90)

Bristles set in a flat metal ferrule with a chisel edge cut at an angle.

***Size*** 6 mm, 12 mm, 18 mm, 25 mm, 31 mm and 37 mm wide.

***Use*** Against a straight-edge, to paint straight lines.

## Coach liner (Fig. 1.91)

Long-haired brushes, usually with ox hair fillings. Available in quill or ferrule.

***Size*** Similar sizes to signwriting brushes.

***Use*** Running thin painted lines.

## Sword striper (Fig. 1.92)

Long, dagger-shaped squirrel hair fillings, either in a metal ferrule or bound to the handle with copper wire.

***Size*** In widths from 8 mm to 13 mm.

***Use*** Running thin painted lines.

## Lining wheel (Fig. 1.93)

The cylindrical body is metal. Paint is poured in at one end when the plunger assembly is removed. The serrated wheel assembly controls the line thickness and is available in a number of wheel widths.

***Size*** Wheel sizes vary from 2 mm to 6 mm wide, and available in double wheel heads.

***Use*** To run thin painted lines, either against a straight-edge, or controlled by the guide which can be adjusted to slide against the edge of a board or panel. Can also be used to run curved lines.

## One-stroke brush (Fig. 1.94)

Metal ferruled brushes which have been squeezed flat to produce a spade-like filling. Commonly available with ox or 'sabox' fillings, but very expensive brushes are available with sable fillings.

**Fig. 1.94** One-stroke brush

**Fig. 1.95** Stencil brush

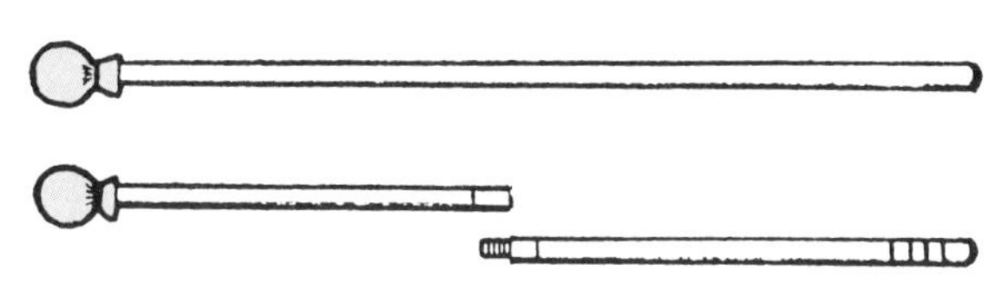

**Fig. 1.96** Mahl stick

***Size*** In widths from 4 mm to 25 mm.

***Use*** (i) Painting types of script letters in one stroke, similar to that obtained with a broad nibbed pen.

(ii) Filling-in large letters or large areas of decorative painting.

(iii) Painting-in heartwood grain when imitating oak, pine and similar woods.

### Stencil brush (Fig. 1.95)

The filling is of short stiff bristles.

***Size*** Various sizes from 10 mm to 38 mm diameter.

***Use*** To obtain decorative effects or lettering by applying paint through a cut-out stencil, or around a template.

### Mahl stick (or rest stick) (Fig. 1.96)

A stiff rod of timber or metal padded at one end and usually covered in chamois leather. Can be in one length, or in two or three lengths which screw or push together.

***Size*** Approximately 500 mm long.

***Use*** To steady the hand when signwriting or decorative painting.

### Palette board or thumb board (Fig. 1.97)

A rectangular timber or plastic board which fits over the thumb of the non-brush hand.

***Size*** Approximately 230 mm × 160 mm.

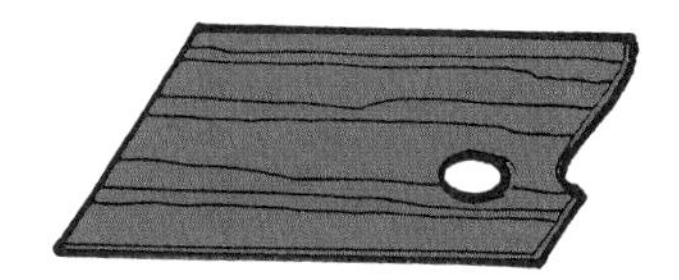

**Fig. 1.97** Palette or thumb board

**Fig. 1.98** Dipper

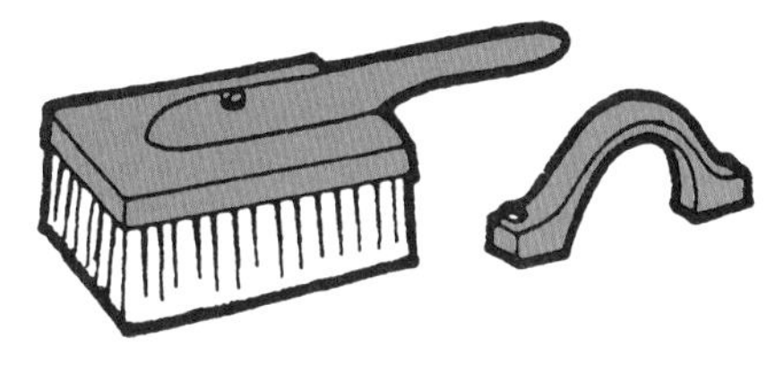

**Fig. 1.99** Stippling brush

***Use*** To hold dippers of paint and upon which to shape the brush when signwriting or decorative painting.

### Dipper (Fig. 1.98)

Small cylindrical metal containers which clip onto the palette board.

***Size*** Various, but most commonly used are 40 mm diameter × 40 mm high.

***Use*** To hold paint and/or thinners when signwriting or decorative painting.

### Stippling brush (Fig. 1.99)

Bristles set in small tufts in a flat stock. The tips of all the bristles are level. The handle may be straight or arched.

***Size*** Various sizes from 100 mm × 75 mm to 200 mm × 150 mm.

***Use*** (i) To remove brush marks from wet scumble to produce a 'stippled' finish.

(ii) To 'blend' wet scumble.

### Two-row stippler (Fig. 1.100)

A narrow version of a stippling brush consisting of only two rows of bristle tufts.

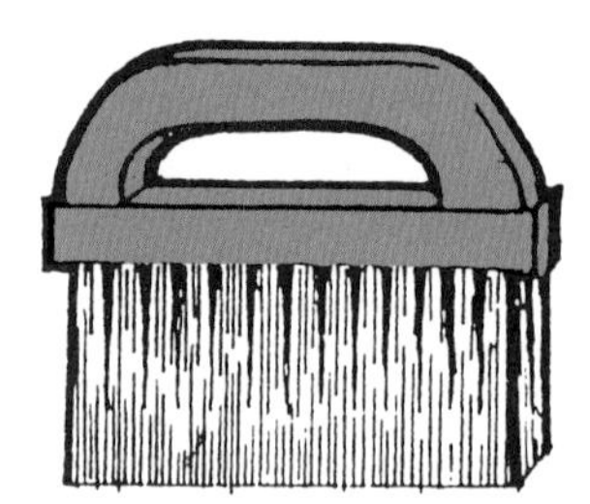

**Fig. 1.100** Two-row stippler

***Size*** 100 mm × 25 mm and 150 mm × 25 mm.
***Use*** In association with a stippling brush for stippling wet scumble in angles, and on narrow features such as mouldings and edges.

### Care and maintenance of signwriting brushes

***Cleaning brushes after use*** Must be carried out immediately after use, otherwise paint that is allowed to harden at the root of the brush may never be removed completely without seriously damaging the hairs. Hardened paint will destroy the shape of the brush. The brush must be rinsed out in the thinners of the paint being used, e.g. white spirit for oil paints, and water for emulsions. When all traces of paint have been removed, the hair should be dried with clean rag to remove the thinners.

The final process is to shape the brush with grease (vaseline). This is done by taking any non-setting grease or oil between the thumb and forefinger, pushing it deep into the root of the filling, and pulling it up to the tip leaving the brush in its natural shape. The greasing protects the filling from damage while being carried around, as well as ensuring that any slight traces of paint which may have been left in the filling will not harden. The grease must be thoroughly washed out with white spirit before re-use.

The use of hot water or strong soaps and detergents to wash soft hairs may result in the hairs losing their springiness.

***Storage*** Lay flat in a tray, drawer or brush box, or bind them to a stiff card using elastic bands. New brushes should be stored with an insecticide to protect them from moth attack which will either destroy the filling completely, or eat away the fine tip rendering the brush useless.

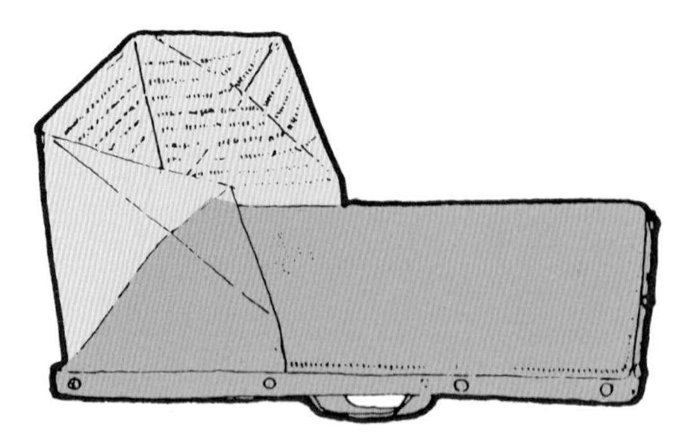

**Fig. 1.101** Gilders' cushion

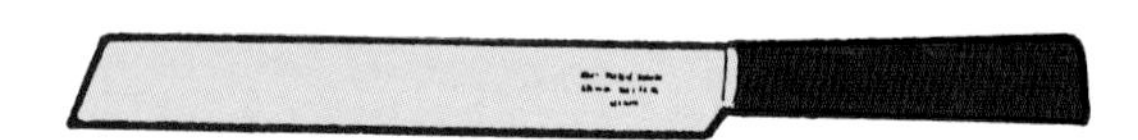

**Fig. 1.102** Gilders' knife

## 1.1.11 Gilding tools and brushes

### Gilders' cushion (Fig. 1.101)

A flat board padded with felt and covered with chamois leather or goat skin. One end has a folded parchment shield to protect the gold leaf from draughts. Underneath the board are two loops, one for the thumb to hold the cushion secure, and the other to sheath the gilders' knife.

***Size*** Approximately 200 mm × 125 mm.
***Use*** Upon which to cut loose gold leaf and other metallic leaves into convenient sizes when gilding relief surfaces.

### Gilders' knife (Fig. 1.102)

Long, flat, bladed knife which is well ground but not sharp.

***Size*** Usually a standard size of 125 mm blade length.
***Use*** To cut loose gold leaf and other metallic leaves on a gilders' cushion.

### Gilders' tip (Fig. 1.103)

Very thin row of squirrel or badger hair set between two pieces of card.

***Size*** Usually 85 mm wide. Available in three lengths of hair: short up to 40 mm; medium up to 50 mm; and long about 60 mm 'length out'.

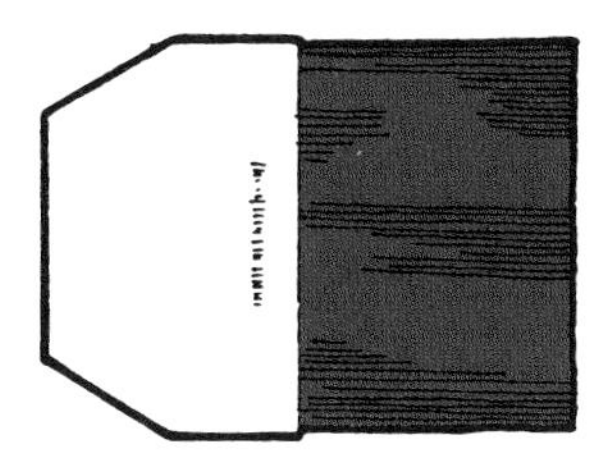

**Fig. 1.103** Gilders' tip

***Use*** There are two schools of thought, when the filling is rubbed against the gilder's hair or cheek, the hairs become sufficiently charged with static electricity to attract the gold leaf from the cushion so that it may be applied to the sized areas, whilst others claim it to be natural body oils coming off the skin.

### Gilders' mop (Fig. 1.104)

Domed shape squirrel hair with filling set in either a metal ferrule or quill.

***Size*** A range of diameters from 6 mm to 25 mm.

***Use*** (i) To push applied gold leaf into crevices when gilding relief surfaces.
(ii) To 'skew' off surplus gold from relief surface.

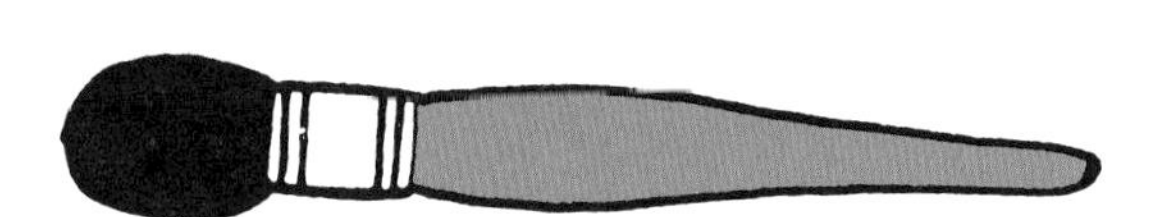

**Fig. 1.104** Gilders' mop

## 1.1.12 Texture-painting tools

### Mixing tool or bumper (Fig. 1.105)

A round plate made from rigid plastic or aluminium with holes near the rim. Takes a standard broom handle.

***Use*** To ensure speedy and thorough mixing of texture paints by plunging up and down during the mixing operation.

### Rubber stippler (Fig. 1.106)

*Type A* – Fine flexible rubber filling set into a flat base with an arched wooden or plastic handle.

***Size*** 150 mm × 100 mm and 200 mm × 150 mm. A very large type measuring 330 mm × 330 mm is available, which can be fitted with an extension handle for ceiling work.

*Type B* – A more rigid stippler completely moulded in rubber with thicker filaments.

***Size*** 150 mm × 60 mm.

***Use*** To produce a range of stippled or swirl effects in texture paint.

### Comb (Fig. 1.107)

Clear plastic flexible blade with 'V' grooves or curves along the edge. A shorter blade is set alongside to ensure even pressure when in use.

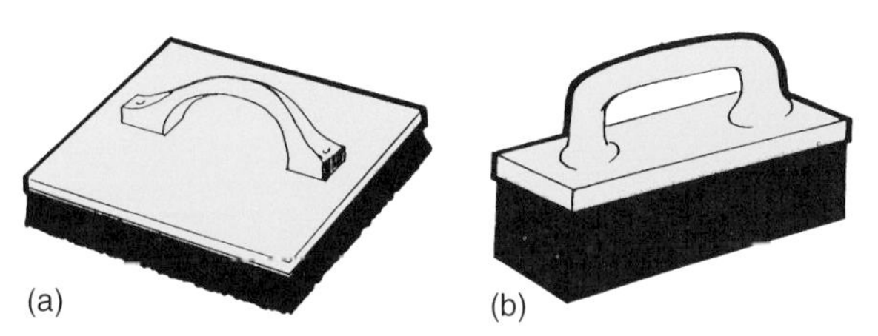

**Fig. 1.106** (a) Rubber stippler Type A; (b) Rubber stippler Type B

**Fig. 1.105** Mixing tool or bumper

**Fig. 1.107** Comb

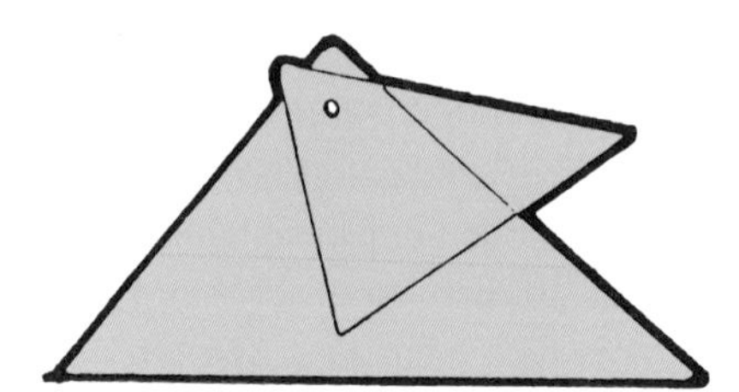

**Fig. 1.108** Lacer

***Size*** 90 mm, 180 mm, 250 mm and 300 mm wide.

***Use*** To produce a range of combed patterns in texture paint.

**Lacer (Fig. 1.108)**

A triangular clear plastic blade with rounded corners, and a smaller blade attached to one corner.

***Use*** Used wet and drawn lightly over partially set texture paint to remove high points and sharp edges or spikes.

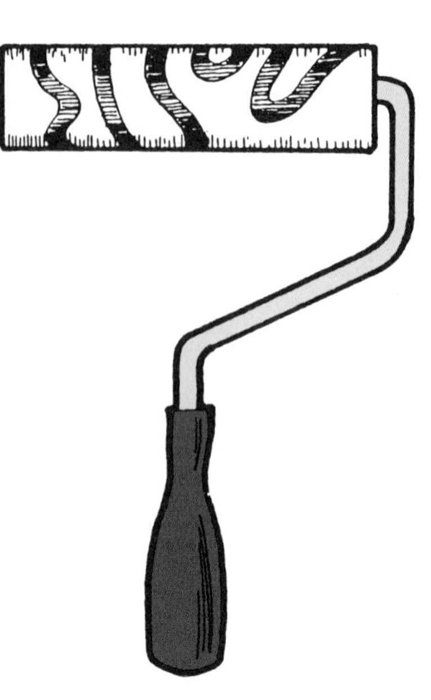

**Fig. 1.109** Textured roller

**Textured roller (Fig. 1.109)**

Synthetic foam rollers with sculptured patterns.

***Size*** Similar to paint rollers approximately 200 mm long.

***Use***
(i) Used over wet texture paint to give a variety of relief textures.
(ii) A smooth plastic type gives a timber bark effect and can also be used as a lacer.

# 1.2 Stripping and cleaning

## 1.2.1 The steam stripper (Fig. 1.110)

There are two types:

(a) heated by liquefied gas (LPG); large, heavy piece of equipment now rarely used, being superseded by (b).
(b) electrically heated; small, compact, easily moveable unit.

Steam strippers consist of a water tank attached to an enclosed burner or heating element. The water is heated and converted into steam which, by means of a flexible hose, is carried to a perforated plate. When the plate is laid against the wall, steam penetrates deeply into the wall covering, softening the paper and adhesive. Paper can then be easily scraped from the surface, leaving it smooth, clean and free from bacteria.

### Use

(i) To assist in the removal of any type of surface covering from ceiling and walls. Particularly useful for materials that are difficult to remove by normal soaking, e.g. Lincrusta, washable, multi-layered papers and coated papers.
(ii) To assist in the removal of water paints, emulsions and some plastic paints.

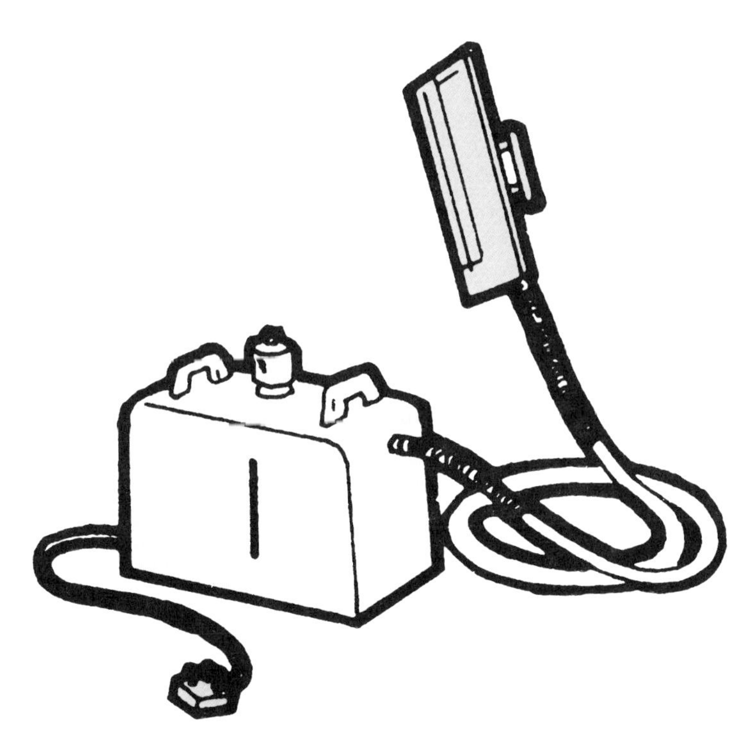

**Fig. 1.110** Electric steam stripper

**Note** (i) If the concentrator is left in one place too long, steam may cause damage to the substrate, e.g. blistering of the top layer of plaster.

(ii) Not recommended for use on old or friable plaster surfaces as the steam penetration will damage the plaster.

### Safety precautions

(i) Never let the water level drop too low.

(ii) Do not use kinked or tightly coiled hose which will block the flow of steam.

## 1.2.2 Flame-cleaning (Fig. 1.111)

Flame-cleaning is a very effective method of removing rust from iron and steel. It will remove millscale also, provided it is not tightly adhering. A flame of extremely high temperature (up to 3000°C) is passed over the surface of the metal. Its effectiveness is due to several factors:

(i) *Differential expansion* When the intense localised heat is applied to partially weathered millscale, the scale, being heated first, expands faster than the base metal and flakes off.

(ii) *Dehydration* Rust is dehydrated by the intense heat. All the moisture is driven off rapidly, and the rust is converted to dry powder which is easily removed by wire brush.

(iii) *Heat penetration* The heat of the flame penetrates the smallest cracks and pits in the metal and dries out all traces of moisture.

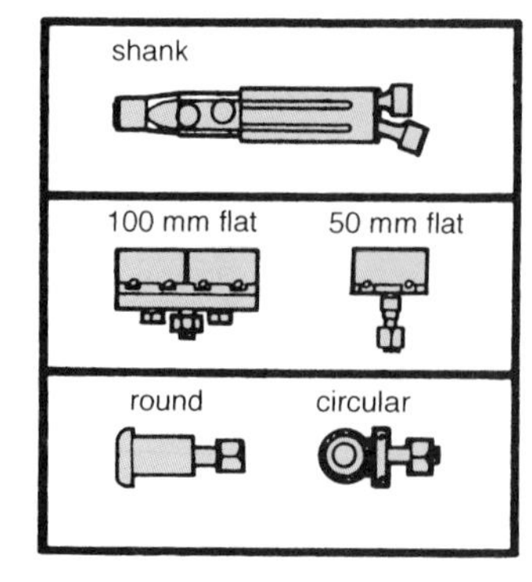


**Fig. 1.111** Flame-cleaning equipment

### Equipment

Oxygen and acetylene gases are used together and burnt in about equal proportions. It is recommended that three cylinders of each gas are coupled together to reduce handling. By using a special propane mixer, propane gas can also be used.

Several different types of multi-jet nozzle are available: round, circular, semi-circular and flat. The flat heads are available from 50 to 200mm wide. This range allows the equipment to be used around bolts, corners and recesses as well as on flat areas.

### Procedure

Normally three men work as a team: *Operative no. 1* flame-cleans the surface, which should appear light grey all over when finished; *Operative no. 2* wire-brushes the surface to move all the dry rust powder; *Operative no. 3* primes the metal while it is still warm (38°C, or the temperature at which the hand can be held comfortably on the steel). The warmth of the plate tends to lower the viscosity of the paint, allowing it to penetrate more easily into small surface irregularities so that an excellent bond is formed between primer and metal.

### Use

The equipment is portable, therefore suitable for use on site – e.g. on bridges, railings, structural steel, gas holders, ships' hulls – where rust and millscale have to be removed.

### Disadvantages

(i) The heat may buckle and distort steel plate less than 6 mm thick. Specialist advice must be obtained if working on thin plate.
(ii) When removing heavy millscale, it is often more economical to chip the scale off before flame-cleaning.
(iii) It may not provide a sufficiently high standard of preparation for special coatings such as two-pack epoxies.

### Safety precautions

(i) All safety precautions necessary when using gases must be observed (see 1.4.3).
(ii) Leather gauntlets and tinted goggles should always be worn.
(iii) A face mask is recommended for protection against fumes when old paint is being removed.

## 1.2.3 Blast cleaning

Blast cleaning, abrasive cleaning, shot-blasting, grit-blasting and abrasive blasting are all terms used to describe a very efficient operation whereby compressed air is used to shoot or blast various types of abrasive on to a surface at very high speeds (up to 450 mph or 725 km/h). The force of impact and the hardness of the abrasive pulverise the rust and roughen the surface of the material.

### Use

(i) To remove rust and millscale from iron and steel.
(ii) To clean surfaces, e.g. stone, concrete, alloy.
(iii) To obtain decorative effects on such materials as stone, concrete and glass.

### Equipment

Two main types of equipment are available, with variations of each produced by different manufacturers.

(a) *Open blasting*
Used for approximately 90–95 per cent of all large-scale cleaning and surface preparation work.
*Unit* (Fig. 1.112) The abrasive is loaded into a hopper and falls into a pressurised pot. It is then fed through a metering valve into the compressed-air stream through the hose to the nozzle.

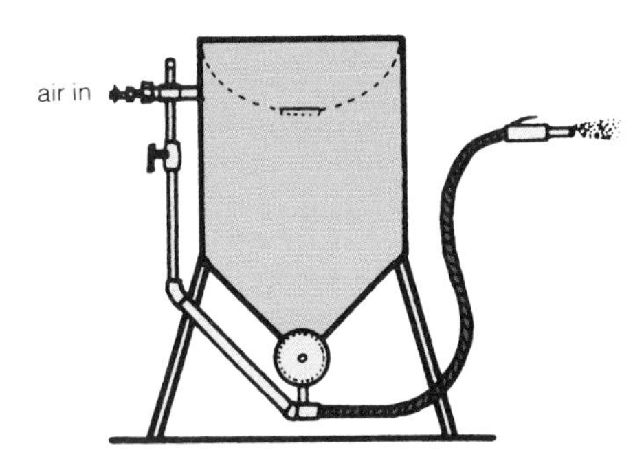


**Fig. 1.112** Open blasting equipment

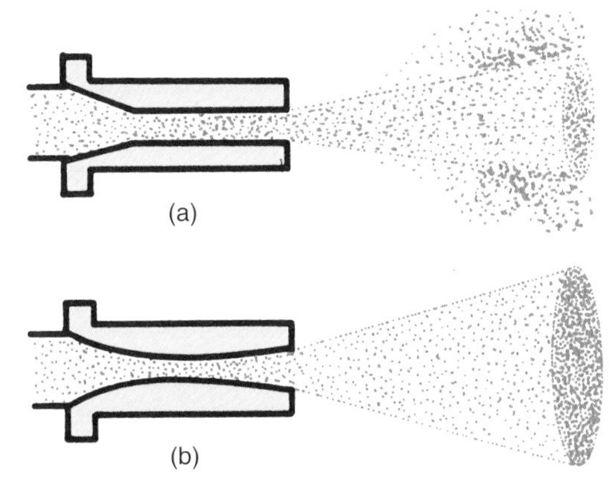


**Fig. 1.113** Nozzle

The equipment is available in various sizes, the choice being mainly dependent on size and type of job.

*Nozzle* (Fig. 1.113) The most important part of the equipment. Two types are available:

(i) Straight cylinder nozzle. Not very efficient and the cleaning pattern tends to be uneven.
(ii) Nozzle with a narrow straight section or throat inside the cylinder, known as a 'venturi tube'. This is much more efficient: velocity increases up to 450 mph (725 km/h) as the abrasive passes through the narrower section; the cleaning pattern is also very even.

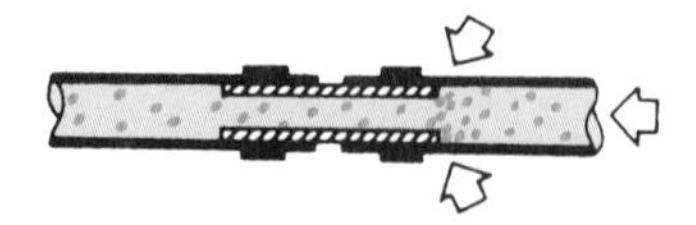

Fig. 1.114 Hose couplings

The inner lining is of tungsten carbide or a special alloy to withstand the severe abrasive action.

Available in various lengths and orifice sizes.

*Hose* The best hose is made of reinforced rubber containing carbon black for protection against static electricity.

*Hose couplings* (Fig. 1.114) Hose couplings which fit internally are not recommended as they restrict the flow of abrasive. Special external couplings are made to allow a straight-through passage for the abrasive.

*Air supply* The force of the abrasive blast is created by compressed air. The higher the air pressure, the greater the force of impact, which in turn produces a better finish. The ideal pressure is 100 psi (6.5 bar) *at the nozzle*. Less pressure than this means slower working and a poorer result.

Air consumption is high. It varies according to the type of equipment, nozzle and air pressure used, but is usually between 80 cfm (for small units) and 330 cfm (for larger plants) (37.8 and 156 litre/s). Quite large compressors are therefore required for this work (see 1.3.3 and 1.4.1).

(b) *Vacuum reclaim process* (Fig. 1.115)

The abrasive is fed from a storage tank to the blasting head, where it cleans the surface and is sucked back by vacuum into the plant. It is then screened and recirculated all in one operation. This is a relatively slow process, used mainly for small work and where cleanliness is of the utmost importance.

Special equipment is available, which combines water or steam with the abrasive to reduce dust. This is effective, but slow and messy.

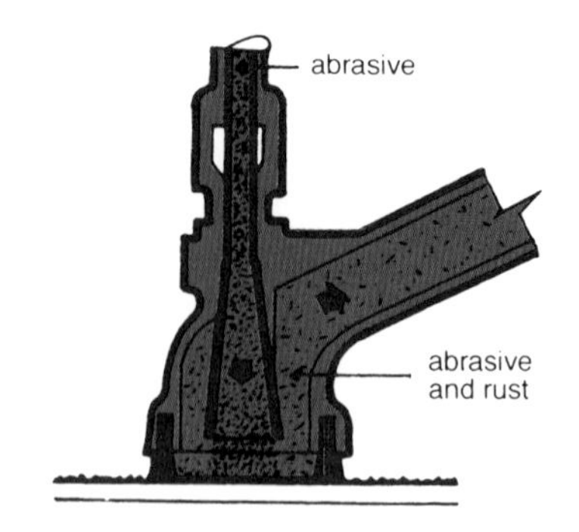


Fig. 1.115 Vacuum reclaim process

## Abrasives

### *Choice of abrasive*

It is very important to choose the correct type, size and shape of abrasive. The choice will depend on the job, and in each case, the following points should be considered:

(i) Type of surface to be cleaned, i.e. metal or stone.
(ii) What has to be removed – heavy millscale, rust, old paint, etc.
(iii) Type of paint system to be used.
(iv) Standards of finish required (see below, Standards).
(v) Cost – whether the abrasive used is or can be reclaimed.

### *Types of abrasive*

*Re-usable abrasives* Where it is possible and economical to reclaim and re-use abrasives (e.g. blast rooms, new steel storage tanks, vacuum units), mineral abrasives are used. These can often be re-used up to 10 times, depending on type. They include:

*cast steel* – angular grit and round shot
*malleable and chilled iron* – angular grit and round shot
*aluminium oxide*
*aluminium*
*brass* } limited specialised use
*copper* }
*glass*

*Expendable abrasives* On most site work, it is difficult and expensive to recover the abrasives; cheaper, expendable types are therefore used once only. Mineral slags, produced in the refining of iron and

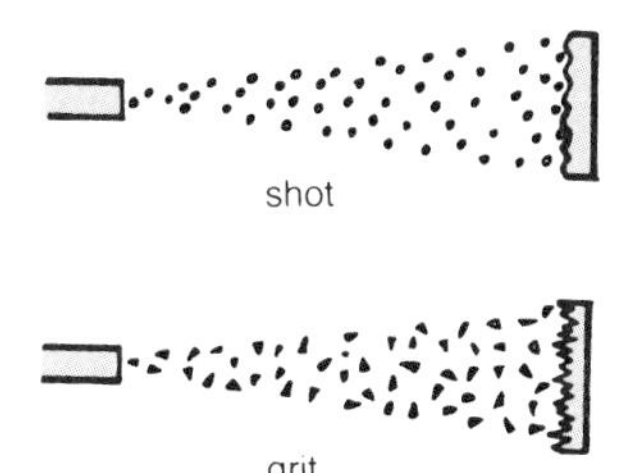


**Fig. 1.116** Abrasive particles

copper, are mainly used. They should be hard and angular to give good cutting power.

***Size of abrasive particles***

Abrasives are available in a wide range of sizes, graded through a British Standard mesh and specified in BS 7079:1990. Selecting the correct size of abrasive for a particular job is very important. Large-particle grit or shot will hammer a surface harder and loosen scale faster than will small particles, but the resulting rough, coarse finish will not usually be suitable for painting. Small-particle abrasives, having more pellets to the kilogram, will produce a finer finish much more quickly, but will have less power of impact. After considering all these factors, for most work it is best to use the smallest size abrasive that will give the desired result for the work in hand.

***Shapes of abrasive particle* (Fig. 1.116)**

Abrasives are obtainable in two shapes:

***Shot*** Small round particles of abrasive which act as tiny hammers on the surface, producing a round profile.

***Grit*** Small angular chips which act as tiny chisels on the surface, producing a sharp angular profile.

## Standards

The standard to which a surface is prepared usually depends on all or some of the following factors:

(i) The type of paint system to be used.
(ii) The state of the metal before treatment.
(iii) The conditions to which the finished job will be subjected.
(iv) *Cost* – the higher the standard, the higher the cost.

Standards for blast cleaning are specified by the following bodies, of which the first two are internationally recognised:

**Table 1.5** Grades of blast cleaning

| *BS 7079* | *SSPC (USA)* | *Swedish* | *Description* |
|---|---|---|---|
| First quality | White metal | SA3 | Perfectly white metal surface all over |
| Second quality | Near white metal | SA2.5 | The whole surface shall show blast cleaning pattern and at least 95% of the surface shall be clean bare steel |
| Third quality | Commercial | SA2 | The whole surface shall show blast cleaning pattern and at least 80% of the surface shall be clean bare steel |

*Note* A fourth, 'Brush off' finish, is sometimes used, but as this produces a much lower standard of preparation, it is not recommended or encouraged as suitable for painting.

The Swedish Standards Organisation (SIS 05 59 00)
The Steel Structures Painting Council, USA
The British Standards Institution (BS 7079:1990)

*Note* The Swedish standards are set by comparing the surface with a set of expertly prepared coloured photographs. The others are only verbal descriptions.

Three grades are commonly used and can be quoted with approximate equivalents in each standard (Table 1.5).

## Profile (Fig. 1.117)

Some surface roughness is a great advantage in providing a physical key for the primer to adhere to; however, the degree of roughness should not be excessive, or peaks of steel may project through the paint film and rust very quickly. As a general guide, the roughness known as 'profile' or 'anchor pattern' should not exceed 50 microns.

## Painting procedure

The surface should be painted as soon as possible after cleaning – never more than 4 hours later, or

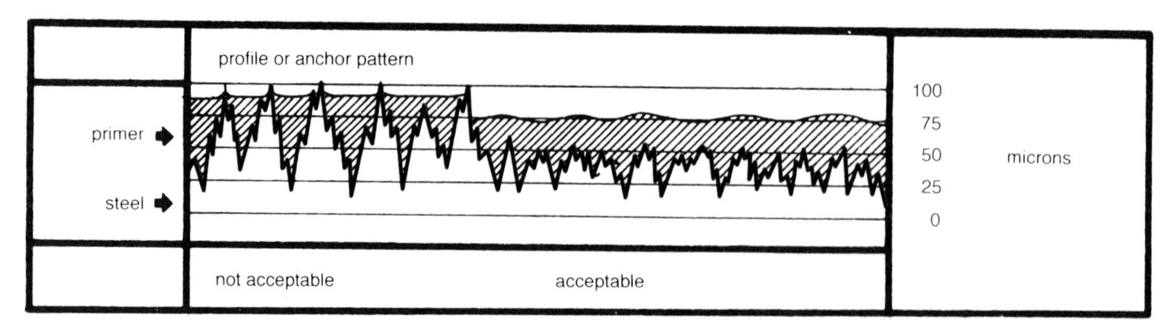


**Fig. 1.117** Profile

further corrosion will start before the surface is primed. (The time interval is dependent on weather and site conditions.)

## Safety precautions

Abrasive cleaning can be dangerous if not carried out correctly. Used correctly, it is perfectly safe.

(i) Always observe the correct working procedure which has been carefully drawn up by the manufacturer of the equipment used.

(ii) Always wear the recommended safety equipment:
Visor or blasting helmet to protect eyes and face. Usually equipped with breathing apparatus.
Abrasive-resistant gloves.
Industrial safety boots.
Rubber or leather apron.

# 1.3 Spraying

## 1.3.1 The spray gun (Fig. 1.118)

The set-up used on a spray gun depends on a number of factors, the first being the way in which the fluid and air leave the gun. Two types of gun are used for general spraying.

### (a) Bleeder gun

A gun without an air valve: the air flows continuously through the gun and cannot be cut off. These guns are used with small compressors which have no air receiver or pressure-controlling device. If the air were not allowed to bleed out, air pressure would build up in the hose and might damage the compressor. The trigger controls only the flow of paint.

### (b) Non-bleeder gun

A gun fitted with an air valve which shuts off the air when the trigger is released. The trigger therefore controls both the air and the flow of paint. Most spray guns are of this type.

### The gun set-up

Although the gun set-up is usually considered to have three units, it can be more easily explained as having two:

(i) The fluid-tip and needle, which must go together and be a perfect fit.
(ii) The air cap or air nozzle – usually considered as a set with the fluid-tip and needle, but some types can be interchanged.

### (i) The fluid-tip and needle

These control the quantity of paint going into the air stream; the size used depends on:

(i) Type and viscosity of paint being sprayed.
(ii) Volume of air available.
(iii) Speed of application required.
(iv) Standard of finish required, fine or coarse atomisation.
(v) Type of air cap being used.

Table 1.6 provides an approximate guide to the range of orifice sizes for use with various types of paint.

### (ii) Air nozzle or air cap

This is the most important part of the spray gun. It directs the air into the paint stream to break it up into a fine vapour (atomisation) and to form the various spray patterns. Choice of type depends on:

(i) Volume of air available.
(ii) Type and viscosity of paint used.
(iii) Type of material container used.
(iv) Size of fluid-tip and needle used.
(v) Size and type of object to be sprayed, e.g. large wall area, small intricate work such as ornamental ironwork.
(vi) Standard of finish required – fine or coarse atomisation.

Two basic types of air cap are available, the internal-mix and external-mix.

***The internal-mix air cap*** **(Fig. 1.119)** The air and fluid mix in a cavity inside the air cap before being released into the atmosphere. Used mainly with low-pressure equipment.

Two types are available (Fig. 1.120):

(a) The paint sprays from a single slot in a fixed fan pattern.
(b) The paint sprays from a single round hole in a round pattern.

***Use*** For work with compressors using $\frac{1}{4}$ or $\frac{1}{2}$ motors producing 2 to 3 cubic feet per minute or cfm (0.95 to 1.3 litre/s) – see 1.4.1.

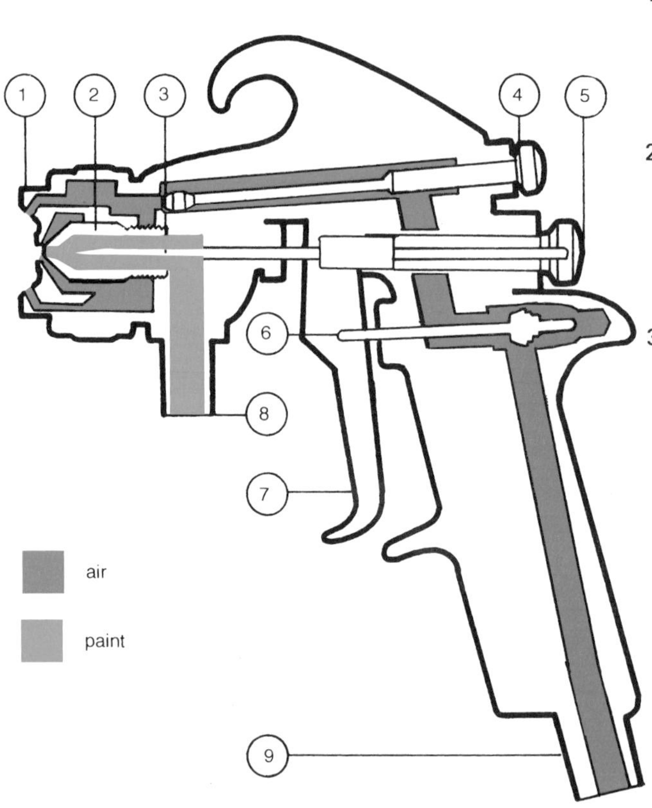


*The spray gun set-up:*
1 *Air cap or air nozzle*
*Directs air into the flow of paint, causing the paint to atomise and form a spray pattern*
2 *Fluid-tip or fluid nozzle*
*Acts as a seating for the needle and controls the amount of paint flowing through*
3 *Fluid needle*
*A spring-loaded needle controlled by the trigger which fits into the fluid-tip and starts and stops the flow of paint*

*Other components:*
4 *Spreader adjustment valve*
*Regulates the amount of air flowing into the air cap, adjusting the fan pattern from a round to a wide fan*
5 *Fluid adjustment valve*
*Controls the adjustment of the fluid needle, regulating the amount of paint flowing through*
6 *Air valve*
7 *Trigger*
*Controls the air inlet valve and the fluid needle*
8 *Material inlet*
*Where the paint enters the gun from either a suction pot or a fluid-hose from a pressure pot*
9 *Air inlet*

**Fig. 1.118** Components of the spray gun

**Table 1.6** Sizes of spray gun orifice

| *Orifice size* | *Use* | *Air consumption (see section 1.4.1)* |
|---|---|---|
| 1.0 to 1.9 mm | Stains, etch primers, cellulose finishes, clear lacquers | 3 to 7 cfm (1.4 to 3.3 litre/s) |
| 1.4 to 2.2 mm | Most types of gloss, undercoat, emulsion and some primers | 5 to 15 cfm (2.3 to 7 litre/s) |
| 2.2 to 2.8 mm | Heavy-bodied paints, some texture paints and heavily pigmented fillers | 7 to 16 cfm (3.3 to 7.5 litre/s) |
| 2.8 to 9.5 mm | Heavy-texture paints, anti-condensation paints and bituminous compounds | 17 to 25 cfm (8 to 11.8 litre/s) |

*Note* The amount of air consumed also depends on the air pressure used.

***Advantages of internal mix***

(i) Requires low volume of air.
(ii) Requires low air pressure.
(iii) Has low overspray.

***Disadvantages of internal mix***

(i) Quick-drying paints tend to clog the exit holes.
(ii) Has coarse atomisation (see 1.4.1).
(iii) No control of the size or shape of the fan pattern.

***The external-mix air cap*** **(Figs. 1.121 and 1.122)** The most commonly used type: the air and paint mix outside the air cap.

There is a central orifice where the paint leaves the gun, surrounded by a circle of air. This effects the first

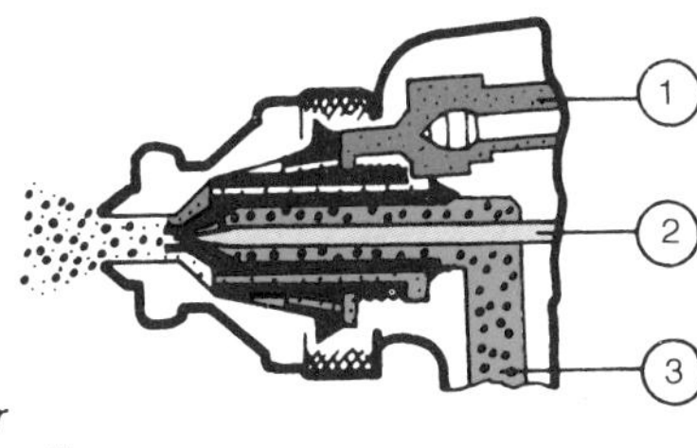


**Fig. 1.119** Internal-mix air cap

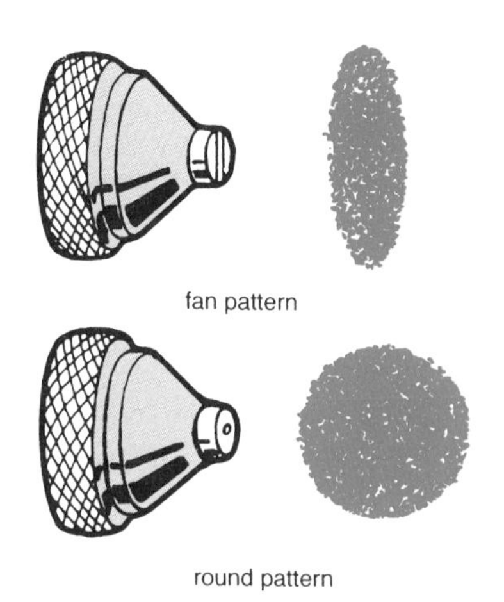


**Fig. 1.120** Types of internal-mix air cap

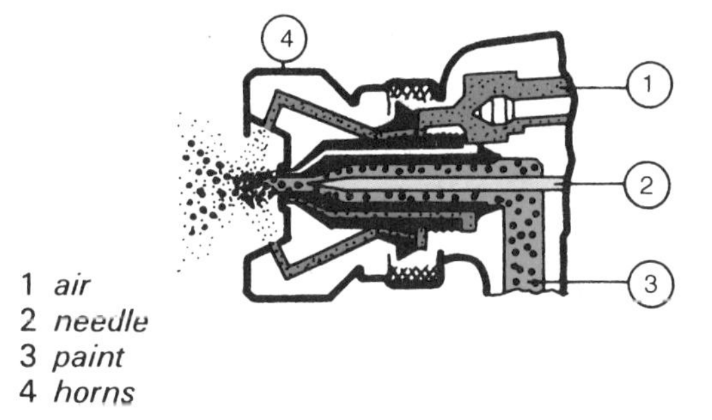


**Fig. 1.121** External-mix air cap

stage of atomisation. Other holes around the central orifice (depending on the type) stop the pattern expanding too fast, help keep the horns clean and provide a further secondary atomisation.

The horns have holes or side port-holes which direct a stream of air from two sides, producing a fan pattern and providing additional atomisation.

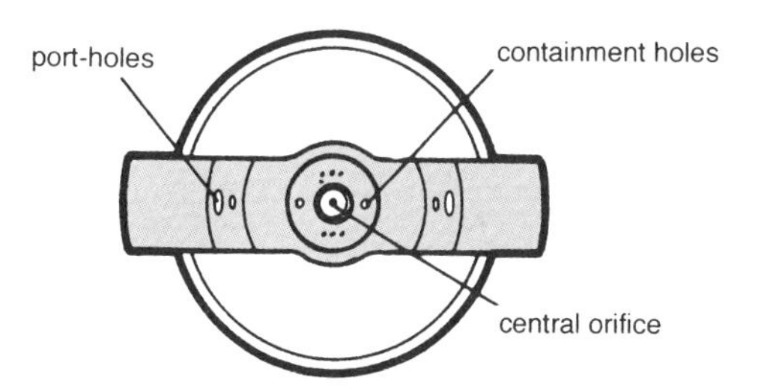


**Fig. 1.122** Multiple-jet air cap

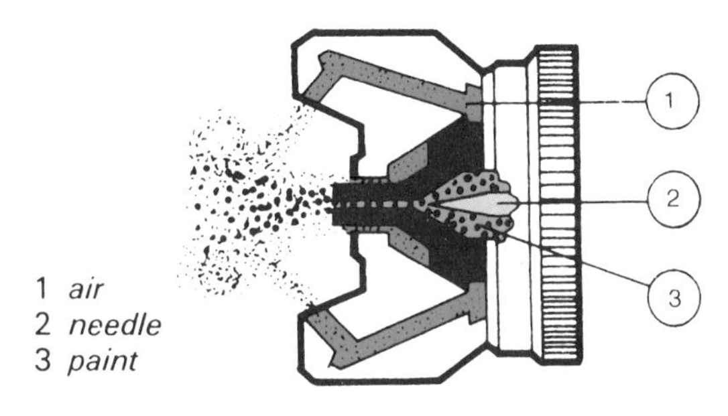


**Fig. 1.123** Syphon-feed air cap

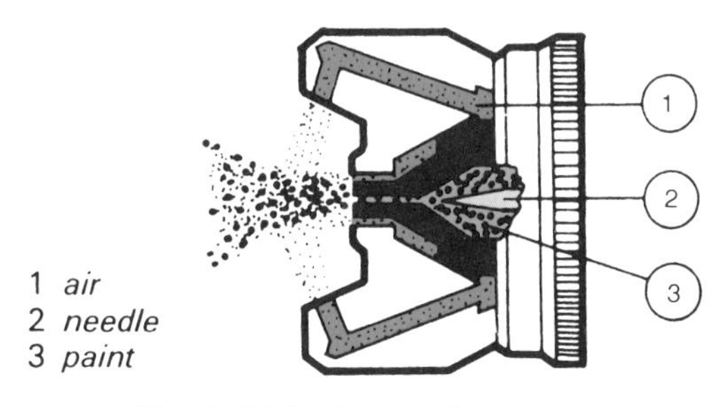


**Fig. 1.124** Pressure-feed air cap

Two types are available:

(a) *Syphon or suction-feed air cap* (Fig. 1.123) The fluid-tip extends beyond the air cap. A partial vacuum is produced as the paint leaves the fluid-tip (see 1.3.2). This type can also be used on gravity and pressure-feed guns.

(b) *Pressure-feed air cap* (Fig. 1.124) The fluid-tip and air cap are flush on this type, as no vacuum is required. The paint is forced to the gun under pressure.

There are between five and fifteen orifices in the air cap. The multiple jet air cap produces better atomisation and a more even fan pattern; good atomisation is

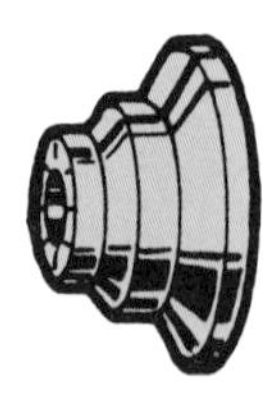

**Fig. 1.125** Veiling

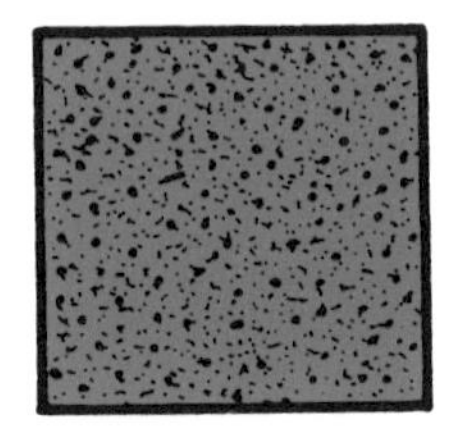
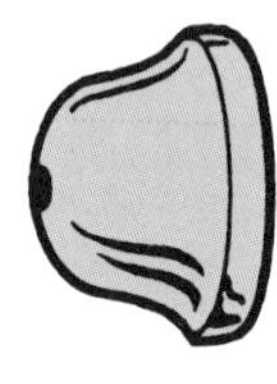

**Fig. 1.126** Spattering

possible with less air pressure, but does require greater volume of air.

***Advantages of external mix***

(i) Very fine atomisation.

(ii) Precise control over spray pattern size.

(iii) Very wide range available.

***Disadvantage of external mix***

Higher volume of air required than with internal mix.

***Air caps for special effects***

Special heads or special set-ups to fit standard guns are available with orifice size and air cap modified to produce decorative effects such as veiling (Fig. 1.125), spattering (Fig. 1.126), flecking and shading or blending.

Although these are used with pressure-feed guns, most special effects require high fluid and low air pressures.

*Note* Spattering effects can be produced with a conventional set-up by increasing the pressure on the fluid and reducing the atomising air pressure. The result is an all-over spot pattern. The particle size may be adjusted by varying the pressures. This type of spraying demands a highly skilled operative to maintain an even, all-over pattern.

## 1.3.2 Containers for spraying

Paint is delivered to a spray gun by one of three main methods:

Gravity feed
Syphon or suction feed
Pressure feed

Many variations of these are available, each equipment manufacturer producing their own shape and size of container with its individual characteristics. The choice of which type to buy or use may depend on several factors.

(i) *Type of job*: large jobs require more paint and large containers need filling less frequently.

(ii) *Speed of application*: large flat surface areas can be sprayed quickly with larger systems whereas the smaller units are quite sufficient for small jobs.

(iii) *Type of compressor available*: the amount of air available will govern the type of equipment used.

(iv) *Type of paint used*: some heavy-bodied paints must be applied under pressure, whereas thinner materials can easily be sprayed using low pressures and low-volume units.

### Gravity-feed container (Fig. 1.127)

A paint container or cup of metal or plastic fitted to the top of the spray gun. Paint is fed into the gun as it falls by the force of gravity.

***Size*** Two cup-sizes are usually available holding either $\frac{1}{4}$ or $\frac{1}{2}$ litre of paint.

***Use*** Can be used for most normal types of paint, but not heavy-bodied material such as plastic texture paints.

The small cup-size and limited angle of working limit its use to staining, preparing patterns and small areas (e.g. radiators and touch-up jobs).

### Syphon- or suction-feed container (Fig. 1.128)

The container or cup is screwed underneath the gun. The air nozzle is designed in such a way (see 1.3.1) that a low vacuum is created in front of the fluid nozzle. This allows atmospheric pressure to force the paint up the siphon tube to the fluid nozzle.

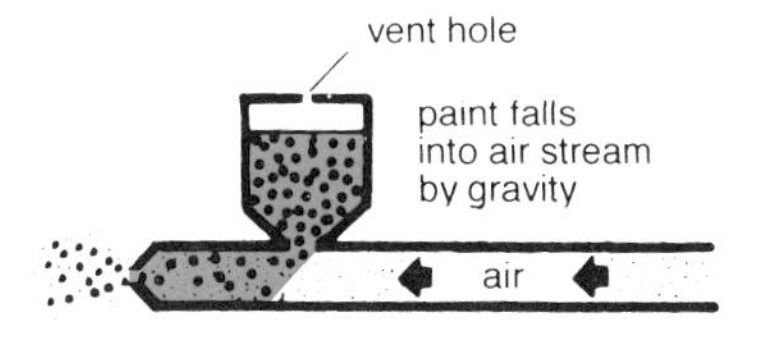


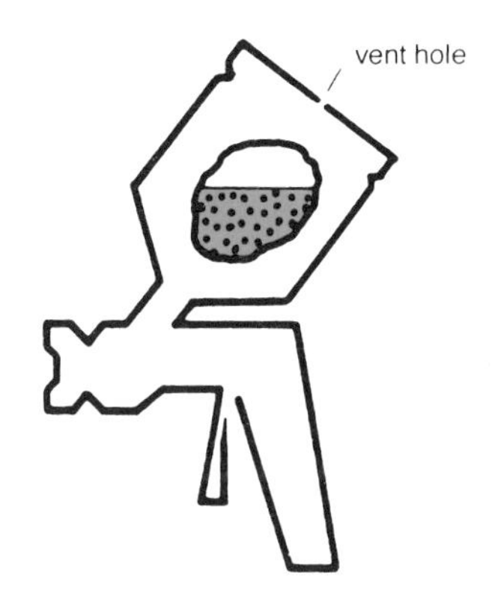


**Fig. 1.127** Gravity-feed container

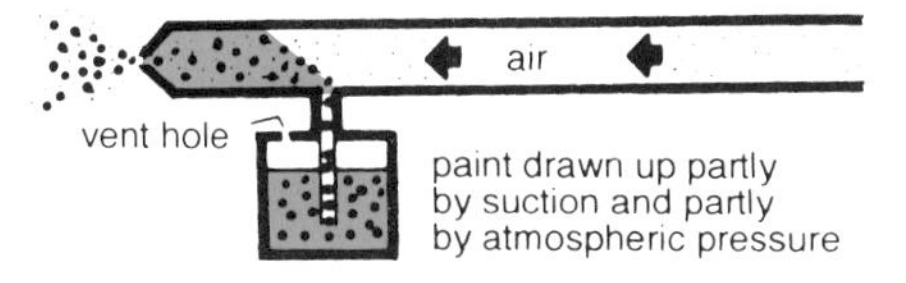


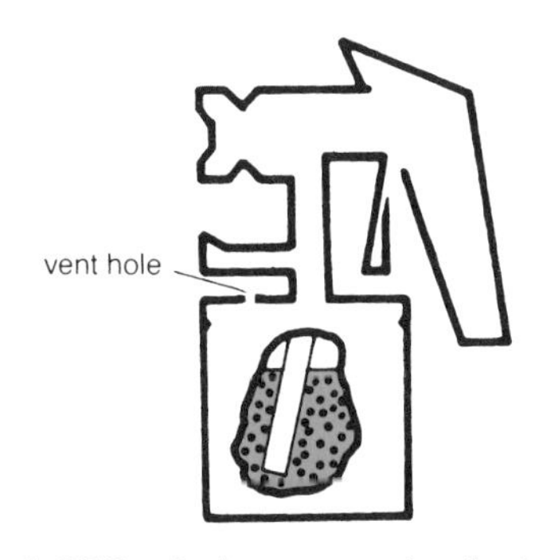


**Fig. 1.128** Syphon- or suction-feed container

***Size*** Cups are available holding either $\frac{1}{2}$ or 1 litre of paint (1 litre is the maximum because of weight).

***Use*** For spraying small quantities of paint or where several colour changes are required. Not suitable for heavy-bodied paints. Many pressure-feed guns can be adapted to take syphon cups.

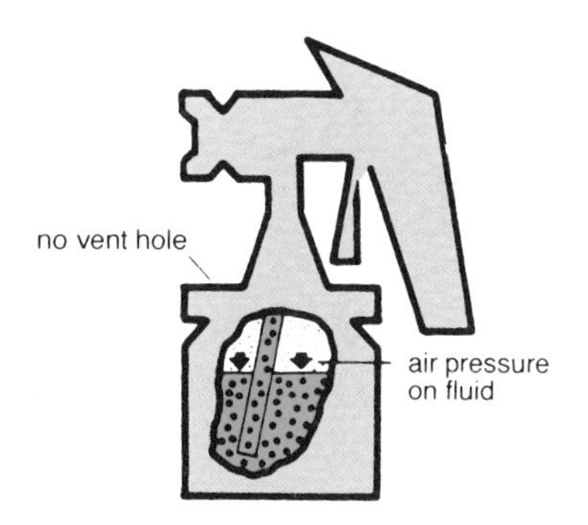


**Fig. 1.129** Pressure-feed cup

### Pressure-feed container

Paint is stored in a closed metal container and forced out to the spray gun by compressed air. The amount of paint fed to the gun is controlled by increasing or decreasing the air pressure in the container. Two main types are available:

Cups – small quantities up to 2 litres.

Tanks – various sizes from 10 to 200 litre capacity.

Being under pressure, they have to be tested and approved for safety. Cups are tested from 40 psi (2.7 bar) to 50 psi (3.4 bar) and tanks to 110 psi (7.5 bar).

***Pressure-feed cup*** **(Fig. 1.129)**

The cup is fitted under the gun as in the syphon-feed container. A small amount of the compressed air going to the bleeder gun is directed into the cup. This forces the paint up the fluid-tube to the fluid nozzle. (Working pressure safe up to approximately 40 psi (2.7 bar).)

***Size*** Usually 1 litre cup.

***Use*** Mainly with a small low-pressure compressor producing approximately 2 cfm (0.95 litre/s), to apply small quantities of most standard decorative paints, but not heavy-bodied materials.

*Note* A special type of pressure-feed-cup gun is available for higher pressure work and for heavy-bodied paints, used mainly for testing paints in laboratories.

***Remote pressure-feed cup*** **(Fig. 1.130)**

A small pressure-feed cup detached from the gun, with all the advantages of the larger pressure-tanks but light enough to be carried in the hand. It has a short hose and is easily controlled as all the controls are in the hands of the operator.

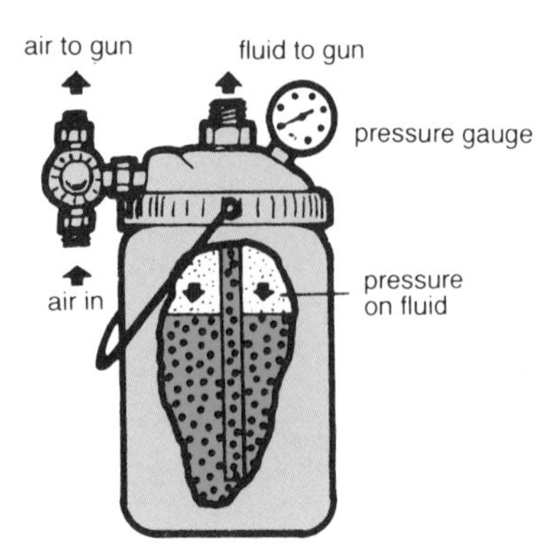

**Fig. 1.130** Remote pressure-feed cup

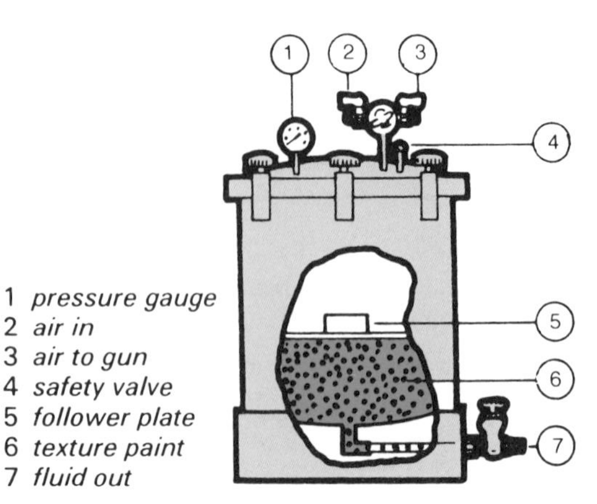

**Fig. 1.132** Pressure-feed tank (bottom outlet)

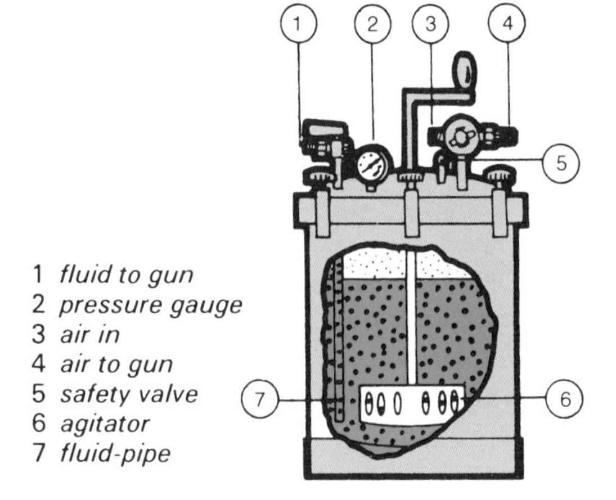

**Fig. 1.131** Pressure-feed tank (top outlet)

Working pressure up to 50 psi (3.4 bar).

***Size*** 2 litre cup capacity.

***Use*** Ideal for applying most types of decorative paint, including multi-colour finishes. Useful when working from ladders or scaffolding and on smaller jobs where 2 litres is a convenient amount to use.

***Pressure-feed tank (top outlet)*** **(Fig. 1.131)**
Made from heavy-gauge steel with a lid which is clamped and screwed down. Some types have a separate container which fits into the pot and is removed for cleaning purposes. Most are fitted with an agitator to allow any heavily pigmented materials which might settle out to be stirred while they are in the tank. Special air-driven agitators are available to allow constant stirring of materials which are likely to settle quickly.

***Size*** From 10 to 200 litre capacity. Tanks holding 25 litres are ideal for larger contract work.

***Use*** Ideal for jobs where large quantities of paint are to be applied. Advantages include:

(i) Time is saved in refilling.
(ii) Application rate is fast.
(iii) Paint delivery is constant.
(iv) Paint cannot be spilt while spraying.
(v) The gun can be tipped to any angle when spraying.

***Pressure-feed tank (bottom outlet)*** **(Fig. 1.132)**
Identical in most ways to the top-outlet type except that the outlet is at the bottom of the tank, so that an insert container for easy cleaning cannot be fitted.

***Size*** Available in various sizes from 10 to 80 litre capacity.

***Use*** For heavy-bodied materials such as texture paints, anti-condensation and heavy bituminous compounds which are too thick to be forced up the normal fluid-pipe in top-outlet tanks.

## 1.3.3 Compressors, receivers, transformers

### Compressor

A machine driven by an electric, petrol or diesel engine and designed to draw in air from the atmosphere and

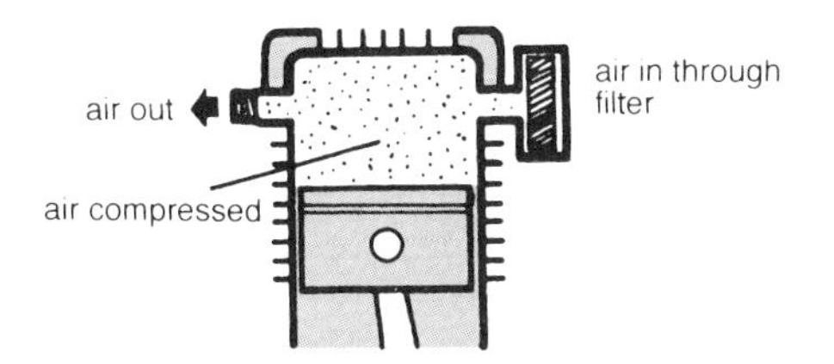


**Fig. 1.133** Piston compressor: single-stage (compressed once)

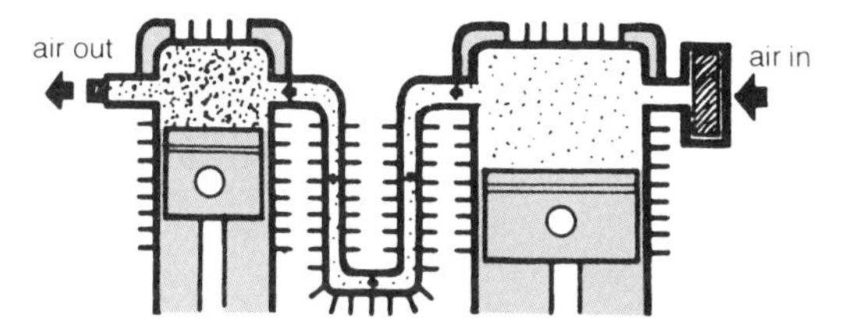


**Fig. 1.134** Piston compressor: two-stage (compressed twice)

compress it by decreasing its volume and increasing its pressure. A continuous flow of air must be supplied at the required pressure and volume to suit the equipment being used.

Many types and sizes are available. Three main types, available as static or portable units, are used by the decorator.

(i) *Piston compressor* (Figs. 1.133 and 1.134)
This is the type of compressor most commonly used. It consists of a piston inside a cylinder, driven by a crankshaft. Air is taken in through filters, compressed by the piston and then stored in an air receiver.

If the required pressure is obtained by a single action, the compressor is described as *single-stage*. This is efficient only up to 100 psi (6.5 bar). Higher pressures required to be maintained at or above 100 psi (6.5 bar) cannot efficiently be produced by a single action. In this case, air discharged from the first piston is fed into another smaller cylinder and compressed still further – *two-stage* compression.

Where air is compressed, there is an increase in temperature, and some form of cooling has to be employed. This is either done by water or, on small units, fins are fitted on the outside of the cylinder.

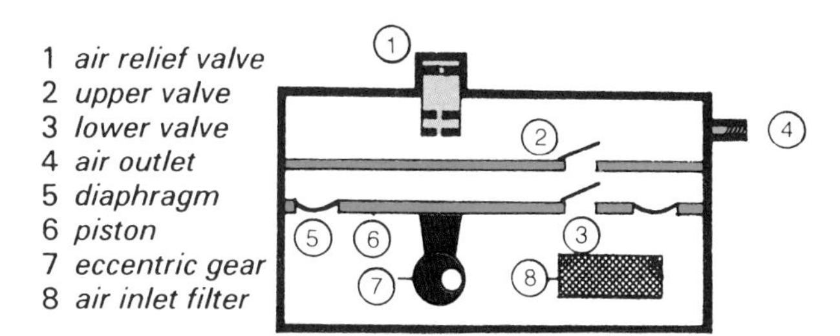


**Fig. 1.135** Diaphragm compressor

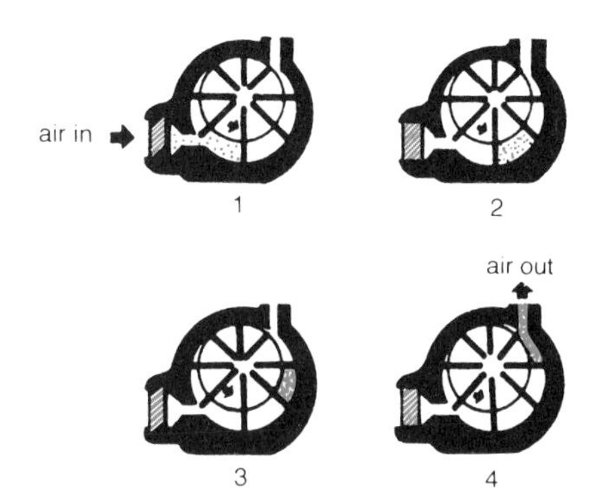


**Fig. 1.136** Rotary compressor

(ii) *Diaphragm compressor* (Fig. 1.135)
Air is compressed by the up-and-down movement of a diaphragm, rather like the action of a bicycle pump. This type has a relatively small capacity and is used for spray units which produce low pressures only.

(iii) *Rotary compressor* (Fig. 1.136)
Air is compressed by a number of vanes like a fan. As they rotate, the air is gradually compressed.

*Static compressors* are installed permanently in sprayshops. They are large units, usually bolted to the floor, with air receivers large enough to supply all the equipment installed in the workshop (Fig. 1.137). They should be placed in such a position as to be able to supply all the air transformers strategically placed around the workshop.

*Portable compressor units* are designed so as to be easily transportable. They are usually fitted with wheels or castors, but some of the smaller plants can be

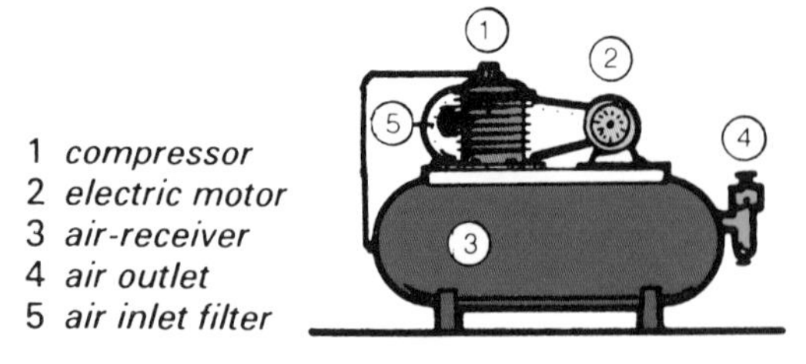


**Fig. 1.137** Static compressor

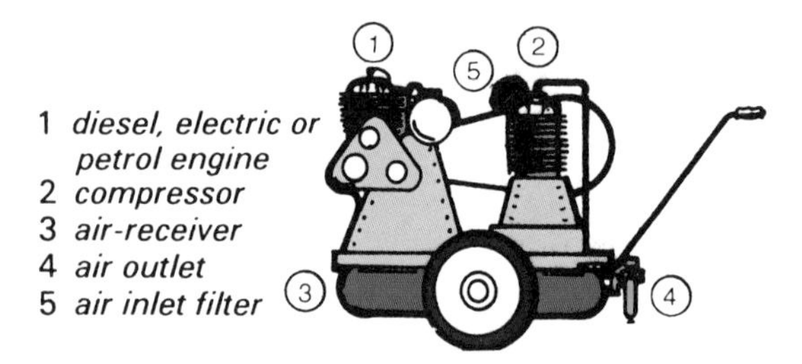


**Fig. 1.138** Portable compressor

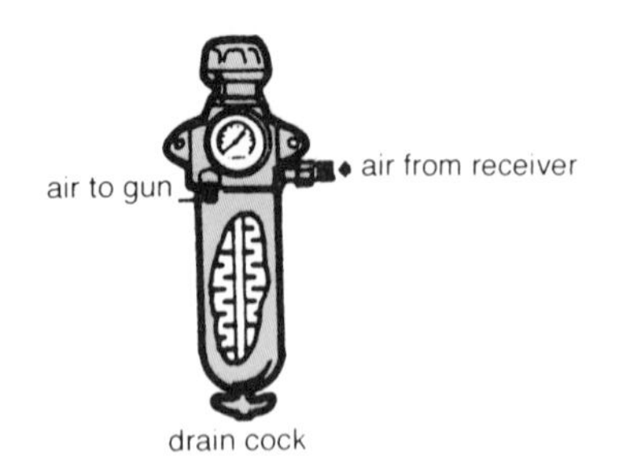


**Fig. 1.139** Air transformer

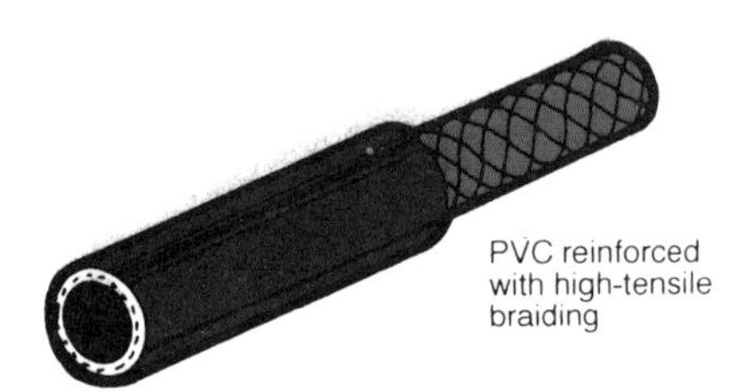


**Fig. 1.140** Air hose

carried easily by one person. They vary in size from small units which can be carried in the boot of a car to large compressors which have to be towed by a vehicle (Fig. 1.138).

### Air receiver

The air receiver is a vertical or horizontal storage tank which stores the compressed air at high pressure. It must be strong and airtight, and must be cleaned and examined at least every 26 months. A solid-drawn receiver must be checked at least every 4 years.

Each receiver must be fitted with:

(i) A safety valve to prevent the tank bursting if too much pressure is built up.
(ii) An accurate pressure gauge.
(iii) A tap at the bottom to drain off any oil or water.
(iv) An inlet and an outlet valve.

### Air transformer (Fig. 1.139)

Apparatus fitted at various points around a spray shop, or on a portable compressor, to which air lines are attached. It strains the air, removing any oil and water in the line, filters any dirt and regulates the amount of air going to the gun.

## 1.3.4 Hoses and connections

### Air hose (Fig. 1.140)

Air hoses are made of high-quality PVC (polyvinyl chloride) reinforced with high-tensile braiding. They are made to withstand maximum working pressures of 210–230 psi (14.5 to 15.8 bars).

***Size*** Available in three sizes:

6 mm ID (internal diameter)
8 mm ID – most commonly used
11 mm ID

***Pressure drop***

When the air pressure at the gun is less than the air pressure at the transformer or compressor, this is known as 'pressure drop'. Although the interior lining of the hose is smooth, it still causes some resistance to the flow of air by friction, resulting in such a drop in air pressure. Therefore (a) the smaller the internal diameter, the higher the drop in air pressure; (b) the longer the air line, the greater the drop in air pressure.

Table 1.7 shows how air pressure can drop in various lengths of 6 mm and 8 mm ID hose.

**Table 1.7** Pressure drop related to hose size

| Air pressure drop at spray gun | | | | |
|---|---|---|---|---|
| ***ID of hose*** | ***3m length*** | ***4.5m length*** | ***6m length*** | ***15m length*** |
| 6mm at 40psi or 2.7bar | 8psi or 0.55bar | 9.5psi or 0.65bar | 11psi or 0.75bar | 24psi or 1.6bar |
| 6mm at 60psi or 4.1bar | 12.5psi or 0.8bar | 14.5psi or 1.0bar | 16.75psi or 1.1bar | 31psi or 2.1bar |
| 6mm at 80psi or 5.5bar | 16.5psi or 1.1bar | 19.5psi or 1.3bar | 22.5psi or 1.6bar | 37psi or 2.5bar |
| 8mm at 40psi or 2.7bar | 2.75psi or 0.2bar | 3.25psi or 0.25bar | 3.5psi or 0.25bar | 8.5psi or 0.6bar |
| 8mm at 60psi or 4.1bar | 4.5psi or 0.3bar | 5psi or 0.35bar | 5.5psi or 0.4bar | 11.5psi or 0.8bar |
| 8mm at 80psi or 5.5bar | 6.25psi or 0.45bar | 7psi or 0.5bar | 8psi or 0.55bar | 14.5psi or 1.0bar |

*Note:* psi = pounds per square inch; the metric equivalents in bars are approximate.

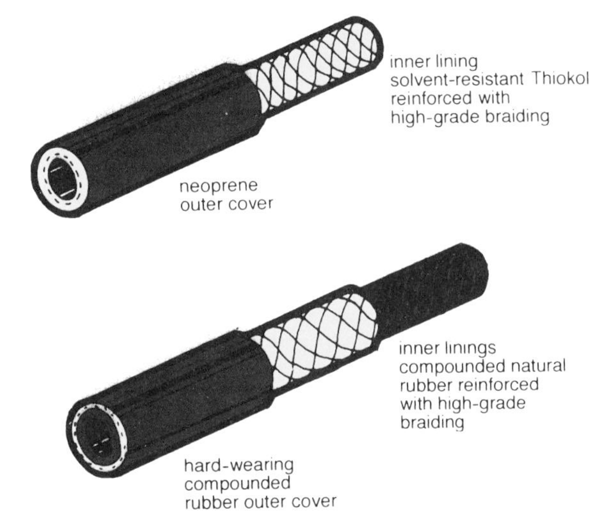


**Fig. 1.141** Fluid hoses

### Fluid hose (Fig. 1.141)

The internal lining of a fluid hose must withstand not only fluid pressure but also water, white spirit, oil, cellulose and synthetic paints and their solvents. The hoses are made from compounded rubber or special plastic materials, mainly Thiokol, with a single or double braid reinforcement.

***Size*** 9mm ID for normal use; 12mm ID for heavy-bodied paints and where lengths longer than 6m are used; 19mm ID special hose for use with bottom-outlet pressure pots to apply heavy-bodied materials such as plastic texture paints.

### Hose for airless spray (Fig. 1.142)

Especially strong high-pressure fluid hoses made with either

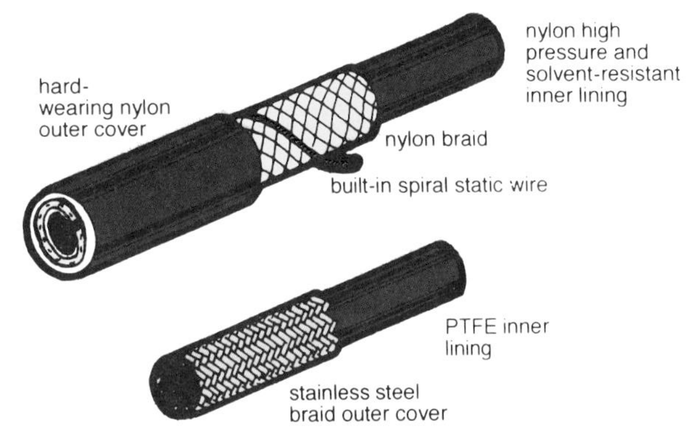


**Fig. 1.142** Hoses for airless spray

(a) A smooth nylon lining reinforced with nylon braid inside a vinyl or nylon cover. This type has a built-in spiral static wire.
(b) A tough plastic (polytetrafluoroethylene, PTFE) lining with a stainless steel braid cover.

***Size*** Available with 3, 6 or 9mm ID.

### Hose connection

Two types are available.

(i) The more commonly used screw-on type for fluid and air-hose connections (Fig. 1.143). Available in several sizes, they vary according to type of hose and manufacturer.
(ii) Quickly detachable or bayonet type which only requires plugging in: self-sealing and can be used on air or fluid hoses (Fig. 1.144). These are useful for work which requires frequent changes of gun, or in factories where equipment is run off a long, fixed air-supply line. (Not used for airless spray.)

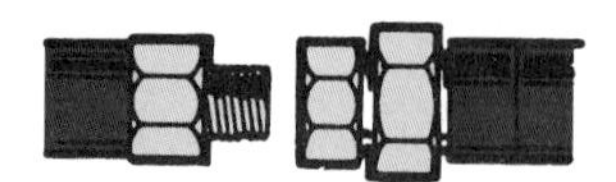

**Fig. 1.143** Screw-on hose connection

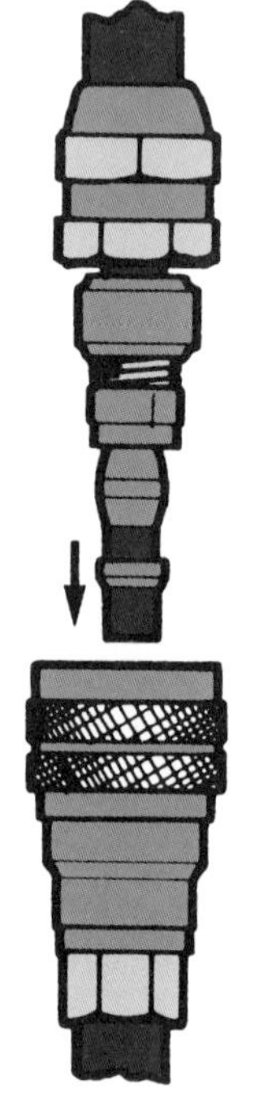

**Fig. 1.144** Bayonet-type hose connection

## 1.3.5 Types of spray system

Many types of spray system are available to suit almost any type of work. Different combinations of the various components make the choice even wider. The four main examples shown are typical of systems which can be used by the painter and decorator.

**Portable lightweight compressor (Fig. 1.145)**
A simple unit which can easily be carried in the boot of a car. It is useful for many small decorating jobs including the application of undercoats, gloss and eggshell finishes, emulsion paint and multi-colour finishes.

**Small portable compressor (Fig. 1.146)**
Used for medium-size decorative, maintenance and industrial painting. Can be used to apply most types of paint.

**Heavy-duty portable compressor (Fig. 1.147)**
Used for larger contract work where two guns or powered hand tools are in constant use.

**Static compressor (Fig. 1.148)**
Large stationary compressor for use in spray shops. Available in various sizes from 1 to 15 hp motor, delivering up to 50 cfm (23.6 litre/s).

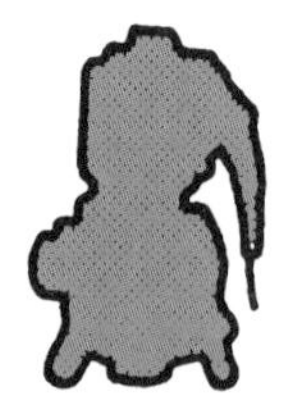

**Fig. 1.145** Spray system: portable lightweight compressor

1 1/2–2 hp petrol, diesel or electrically
driven compressor delivering
approx 7 cfm (3.3 litre/s)
at 60–70 psi (4–4.7 bar)
weight approx 125 kg

1 suction-feed spray gun
requiring 5–7 cfm
(2.4–3.3 litre/s)
at pressures of
30–50 psi (2–3.4 bar)
or

1 pressure-feed gun with
pressure pot for faster production
requiring 5–7 cfm
(2.4–3.3 litre/s)
at pressures of
45–65 psi (3–4.43 bar)
or

**Fig. 1.146** Spray system: small portable compressor

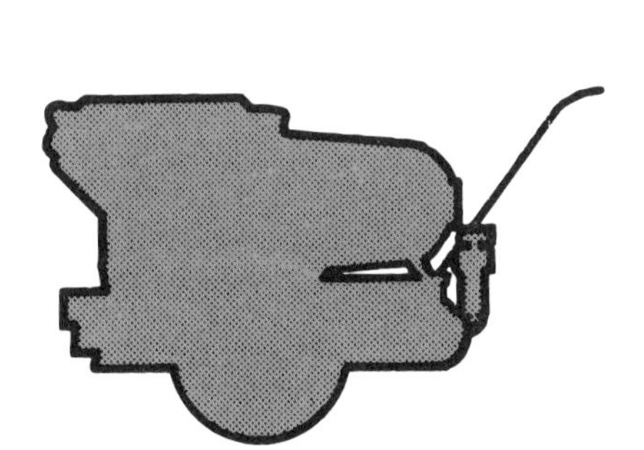

5–7 hp petrol, diesel or electrically
driven compressor delivering
approx 18–20 cfm (8.5–9.45 litre/s)
at 60–70 psi (4–6.8 bar)
according to load
weight approx 375 kg


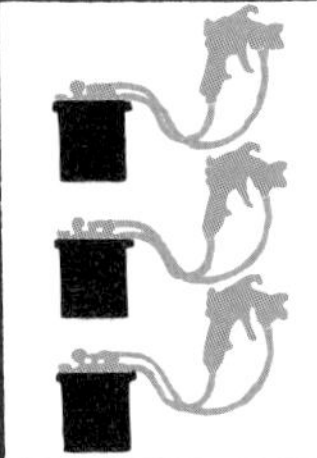

3 pressure-feed guns with
pressure pots spraying
standard decorative paints,
each gun requiring approx
5 cfm (2.36 litre/s)
at 60–70 psi (4.1–4.7 bar)
or

1 bottom-outlet pressure pot
and heavy-duty gun to spray
texture paints, requiring
12–17 cfm (5.6–8.1 litre/s)
at 40–50 psi (2.75–3.4 bar)
or

1 descaling pistol requiring
10 cfm (4.72 litre/s)
at 80–90 psi (5.45–6.1 bar)
also
1 pressure-feed spray gun
and pressure pot


**Fig. 1.147** Spray system: heavy-duty portable compressor

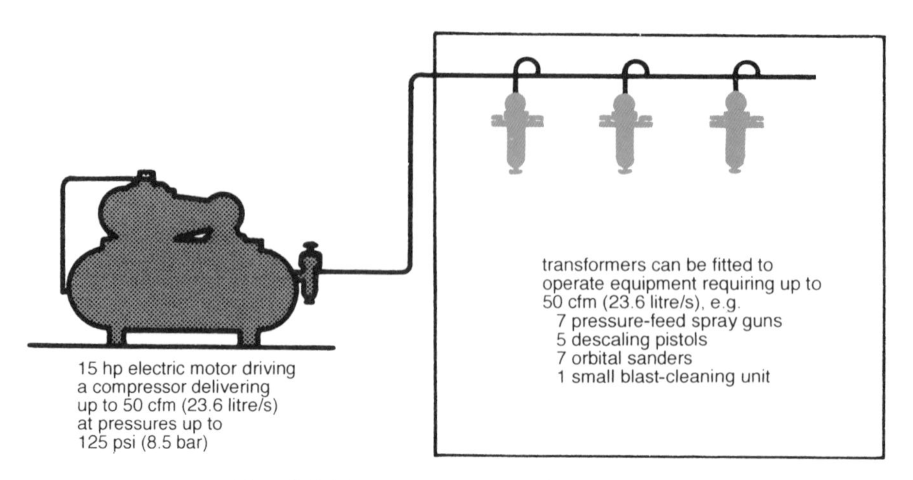


**Fig. 1.148** Spray system: static compressor

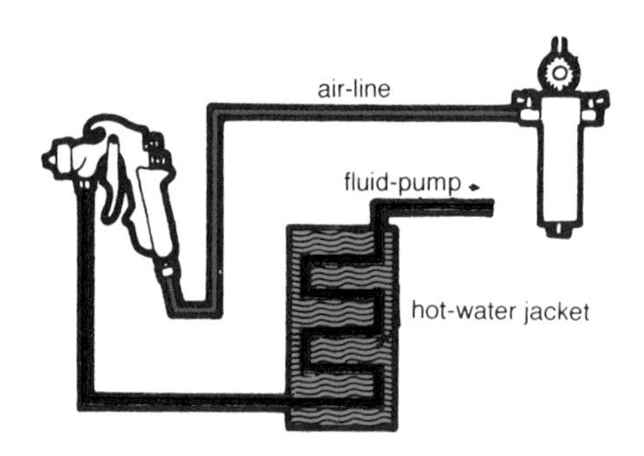


**Fig. 1.149** Spray system: hot spray

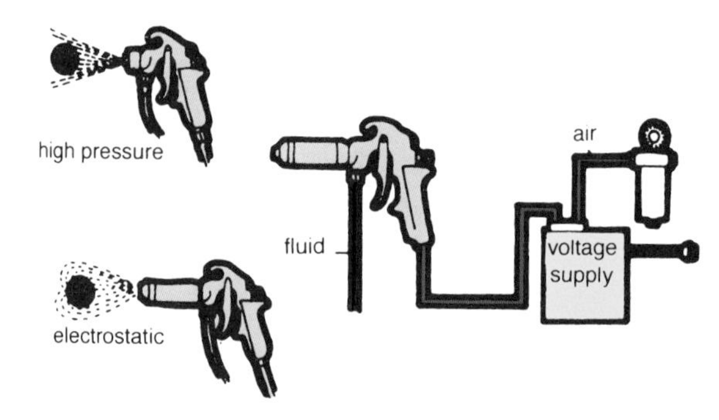


**Fig. 1.150** Electrostatic spray equipment

## Hot spray (Fig. 1.149)

Thick oil or synthetic paints are difficult to apply. They can be thinned in two ways:

(i) Add thinner, which must evaporate before the paint can dry and results in a thinner paint film.

(ii) Heat the paint, whereby it becomes thinner, easier to apply, flows out better and dries faster. By incorporating a heating unit into a normal spray-painting system, these properties can all be used to advantage.

***Equipment***

A conventional air compressor, spray gun, container and hose are used, but with a heater fitted between the container and gun. A specially jacketed fluid hose keeps the paint warm.

***Use*** Mainly for factory production spraying but can be used on site to spray oil or synthetic paints on large work.

***Advantages***

(i) Saving in thinners.

(ii) Paint can be sprayed at lower air pressures, reducing overspray.

(iii) The equivalent of two coats of paint can be applied in one application.

(iv) The warm paint flows out better and gives a better finish.

(v) The solvents that are used evaporate quickly and the paint dries faster.

***Disadvantages***

(i) Not suitable for water-thinned materials.

(ii) The equipment is costly and not easily manoeuvrable.

## Electrostatic hand gun (Fig. 1.150)

A very specialised method of spraying which incorporates the basic law of electricity that 'unlike electrical charges attract each other'.

Paint is sprayed from a pressure-feed tank through a fluid hose to the gun in the usual way. At the same time, the paint is given a negative electrical charge. As the electrically charged paint leaves the gun, it atomises and each particle is attracted to any earthed object in its path – i.e. the article being sprayed. This causes the paint not only to cover the face of the object but to wrap round behind to get to earth. The result is that some articles, such as tubes, can be coated all round when sprayed from only one side.

***Equipment***

An air compressor, paint container, electrostatic voltage supply and special spray gun.

The voltage supply usually comes from the mains, but small, mobile battery-run units are available which can be used on open sites.

***Use*** Many industrial applications; the mobile units are especially useful for painting large areas of railing and chain-link fencing.

***Advantages***

(i) No overspray. Very clean to use.
(ii) Great saving in paint.

***Disadvantages***

(i) Not suitable for water-thinned paints or heavily pigmented metallic coatings.
(ii) The equipment is costly and not easily manoeuvrable.

## 1.3.6 Airless spray

Airless spray is created by forcing paint at extremely high pressures through an accurately designed small hole or orifice. As the paint leaves the gun and meets the atmosphere it expands rapidly. These two factors cause the paint to break up into an extremely fine, very even spray pattern. No air is used to atomise the paint, hence the description 'airless'. Owing to the high pressures and potential danger of the equipment, many contractors/local authorities insist on operatives holding evidence of training.

### Equipment

***Fluid-pump***

This sucks in paint from an unpressurised container and forces it out at very high pressures through a fluid-line to the gun. There are three methods of forcing out the paint:

(i) *Air motor* Compressed air is supplied from an independent air compressor. Pressures between 60 and 100 psi (4 to 7 bar) are required with a minimum consumption of 3.5 cfm (1.7 litre/s). The pumps have ratios of from 23:1 to 48:1 according to type and make: i.e. the pump multiplies every pound or bar of air taken in by the ratio number. A 30:1 ratio pump with an air-intake pressure of 100 psi (6.8 bar) will produce spraying pressures of 3000 psi (204 bar) – i.e. $30 \times 100 = 3000$. (Fig. 1.151).
(ii) *Electrically driven hydraulic pump* Hydraulic fluid pressure is used to drive the fluid-pump. This is more efficient and requires no air pressure, having only to be plugged into the main electricity supply. Capable of producing spraying pressures of 3000 psi (204 bar).

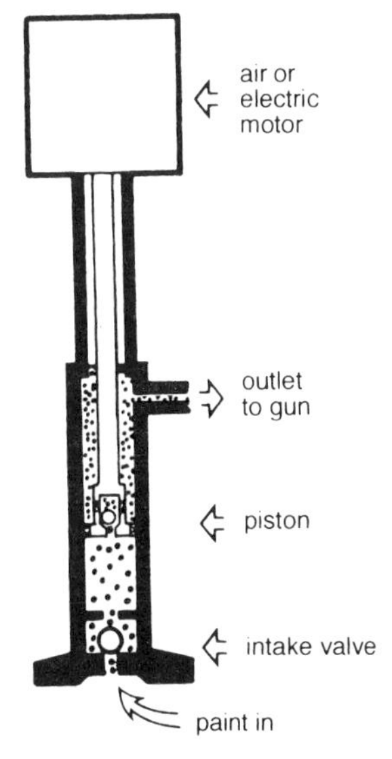


**Fig. 1.151** Fluid-pump

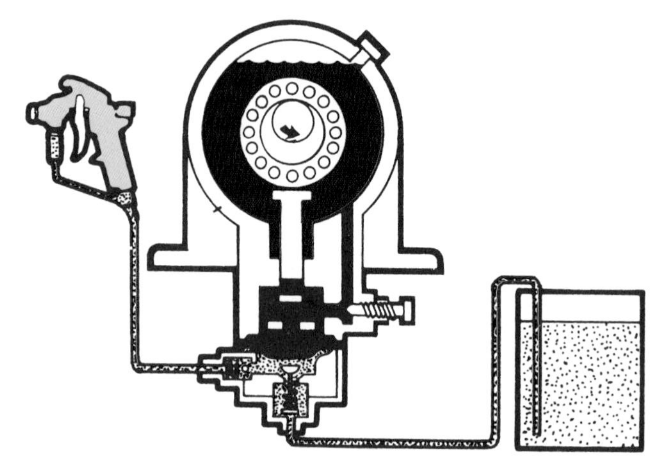

Fig. 1.152 Diaphragm pump

(iii) *Diaphragm pump* An electrically driven motor operating a diaphragm-type hydraulic pump. It has a low pump capacity but works at much greater speeds than a normal pump. This produces pressures of up to 3000 psi (204 bar) delivering a surge-free paint flow of up to 3 litres per minute. It has fewer working parts, is smooth-running and quiet to operate, has no friction or heat wear and is therefore easier and cheaper to maintain (Fig. 1.152).

***Fluid-line***

Special high-pressure, solvent-resistant small bore hose (see 1.3.4).

**Gun (Fig. 1.153)**

Specially designed, well-balanced gun with a trigger but no other controls. Fan pattern and material flow are controlled by the fluid-tip. A safety catch is fitted to prevent accidental spraying. The needle, tip and seating are made of tungsten carbide to withstand the high pressures and abrasive action of some paints.

***Fluid-tip*** Made from tungsten carbide for long life. The orifice size controls the quantity of paint passing through. The angle of the tip controls the size of the fan pattern. Tips have to be changed when adjustments are required. A wide range is available (Table 1.8).

***Reversible head*** As the orifice size of the fluid-tip is so small, it can easily be blocked by small particles of

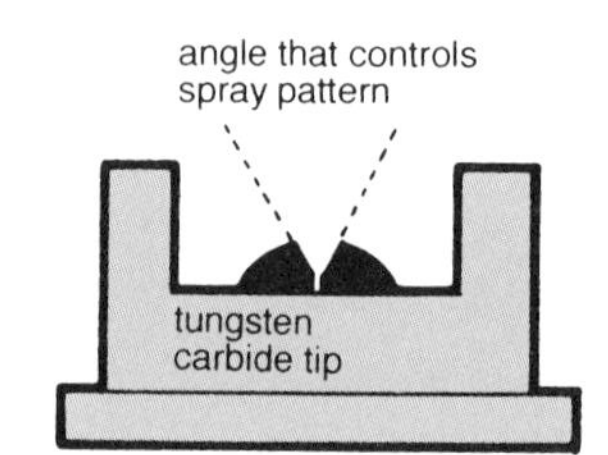


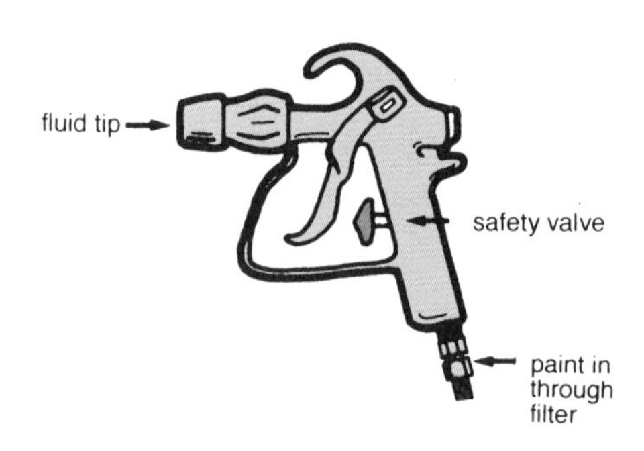


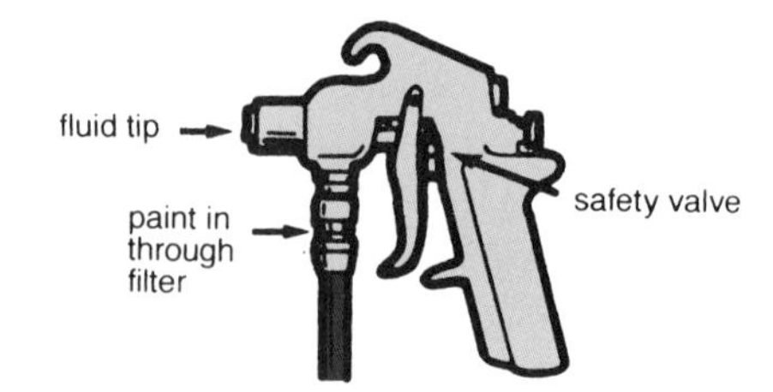


Fig. 1.153 Airless spray gun

**Table 1.8** Fluid-tips in common use with decorative paints

| *Type of material* | *Orifice size* | *Orifice angle* | *Approximate spray pattern* |
|---|---|---|---|
| Lacquers | 0.225 mm | 40° | 180 × 75 mm |
| | 0.275 mm | 25° | 275 × 50 mm |
| Thin primers | 0.275 mm | 40° | 300 × 75 mm |
| Gloss paint | | | |
| Undercoat | As above, also: | | |
| Eggshell | 0.325 mm | 25° | 180 × 50 mm |
| Emulsion | 0.325 mm | 50° | 350 × 50 mm |
| Fillers | 0.375 mm | 40° | 275 × 50 mm |
| Heavy primer surfacers | 0.450 mm | 65° | 375 × 50 mm |

*Note:* In some orifice sizes, angles of 5°, 15°, 80° and 95° are also available.

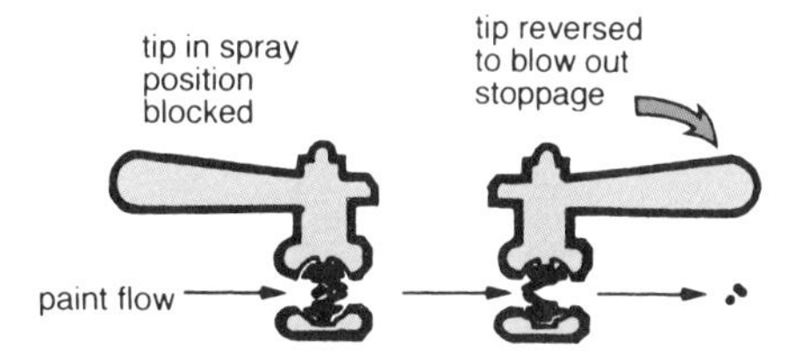


**Fig. 1.154** Reversible head

dirt. To save removing, cleaning and replacing the tip each time this occurs, special heads have been designed to allow the tip to be reversed so that the dirt can be blown out and the tip then returned into position. There are several types availablo inoluding those where the tip is reversed by a single 180-degree turn of a lever (Fig. 1.154).

### Filtering (Fig. 1.155)

Fine-gauge filters are fitted over the pump paint intake and under the gun fluid-feed connection, but these can easily become clogged if unstrained paint is used. Vacuum-assisted paint strainers are designed to strain paint quickly and efficiently before use. They consist of a large circular funnel with detachable wire gauze

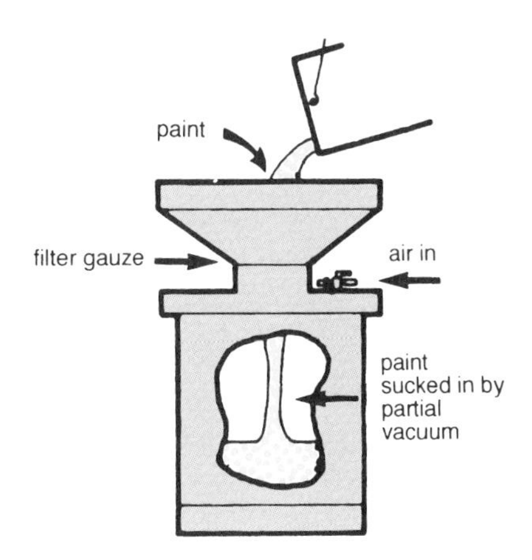


**Fig. 1.155** Vacuum-assisted paint-strainer

fitted to a lid. This is fitted over a container. Compressed air is then blown through a special jet causing a partial vacuum in the container. This draws the paint through very quickly.

### Advantages of airless spray

(i) Airless spray applies most types of paint much faster than any other manually operated method of paint application.

(ii) As the paint container is not under pressure, the pump can operate direct from the manufacturer's can.

(iii) Because no air is used, there is no overspray. A degree of paint 'drop-out' occurs but this can be reduced by pressure control.

(iv) A uniform thick coating is produced, reducing the number of coats required.

(v) A very wet coating is applied, ensuring good adhesion and flow-out.

(vi) Most paints can be sprayed with very little thinning; with less solvent the paint sets faster.

(vii) No thin edges on the spray pattern means no overlapping on each stroke. With wide fan patterns this makes airless spray much faster than conventional spray.

(viii) Because of the even spray pattern and clean edge, cutting in at angles can be achieved.

(ix) The single hose connection to the gun makes it easy to handle.

## 1.3.7 High-volume low-pressure spray (HVLP)

HVLP can be split into two groups: HVLP turbine and HVLP powered by compressed air. To comply with the specification of HVLP it is necessary to guarantee a pressure at the air cap to be no more than 10 psi; if it exceeds this, then it ceases to be HVLP.

HVLP turbine uses air supplied from a low-pressure turbine similar to those used in a vacuum cleaner.

HVLP powered by compressed air came about following The Environmental Protection Act (EPA), introduced to reduce the emissions of volatile organic compounds (solvents). The introduction of 'compliant' technology spray guns has now replaced these. But the principle as used in the HVLP turbine has not been superseded and provides an efficient method of applying coatings for the painter and decorator.

### Equipment

***The turbine***

Produces high volumes of low-pressure air 70–90 cfm at less than 10 psi, compared to conventional spray where 8–12 cfm is used at about 30–50 psi.

***The gun***

HVLP cup guns feed material to the nozzle by pressurising the cup. Using a separate small hose from the incoming air diverted through the cup lid onto the surface of the material results in the fluid being forced up to the fluid nozzle. Note: *the cup gasket must be airtight.* Most air-spray cup guns other than HVLP, for example 'siphon or suction', send fluid to the spray nozzle by creating a vortex in front of the nozzle. Air-spray guns work even if the cup gasket seal is not airtight. They are dependent on atmospheric pressure forcing the fluid to the nozzle by means of a vent hole in the cup lid.

### Material/coating preparation

When using HVLP turbine, most materials will need thinning in order to be spray applied. Thin materials such as wood stain will probably spray apply without added thinners. Always add the manufacturer's recommended thinner and measure the viscosity accurately. A viscosity cup is used to measure a material's viscosity, typically Ford No. 4 or BS 4; flow rate through the cup is measured in seconds, the thinner the material, the quicker it will empty, i.e. paint taking 25 seconds to pass through the cup will be substantially thinner (lower in viscosity) than those passing through in 40 seconds. If the material is too viscose (thick), it may result in poor atomisation which may result in the finish being 'lumpy' or 'orange peel'. *Note:* Temperature will also affect viscosity.

### Needle/nozzle choice

Good results depend upon the right choice of needle and nozzle. Nozzle size also affects film thickness. If the correct needle/nozzle is used, there should be no need for restricting the needle movement to control fluid flow. Needle travel should only be restricted if the fan pattern is reduced.

| Nozzle diameter | Nozzle number |
|---|---|
| 0.8 mm | No. 2 |
| 1.3 mm | No. 3 |
| 1.8 mm | No. 4 |
| 2.2 mm | No. 5 |
| 2.5 mm | No. 6 |
| 2.8 mm | No. 7 |

The following table is a guide only – manufacturer, temperature and required finish may influence viscosity.

| Coating | Amount thinned | Nozzle number |
|---|---|---|
| Water-based primer/undercoat | Thinned 10–15% | No. 4 |
| Water-based eggshell | Thinned 10–15% | No. 4 |
| Water-based gloss | Thinned 10–15% | No. 4 |
| Fine finish enamel | 25–35 seconds | No. 3 |
| Cellulose lacquer | 25–35 seconds | No. 3 |
| Polyurethane | 25–35 seconds | No. 3 |
| Stain | Un-thinned | No. 2 |
| Epoxy | 30–35 seconds | No. 3 |
| Emulsion (detailed work) | Thinned 10–15% | No. 3 |

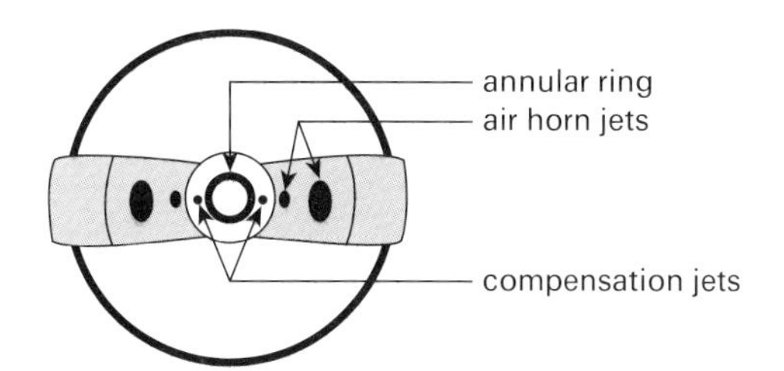


**Fig. 1.156** HVLP spray gun air cap

### Air volume setting

The air volume delivered to the spray gun can be controlled in three different ways:

- The ball valve situated at the gun end of the spray hose.
- The electrical high and low speed on the turbine.
- Airflow control on the spray gun.

### Air cap

Fig. 1.156 shows the HVLP spray gun air cap indicating each air jet.

### Annular ring

Atomising air exits through the gap between the air cap and the fluid tip/nozzle. It is responsible for atomisation and does not play a part in shaping the fan.

### Air horns

The protrusions on the air cap are called air horns. These holes help to shape the fan by flattening the atomised particles. The principal requirement of the air horns is to shape the fan and not to atomise the material.

### Compensation jets

The smaller holes between the annular ring and the air horns help shape the fan.

### Fan control

The fan control regulates the volume of air to the air horns without affecting the atomising air to the annular ring and compensation jets. Air exiting from the air horns flattens the spray pattern and therefore produces the 'fan' pattern. Reducing air volume to the air horns reduces the fan pattern width. When the fan pattern is completely closed, the spray pattern is round. Changing from flat to round spray pattern may require reduced needle travel to avoid sags and runs due to the concentration of material in a smaller area.

*Note:* If you reduce the fan pattern, test the atomisation on a test card before recommencing work.

### Advantages of HVLP

(i) Complete system is lighter and more portable
(ii) Due to low pressures involved, condensation is eliminated
(iii) Reduced overspray
(iv) Low power consumption
(v) Works well on site voltage (110 V)

### Limitations when compared with air spray

(i) Materials may need more thinning
(ii) Some materials may slow production
(iii) Gun and hose are more bulky

## 1.3.8 Cleaning, maintenance and storage of spray equipment

### Cleaning

Perfectly clean spray equipment is essential to trouble-free spraying. Dirty and badly maintained equipment are the main causes of breakdown. Remember:

(i) It is much easier to clean paint from equipment immediately after spraying, while it is still wet. Dry, hard paint is very difficult to remove.
(ii) Use the solvent recommended for the type of paint that has been sprayed:
Water for emulsion paints
White spirit for oil and oil/resinous paints
Cellulose thinners for cellulose paints and lacquers
Special solvents as supplied for special paints
Never use caustic solutions as they will attack the aluminium used for some parts of the equipment.

### Cleaning air spray equipment

***Cleaning the paint container***

The method will vary slightly according to the type of container used, but the sequence is:

(i) Remove all remaining paint from container.
(ii) Where necessary, blow back to remove paint from the fluid hose.
(iii) Spray several small amounts of clean solvent through the system until it is perfectly clean.
(iv) Wipe the container outside and inside, including lid gaskets, with solvent until perfectly clean.
(v) Ensure that vent holes in gravity and suction feed pots are not blocked.

***Cleaning the fluid hose***

Provided it is carried out efficiently, the above process will also clean out the fluid hose. It is essential that all paint is removed as (a) paint left inside will slowly build up and reduce the bore, and (b) dried paint will break away and contaminate fresh paint, causing gun blockages.

Hoses should be blown through with air to dry them, or hung over a peg to drain and dry.

Ensure also that the connections are clean and not damaged.

***Cleaning the spray gun***

Correctly carried out, the above process will also remove paint from inside the gun.

Following this, the air cap should be removed, scrubbed clean, and any paint in the holes removed with a thin stick or fibre (never metal). At the same time, make sure that the outside of the fluid-tip and all the outer gun-casing is wiped clean.

Never leave the gun to soak in solvent. This destroys the grease around the fluid-needle spring and dries out the valve packings.

### Cleaning airless units

(i) Reduce or cut off pressure.
(ii) Remove fluid-tip and clean.
(iii) Remove paint from equipment or pump from paint.
(iv) Drain off paint from fluid hose.
(v) Place solvent in fluid container or place pump in solvent container.
(vi) Set pump at low pressure and circulate solvent until system is clean (Fig. 1.157).
(vii) Wipe all units clean ready for storage.

### Maintenance

When the equipment is clean and dry, the following simple maintenance operations should be carried out.

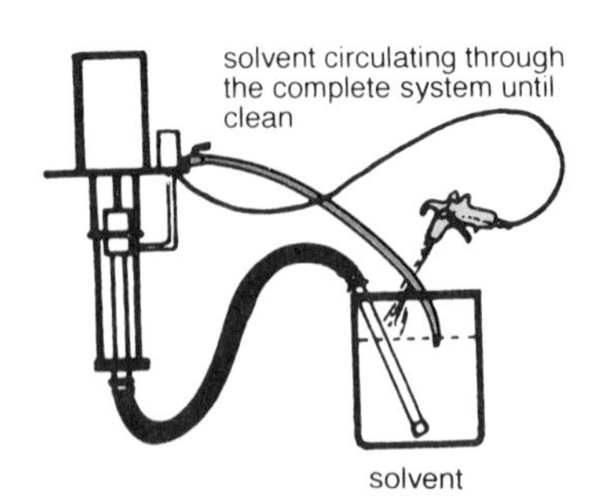


**Fig. 1.157** Cleaning the airless spray unit

(i) The guns should be lubricated and greased to reduce wear and ensure free movement of all working parts (Fig. 1.158).
(ii) Coil up air and fluid hoses without twists or kinks.
(iii) Open the drain cocks on the receivers and air transformers to let out any moisture and oil droplets. Leave open until used again.
(iv) Electric cables should be coiled without kinks or twists.

### Storage

*Compressors and plants* should be kept in a dry store where they cannot be knocked or damaged by exposure to weather.

*Hoses* should be coiled, tied and hung over two pegs in a dry store.

*Guns* are safest hung in a cupboard or laid in a drawer in a warm, dry store.

## 1.3.9 Spray guns for special purposes

### Air brush (Fig. 1.159)

A range of small, low-pressure spray guns, capable of very fine adjustments to the air and fluid controls. This enables a spray pattern to be produced ranging from a fine hair-line to a broad, even spray.

Usually fitted with an internal-mix air cap. A spatter effect can also be produced with a special air cap. Air brushes are fitted with either a gravity-feed or a suction-feed fluid container. The fluid containers are small, and range from 5 ml to 34 ml in capacity.

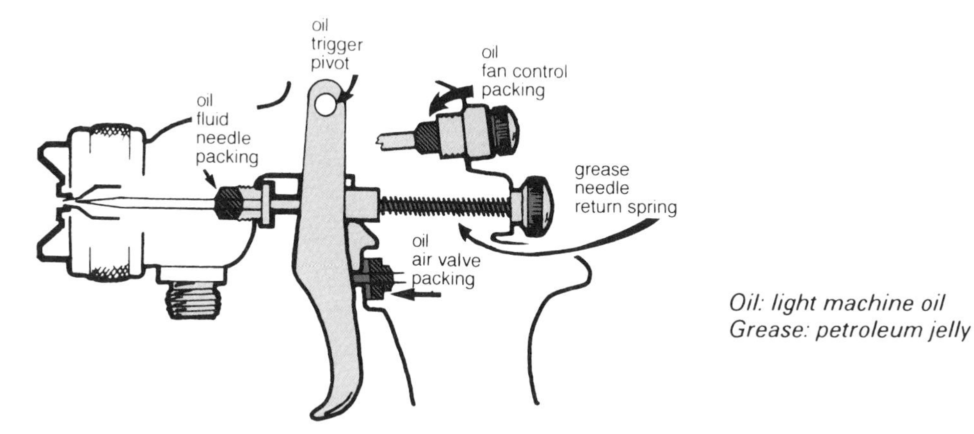


**Fig. 1.158** Lubrication of spray equipment

**Fig. 1.159** Air brush

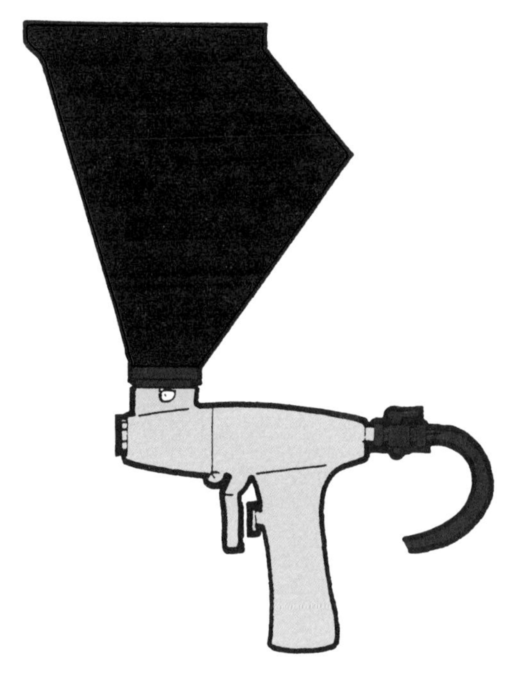

**Fig. 1.160** Hopper feed gun

The guns operate on compressed air supplied by either small, portable air compressors containing an air receiver, or disposable cans of aerosol propellant.

***Use*** Vehicle customising, edge-sprayed motifs, spray stencilling, decorative shading and spattering.

### Hopper feed gun (Fig. 1.160)

A range of gravity-feed spray guns specially designed for the application of heavy texture coatings.

The guns operate by allowing heavy texture material to fall by gravity into a stream of compressed air. The air pushes the material through a small aperture in a nozzle and expands it into a wide but coarse spray.

The material is carried in a lightweight plastic hopper fixed above the gun. Hoppers have a capacity of approximately 4.5 litres (1 gallon). The spray nozzles have apertures ranging from 5 mm to 7 mm diameter. Also available is an elbow adaptor for spraying floors and ceilings.

**Fig. 1.161** Flocking gun

The guns consume about 200 litres/second (7 cfm), at a pressure of 2.75 bar (40 psi), and can be operated with small, portable air compressors containing an air receiver, and delivering a minimum of 230 litres/second (8 cfm), at a pressure of 3 bar (45 psi).

***Use*** For the rapid and economical application of texture coatings such as cement paints, texture relief coatings, non-slip deck paint, stone paints, masonry paints, emulsion paints.

**Flocking gun (Fig. 1.161)**

A range of spray guns specially designed for the application of flock to a surface which has been previously coated with a suitable adhesive. The flock adheres to the surface and produces a soft, textured finish similar to felt.

The guns operate by directing a low-pressure stream of compressed air into a container of flock attached beneath the gun. The stream of air disturbs the flock, and blows it through a simple nozzle, which directs the flock onto the sticky adhesive.

The guns can be operated with any air compressor containing an air receiver.

***Flock*** Cotton or nylon fibres available in a range of colours. The fibres are very fine (2 denier) and approximately 2 mm long.

***Use*** Decorative motifs, stencils, vehicle customising.

# 1.4 Power sources

## 1.4.1 Compressed air

Air can be used to drive certain types of air-operated mechanical hand tool and is used in spray painting. To be useful, it must have pressure above normal atmospheric pressure (approximately 15 psi or 1 bar). Air pressure is increased by compressing the air into a smaller volume: the more a given quantity of air is compressed, the greater the pressure.

**Example (Fig. 1.162)**

1 cubic metre of air at 15 psi or 1 bar; or $\frac{1}{2}$ cubic metre of air at 30 psi or 2 bar; or $\frac{1}{4}$ cubic metre of air at 60 psi or 4 bar.

Air is compressed and raised to the required pressure by mechanical means in compressors.

When choosing air compressors and spray equipment, it is important to understand the difference between air volume, air pressure, air displacement and air delivery.

*Air volume* is the amount in cubic feet or cubic metres of air at any pressure, i.e. the volume of air in the above example is 1 cubic metre, $\frac{1}{2}$ cubic metre and $\frac{1}{4}$ cubic metre. (The bigger the orifice of a spray gun, the more air it lets out, and the greater the volume of air it consumes.)

*Air pressure* is the pressure in pounds per square inch (psi) or bar (1 bar = approximately 15 psi) exerted by air which has been compressed. The greater the pressure, the greater the force: it is the air pressure that atomises the paint when it is sprayed.

*Air displacement* is the amount of air that can theoretically be produced by a compressor in a given period of time. It is measured in either (a) cubic feet per minute (cfm) or (b) litres per second (litre/s).

*Air delivery* is the actual amount of air delivered to the gun or receiver. No compressor is 100 percent efficient (usually 60–80 percent). Therefore, although an air displacement figure is quoted for each compressor, this does not represent the amount of air it can actually deliver to the air receiver or the gun. For instance, if an air compressor has a displacement of 10 cfm (4720 litre/s) and it is only 70 percent efficient, the delivery or amount of air actually available for use will be only 7 cfm (3.304 litre/s).

The above definitions are important as, to run each gun or piece of equipment, a specific volume of air within a given time is required: for maximum efficiency, a specific air pressure is also required (see Table 1.9).

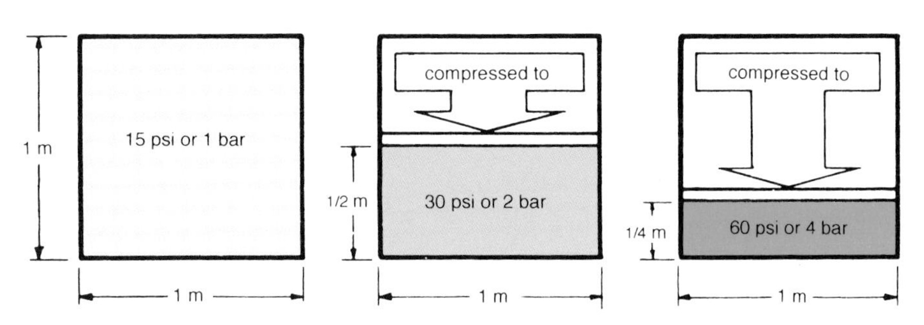


**Fig. 1.162** Compression of air

**Table 1.9** Air delivery/air pressure requirement of equipment

| Equipment | Volume of air required (air delivery) | Air pressure required |
|---|---|---|
| Spraygun (depending on type and size) | 2–10 cfm or 0.944–4.720 litre/s | Average 50–70 psi or 3.3–4.6 bar |
| Descaling pistol | 5–10 cfm or 2.360–4.720 litre/s | 100 psi or 6.5 bar |
| Large abrasive blasting equipment | 330 cfm or 155 litre/s | 100 psi or 6.5 bar |

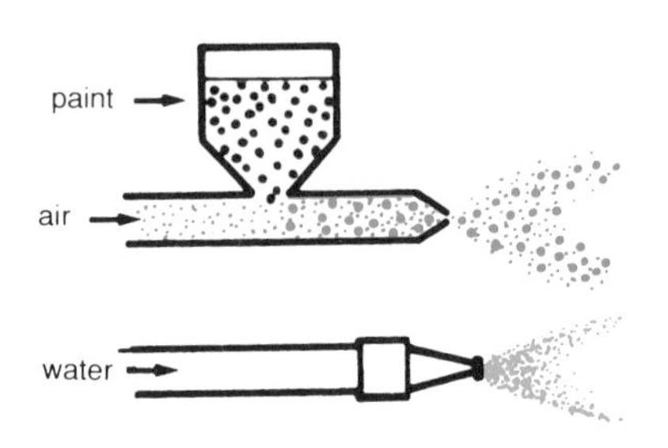


**Fig. 1.163** Atomisation

*Atomisation* is the breaking up of a liquid into small droplets, usually in a high-speed jet of air or by forcing it through a very small orifice at high pressure, as in a hose pipe (Fig. 1.163).

## 1.4.2 Electricity

Electricity is composed of small particles, known as electrons, which flow through conductors such as wires. It flows only when it can complete a circuit, i.e. from the source, along a wire and back to the source. Domestic electric cables contain at least two wires which are colour-coded: electricity flows from the source by way of the *live* brown-sheathed wire to the appliance and back along the *neutral* blue-sheathed wire (Fig. 1.164 a). If the wires are joined or in some way are able to touch each other, a short circuit occurs; this blows the safety fuse (Fig. 1.164 b).

In power stations, electricity is produced at very high voltages; these are reduced by a series of transformers to 240 V for domestic use.

### Units of measurement

***Volt (V)*** Electricity flows along a conductor as a result of electrical *pressure* which is measured in volts.

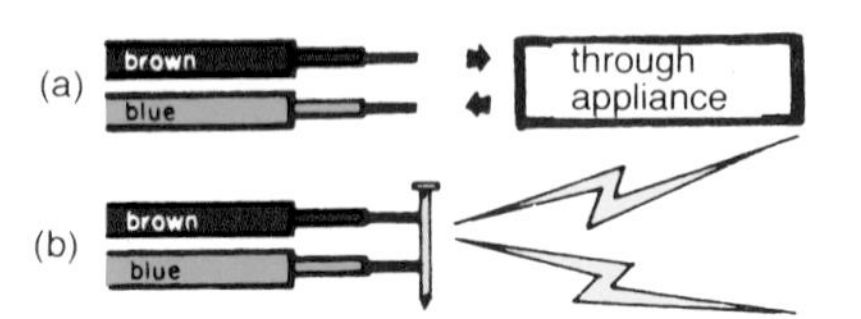


**Fig. 1.164** Electrical circuit

Most electrical appliances in the UK are designed for 240 V, or 110 V for equipment used on construction sites. The required voltage is usually stated on the appliance.

***Ampere or amp (A)*** This is the measure of the *quantity* of electricity which flows along the conductor or wire. Fuses, cables and plugs are all rated according to the quantity of electricity they can carry. To calculate amps, divide wattage by voltage.

***Watt (W)*** The watt is the unit of the *power* of electricity; all appliances state how much power is required to operate them, e.g. a 100 W bulb. Wattage can be calculated by multiplying volts by amps. As a watt is a relatively small unit, the kilowatt is often used:

1000 W = 1 kW.

***Example***

Most modern houses have a 13 amp ring-main system with a voltage input of 240 V, i.e. 13 × 240 = 3120 W. If a number of appliances with a total wattage above 3120 W are all switched on together, the system will be overloaded and the fuses will blow.

### AC/DC

All mains supplies are alternating current (AC) whereas local generators are direct current (DC). All power tools are supplied for AC, some for both AC and DC.

## Transformer

A portable apparatus for increasing or reducing voltage: its main use is to reduce 240 V mains supply to permit the use of 110 V appliances.

## Insulators and conductors

A conductor is a material which allows electricity to pass through it. Most metals are good conductors of electricity, the best being copper and silver.

An insulator is a material which will not allow electricity to pass through it. Most non-metallic materials are good insulators, the best being rubber, glass, ceramics and plastics.

## Single-phase and three-phase electricity

All power hand tools, e.g. drills, sanders, angle grinders, are powered by single-phase electric motors. Single-phase motors operate from either the normal domestic electrical installation of 220/240 V, or this domestic voltage transformed to a safer level of 110 V.

For producing higher levels of power than that required by hand tools, three-phase motors are used. Factory machinery and some air compressors use three-phase motors. Special electrical installations are necessary to operate these motors, and they cannot be connected to the normal domestic supply. The connecting plug contains at least five pins, and the voltage measured across these pins can be in excess of 400 V.

Because three-phase supply is rare on most sites on which a painter may work, a portable three-phase electric air compressor is of little value. They are ideal for static equipment in a properly installed workshop where their greater power and efficiency can be used to their best advantage.

## Flex and cables

The word 'flex' is an abbreviation of 'flexible cord'. Flex is used to connect electrical equipment to a power point. Each wire or conductor is made up of a number of thin strands, which gives it flexibility.

Cable is used for mains wiring. It is more rigid than flex, the conductor being a single thick wire strand.

## Fuses

There is a limit to the amount of electricity that can safely pass through a wire. If the system is loaded above this limit, heat develops, causing the insulation to break down and a short circuit to occur. To prevent this happening, fuses are fitted into the circuit or appliance: these are weak spots in the system, designed to fail as soon as the circuit is overloaded. It is therefore vital that the correct fuse should be fitted since, if a higher-grade fuse or piece of metal is used, it may allow too much overheating resulting in damage to the appliance or even in fire.

There are three types of fuse:

(i) The rewirable type, where various grades of wire can be used according to the quantity of electricity being used in the circuit.

(ii) The cartridge type, where the required grade of wire is enclosed in a small cartridge (Fig. 1.165 a). This is used in all fused plugs, some appliances and switch fuses.

(iii) The miniature circuit breaker (MCB), a device that protects against over-current conditions. The modern equivalent to a fuse, the advantage is that it can simply be reset (usually by a button or switch) and does not require charging.

(a)

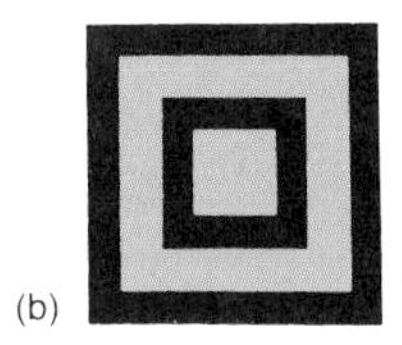

(b)

a *Cartridge fuse*
b *'Double insulated' symbol*

**Fig. 1.165**

## Earthing

All exposed metalwork on meters and appliances must be earthed. This ensures that if any appliance becomes faulty and the metal framework becomes live, the current will flow to earth, blow the fuse and prevent any risk of electric shock.

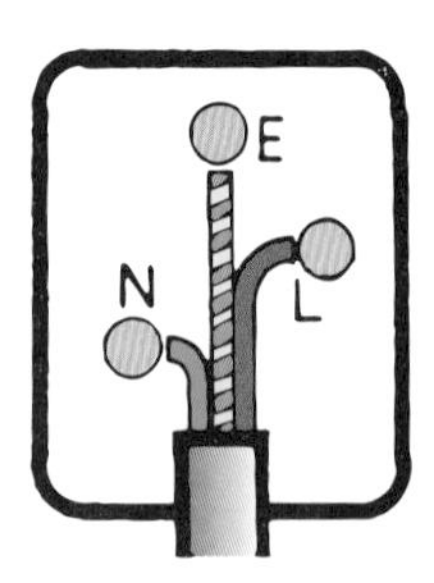

**Fig. 1.166** Wiring

Only special appliances that are completely insulated from the user need not be earthed. These are marked 'double insulated' (Fig. 1.165 b).

### Residual current circuit breaker (RCCB)

A small piece of apparatus connected to a socket outlet or fuse box which immediately cuts off the electrical power if a fault occurs, such as a cut flex, so protecting the user from a possibly fatal electric shock.

### The international wiring code (Fig. 1.166)

E or ⏚ is earth and is shown by green and yellow bands.

L is live and is coloured brown.

N is neutral and is coloured blue.

### Electric shocks

An electric shock occurs when the body becomes part of an electric circuit. This can happen when the body comes between either (a) neutral and earth or (b) live and earth.

The resistance of a human body to electric shock varies according to the condition of the body, age, situation and the weather. A shock of 40 V may be dangerous, and 240 V can cause death or severe burning. Because of this, most new building sites in this country have the voltage reduced to 110 V during construction. This greatly reduces the chance of a fatal accident.

Special equipment is necessary for 110 V: 240 V equipment *is not suitable* unless a transformer is used.

For treatment for electric shock, see 6.8.2.

### Safety precautions

Electricity is an unseen hazard. All electrical gear is a potential source of danger.

***Never*** use too many appliances on one circuit: overloading is dangerous.

***Never*** tape joints: use junction boxes or insulated connectors.

***Never*** drag cables through pools of water.

***Never*** fix wires into sockets with matchsticks or pieces of wood.

***Never*** remove plugs from sockets by pulling the lead.

***Never*** coil cables incorrectly: they may kink and break.

***Never*** drag tools by their cable: their connections may break.

***Never*** use a machine or piece of equipment for a purpose for which it is not designed.

***Always*** wire up correctly: use the correct socket and plug for the appliance or get them wired up by a competent electrician.

***Always*** ensure that the use and control of the equipment are understood before use.

***Always*** check the voltage marked on the equipment and the voltage on the supply. They must be the same.

***Always*** check for loose or missing screws and for damaged leads.

***Always*** attach earth wire when wiring up.

***Always*** have tools regularly checked for wear and have damaged or worn parts replaced.

***Always*** report defects or damage to the foreman or supervisor.

*Note:* The Construction (Health, Safety and Welfare) Regulations 1996 state that any electrically charged overhead cable or apparatus must be made safe or sealed off. This is to prevent accidents when carrying or moving equipment.

## 1.4.3 Gases

Two main types of liquefied petroleum gas (LPG) are used in the building industry: butane and propane. They are produced by the distillation of crude oil. The gases are purified, have no odour and are non-toxic. To comply with international regulations on safety of gases, an agent is added to produce a smell which can

**Table 1.10** Properties and uses of butane and propane gases

| | *Butane* | *Propane* |
|---|---|---|
| Ratio gas : liquid (in volume) | approximately 240:1 | approximately 275:1 |
| Amount of gas from 0.45 kg liquid | 6.7 $ft^3$ (0.190 $m^3$) | 8.65 $ft^3$ (0.245 $m^3$) |
| Temperature at which liquid vaporises | 0°C<br>Cannot be used below freezing point | −44°C<br>Usable in all conditions |
| Thread on cylinder fittings | Left-hand (female) | Left-hand (male) |
| Vapour pressure | At 15°C = 25 psi<br>At 45°C = 85 psi | At 15°C = 100 psi<br>At 45°C = 285 psi |
| Use | Mainly domestic, also used by painters for burning off and with steam strippers | Mainly in industry for heating and lighting |

easily be detected. Both gases are heavier than air and are available in bottles and cylinders containing from 1 to 82 lb (0.45–36 kg).

At normal air temperatures propane and butane are gases, but they will liquefy if cooled and will remain liquid at normal temperatures if kept under pressure in cylinders. Once the pressure is released, they vaporise and revert to gas. Blow torches used by painters require approximately 10 psi (0.714 bar). This is controlled by using a pressure regulator fixed to the cylinder.

### Gas level

The only way to check the amount of gas remaining in a container is to weigh the cylinder and subtract the tare weight, which is always marked on the outside. (Tare weight is the weight of the empty container.)

### Safety precautions

**Never** look for leaks with a naked flame: use soapy water.

**Never** repair faulty hoses: replace them.

**Always** use special reinforced hoses as specified by the manufacturer.

**Always** keep cylinders away from heat: heated cylinders may explode.

**Always** keep gas cylinders upright: liquid gas may get through the regulator, causing high pressure at the torch.

**Always** use a pressure regulator to control pressure and prevent blow-back.

**Always** ensure plenty of ventilation when using gases indoors.

### Other gases

*Acetylene* is a combustible gas and has an unmistakable smell. It is supplied in maroon cylinders with left-hand thread fittings. Only red hoses should be used.

*Oxygen* is a non-combustible gas but is essential for burning acetylene gas. It has no smell and is supplied in black cylinders. Fittings have a right-hand thread and hoses are black or blue.

*Never* allow oil or grease to come into contact with oxygen as the two form a mixture liable to spontaneous combustion.

**Use** Oxygen/acetylene gases are mixed to produce the intense heat required for flame cleaning.

For *storage* of gases see 6.2 and 6.3.

## 1.4.4 The internal combustion engine

In an internal combustion engine, a mixture of fuel and air is burned in such a way that the hot gases produced from the combustion exert a force on working parts of the engine, thus generating power.

Internal combustion engines can be divided into two main groups:

(i) Petrol engines, in which a mixture of petrol and air is compressed and ignited by means of a spark

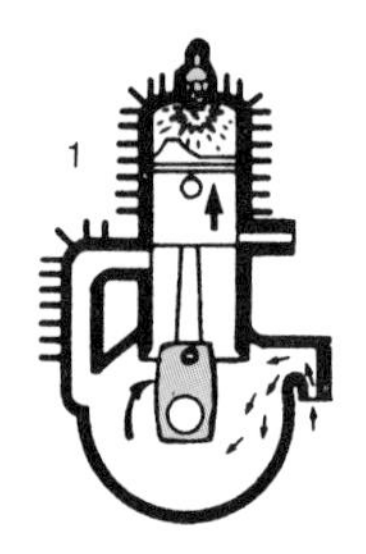


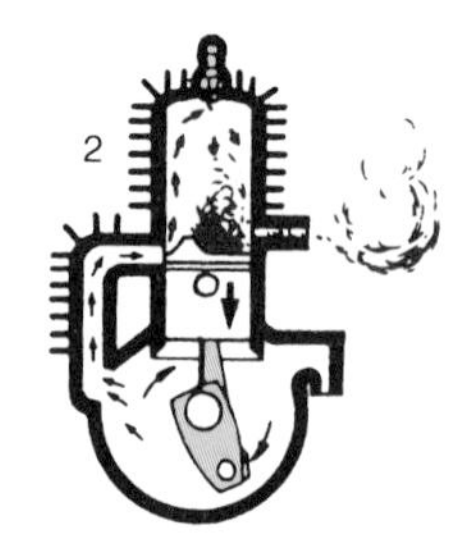


Fig. 1.167 2-stroke engine

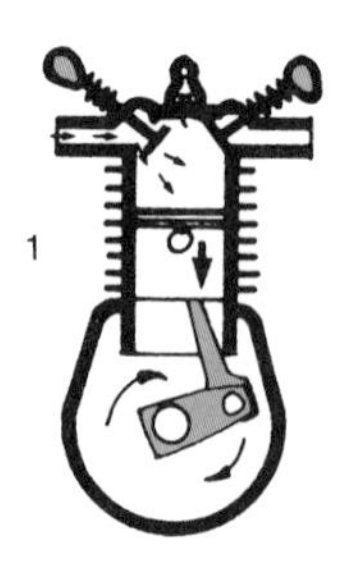


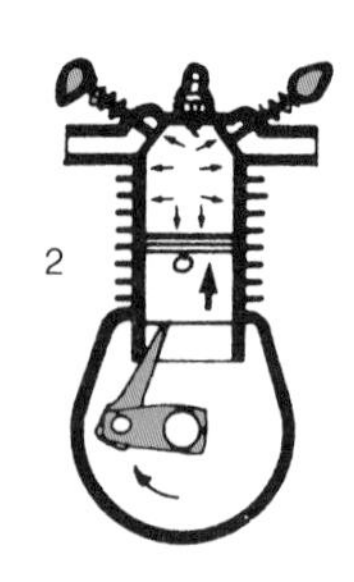


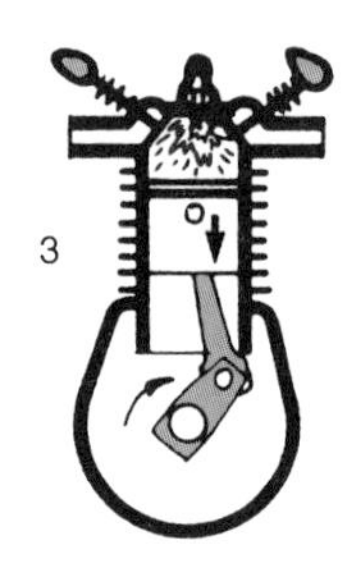


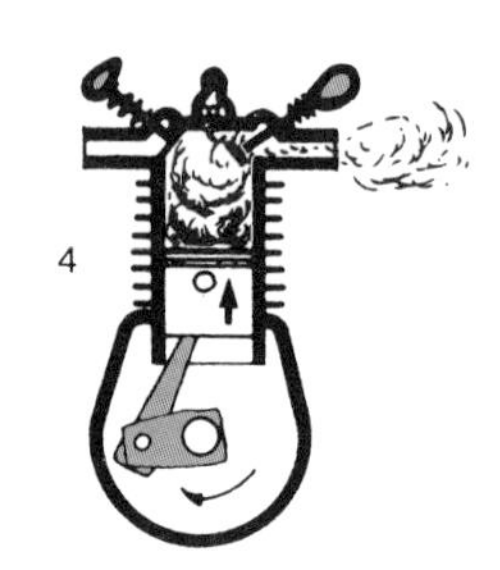


Fig. 1.168 4-stroke engine

from a sparking plug. The hot gases produced exert a force and drive the engine.

(ii) Diesel engines, where air is compressed into a small space in the top of a cylinder. This creates a very high temperature, hot enough to ignite a low-grade fuel (diesel oil) when it is injected into the cylinder, without a spark.

The cycle of operation of petrol and diesel engines is similar, the only difference being that diesel engines have no sparking plug and run on a lower-grade fuel. Both work on either a 2-stroke or 4-stroke cycle (Figs. 1.167–1.169).

### 2-stroke engine (Fig. 1.167)

The cycle is completed in a one-up and one-down stroke of the piston, or one turn of the crankshaft.

(i) Compression: in the first stroke, the piston rises and compresses the fuel in the top of the cylinder while at the same time fresh fuel is taken in under the piston. At the top of the stroke, the mixture is fired by a sparking plug and an explosion occurs.

(ii) Power: as the piston is forced down, it drives the crankshaft; as the piston drops, the exhaust fumes escape, and fresh fuel is forced into the top of the cylinder. The cycle is then complete.

### 4-stroke engine (Fig. 1.168)

The cycle is completed in two upward and two downward strokes of the piston and two full turns of the crankshaft.

(i) Intake: the piston goes down and a petrol/air mixture is let in through the inlet valve.

(ii) Compression: the piston then rises and the mixture is compressed in the top of the cylinder.

(iii) Power: the mixture is ignited, an explosion occurs and the piston is forced down – 'the power stroke'.

(iv) Exhaust: the piston rises again, forcing all the burnt gases out through the exhaust valve which opens with the fourth upward stroke. The cycle is then complete.

### Simple maintenance checks

(i) Always make sure there is fuel in the tank. If it is allowed to become empty, air and dirt may enter the pipes and make restarting difficult.

(ii) Use the correct fuel, i.e. petrol, diesel oil or petrol/oil mixture for 2-stroke petrol engines.

(iii) All oil levels must be regularly checked and topped up to prevent the engine seizing up.

(iv) Where applicable, always check the water level in the cooling system – overheating can cause the engine to seize up.

(v) Remember: regular servicing and correct use prevent breakdown, reduce running costs and increase the life of any engine.

### Safety precautions

(i) All exhaust fumes must be piped into the open air as they contain a poisonous gas (carbon

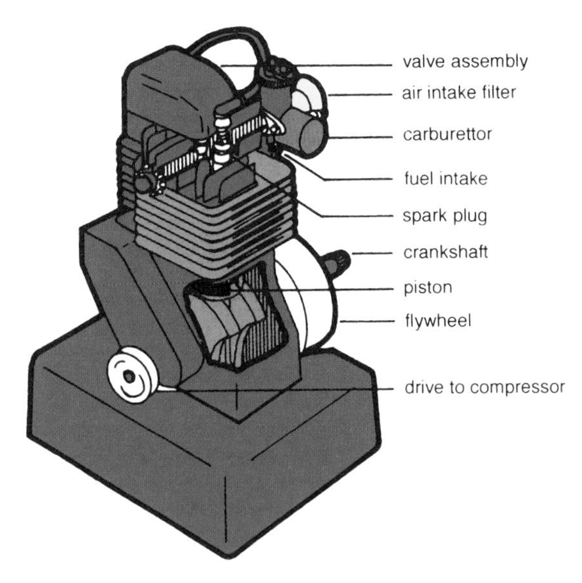


**Fig. 1.169** Parts of the internal combustion engine

monoxide) which can cause headaches, drowsiness and death if inhaled in large enough doses (see also 6.6).

(ii) Always keep the top on the fuel tank; fuel spilled on to a hot engine may cause fire or an explosion.

(iii) Always make sure that portable engines are on firm ground and wedged to prevent them running away with the vibration, when they may be damaged, causing injury, fire or explosion.

In this section, bars and litres per second have been used to express volumetric pressure because both units are commonly used by equipment manufacturers and the construction industry.

# PART 2

# Surface Coatings and Specialist Materials

# 2.1 Introduction

## 2.1.1 Volatile Organic Compounds (VOCs) – 2010 Legislation

The above is the condensed version for 'The Volatile Organic Compounds in Paints, Varnishes and Vehicle Refinish Products Regulations 2005'.

VOCs are chemicals that evaporate into the atmosphere. They can be found in cleaning products, car exhausts, aerosols and, of course, paint, and they do affect the air we breathe and are thought to contribute to global warming. Even though in the UK this amounts to less than 1 per cent from paint, the industry has agreed to reduce the VOC emissions, hence the new regulations. You will be able to identify compliant products by one of two methods: either the text box panel or the globe label (introduced by B&Q, Fig. 2.1) as indicated below:

**EU limit value for this product (cat:A/d) 75 g/l (2007)/30g/l (2010)**
**This Product contains max. 30 g/l VOC**

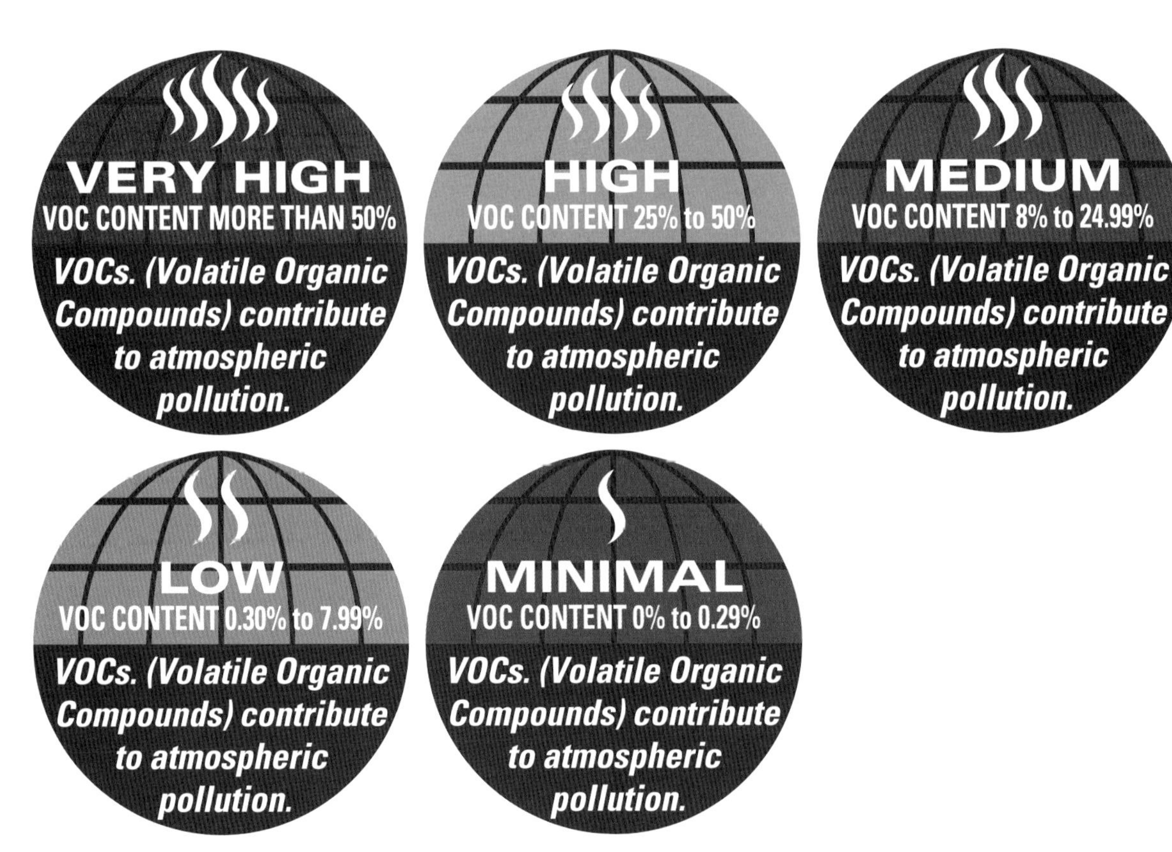


**Fig. 2.1** Globe labels for VOC emissions (courtesy of B&Q)

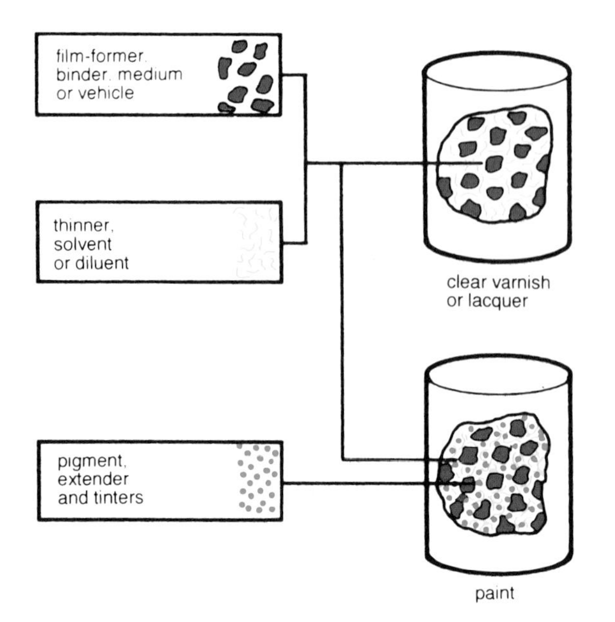


**Fig. 2.2** Composition of surface coatings

Cat A/d refers to section A of schedule 2 of the regulations and to category (d) (interior/exterior trim & cladding paints for wood, metal, plastic). From this, you can identify that the limit for this product in 2007 was 75 grams per litre, which was reduced in 2010 to 30 grams per litre and, finally, the actual VOC content of the product.

## 2.1.2 Composition (Fig. 2.2)

All surface coatings have two principal components:

(i) A liquid *film-former*, also known as medium, vehicle or binder, whose purpose is
  (a) to convert from the liquid coating to a solid dry film;
  (b) to bind the pigment particles together;
  (c) to control the gloss of the coating;
  (d) to make the coating adhere to the surface;
  (e) to give elasticity to the film;
  (f) to provide resistance to water, chemicals and abrasion;
  (g) to hold the pigment in suspension.

The film-formers most commonly used are:

*Drying oils* – linseed oil; soya bean oil; tung or Chinese wood oil; dehydrated castor oil; oiticica oil.

*Resins* – copal, rubber, lac, alkyd, coumarone, phenolic, epoxy, polyurethane, polyvinyl acetate (PVA), acrylic, silicone. Water-based resins like most acrylates are often referred to as latex.

*Oil-modified resins* – natural or synthetic resins mixed with oils, i.e. coumarone resin/tung oil, alkyd resin/ linseed oil.

(ii) *A thinner* or solvent (sometimes called a diluent), whose purpose is to make the coating liquid enough to be easily and evenly applied by any method.

It evaporates completely once the coating has been applied.

The thinners most commonly used are water, white spirit, solvent naphtha, methylated spirit hydrocarbons and mixtures of special solvents.

The combination of a film-former and a compatible thinner produces a transparent coating which may be called a varnish, lacquer or clear finish. This type of coating has a limited use.

The most common type of coating is *paint*, which is an opaque, usually coloured material; pigments and extenders provide these qualities.

(iii) *Pigments*, according to type, may provide one or more of the following:

- Opacity or hiding/covering power
- Colour
- Aid the film-former in protecting the surface, i.e. from UV radiation
- Magnetic or other functional properties

The most common pigment is titanium dioxide as the only practically usable white pigment. This is often used in conjunction with other coloured pigments or tinters.

(iv) *Extenders* are usually white/off white and 'transparent'. They provide build and aid film thickness – the reason why they are sometimes also called fillers.

Further functions are:

- Promote adhesion to surfaces

- Help with easiness of application with some paints
- Help to control sheen, particularly with matt and eggshell paints
- Lower the cost of the paint
- Can contribute slightly to providing opacity

Modern surface coatings are complex materials which also contain, depending upon the type of film-former used or the use to which the coating is put:

(v) *Drier, hardener or catalyst* ensures that the coating converts from the liquid to the solid state: driers are added to paints containing drying oils to speed up the drying process; curing agents, catalysts or hardeners are used to convert epoxy and polyurethane resins into hard films.

(vi) *Plasticiser* is required by some coatings to prevent the film being too brittle.

(vii) *Stabiliser* is required by some coatings to ensure that the complex film-former remains intact.

(viii) *Anti-skinning agents* are added to some oil-based paints to prevent skinning in the tin.

(ix) A *thixotropic agent* gives a jelly-like structure to the paint, providing anti-sag and high build properties.

## Types of coating

Most coatings used by the painter and decorator can be classified as either (a) *water-borne* – those whose film-former is either diluted in water, or is of a nature that allows it to be suspended in water, or (b) *solvent-borne* – those whose film-former is soluble in a material other than water.

## Drying (Fig. 2.3)

The process that converts the liquid material into the solid film is also complex. Most coatings undergo one or a combination of any of the following four processes:

***Evaporation*** This occurs soon after the coating is applied and is the changing of the thinner into a vapour which is absorbed into the atmosphere. Every material goes through this process, which is recognised by the paint *setting up*, so that it can no longer be brushed or

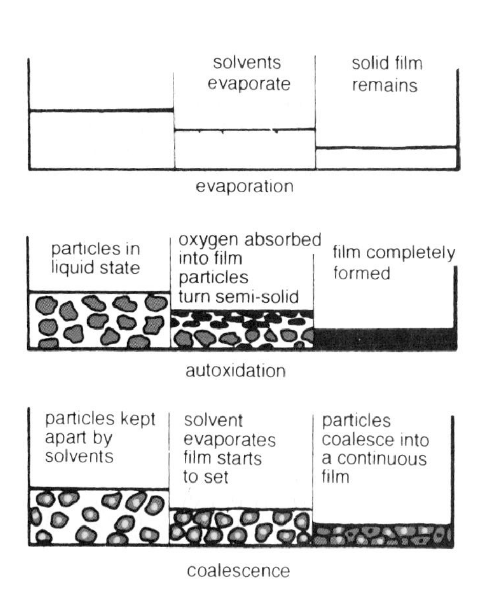


**Fig. 2.3** Drying processes

rolled and its flowing property has gone. A few coatings dry solely by this process.

***Oxidation/autoxidation*** When drying oils are included in the film-former they need to combine with oxygen to complete their chemical change from liquid to solid. Once this has been completed the original film-former has altered its structure and the process cannot be reversed.

***Polymerisation*** Certain elements in the film-former are combined to produce a solid state: this is a reasonably fast process when applied to spirit-borne coatings which contain no drying oil, but is very slow after initial oxidation has occurred in oil-based paints.

***Coalescence*** A thinner may have a secondary function, that of keeping apart particles of resin which, if they came into contact, would form a solid film. In this drying process the coalescing, or coming-together, occurs as soon as the thinner has evaporated.

***Drying conditions*** All coatings used by the painter and decorator are classified as air-drying, meaning that exposure to the atmosphere is essential to start the drying process. Air is necessary to ensure a satisfactory drying action; an absence or shortage of air during the drying process may prevent or retard the forming of a surface coating film. Temperature and moisture also affect the speed of drying.

### Types of film

(a) *Convertible coating (non-reversible film)*
A coating which goes through a chemical change upon drying, so producing a material chemically different from its liquid state. It cannot be converted back to its liquid state by the use of its original thinner.

(b) *Non-convertible coating (reversible film)*
Some coatings are formed by dissolving a resin in a solvent. When the material is applied, the solvent evaporates, leaving the resin in its original form spread evenly over the surface. The change has been physical only and the film can be reversed to the liquid state by the application of its original solvent.

### Types of finish

When surface coatings dry, they produce films with varying grades of sheen. The range extends from flat or matt finishes, which have no sheen, through increasing degrees of lustre to high gloss finishes. The degree of sheen is dependent upon the angle at which light is reflected from the surface. A smooth surface gives a uniform angle of reflectance and appears glossy. A rough surface causes a scattering of reflected light waves and appears less glossy or even matt.

Smooth surfaces are produced by finely ground pigments in small proportion to film-former. Rough surfaces are produced by slightly larger extruder particles used in a higher proportion to film-former (Fig. 2.4).

Alkyd gloss, matt and eggshell finishes contain similar materials, their difference lying in the sheen resulting from the type, size and quantity of pigment used. Matt paints have a high proportion of pigment to film-former, whereas gloss paints have less pigment and more film-former. This is known as the pigment : binder ratio.

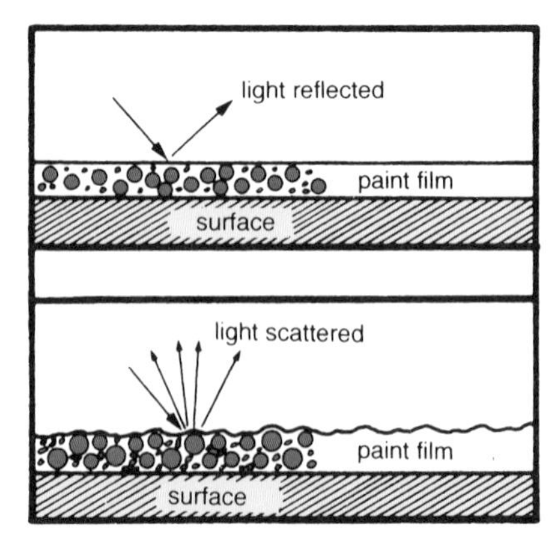


**Fig. 2.4** Types of finish

## 2.1.3 Pigments and extenders

### Pigment

Pigment is a solid component of paint. A vast number of pigments are used in paint manufacture, obtained from numerous sources which can broadly be classified as organic and inorganic.

*Organic* matter is that which at some time has been living. The principal sources of organic pigment are crude petroleum and coal-tar oil. The pigments are produced as dyestuffs (sometimes referred to as lakes) in a wide range of colours; because of their fineness, many are stuck on to bases of extenders to give them substance. Vandyke brown is one organic pigment which is not a dyestuff; it is obtained direct from peaty earth.

*Inorganic* pigments may be derived from:

(a) coloured earths, sometimes known as earth or natural pigments, which are mixed and ground to a fine powder, e.g. yellow ochre, raw umber;
(b) metals chemically treated to render them into fine powders, e.g. titanium white, white lead, iron oxide;
(c) combinations or heat treatments of chemical compounds, e.g. Prussian blue, zinc chromate, lithopone.

***use*** (i) To give body and covering power to paint: most widely used for this purpose is titanium dioxide (white).
(ii) To colour the paint: a property which all pigments possess.
(iii) To aid the film-former in protecting the substrate; those which do this include red lead, zinc chromate, calcium plumbate.

***Stainers/tinters (colourants)***

The painter and decorator usually obtains pigments in one of the following forms for use in matching colours or staining glazes.

(a) Oil tinters, which are ground in an oil medium to a thick paste and available in pack or tube form.
(b) Universal tinters, which are generally thinner than oil tinters but have the advantage of being compatible with any type of paint.

### Classification

Table 2.1 classifies the most common pigments according to colour. Beside each pigment is the following information:

Column 1 indicates its source.

Column 2 states its specific gravity (SG). This is a measure of the pigment's density compared with an equal volume of water: water's SG is 1. Pigments with an SG of 2 are not very dense; pigments with an SG of 8 are very dense and would cause paints to be heavy, and would also tend to settle out of the paint in storage.

Column 3 gives a guide to the pigment's ability to obliterate a surface. Those with a rating of 1 have very good opacity, and those with 5 have poor opacity.

Column 4 gives an indication of the power of a pigment to stain a white paint, or of a white pigment to resist staining: 1 = excellent staining power or reducing power; 5 = poor.

Column 5 indicates the pigment's ability to resist the three most common elements which may affect the colour of pigments: 1 = very good resistance; 5 = no resistance.

### Extenders **(Table 2.2)**

Extenders are also solid constituents of paint and are used in many types of surface coating. They are obtained from natural materials which are mined and ground to a fine powder. They are generally white in colour, and are transparent in oil, therefore having no influence on the colour of oil paints. They are sometimes referred to as 'mineral whites'.

***Properties***

(i) Allow certain paints to be applied with ease.
(ii) Give greater adhesive properties to undercoats.
(iii) Provide a 'roughness' in the film of undercoats to improve adhesion of the finishing paint.
(iv) Prevent certain dense pigments settling out of the paint.
(v) Give bulk to certain paints which do not contain sufficient pigment content, e.g. black, or bright colours which do not contain white.
(vi) Provide added hardness to paint films.
(vii) Improve the flowing properties of certain paints.

## 2.1.4 Film-formers

The film-formers used in modern surface coating can be broadly divided into four groups: drying oils, resins, oil-modified resins and other film-forming substances.

### Drying oils **(Table 2.3)**

Drying oils are vegetable oils which, when exposed to air, convert from liquid to solid. The conversion (they are known as convertible coatings) is a slow process: the oil absorbs oxygen and slowly thickens until it dries to a solid film. This drying process is called oxidation. Drying oils are occasionally used as the sole film-former but more often are mixed with resins (see resins and oil-modified resins).

### Resins

Resins form the basis of most film-formers and provide the properties required by the range of coatings described in Sections 2.2–2.5. They are obtained from two main sources.

***Natural resins*** (Table 2.4) are obtained from trees and may be either (a) extracted from living trees or (b) fossil resins from the remains of trees, mostly tropical, long extinct and lying fossilised in swampy areas in many parts of the world. Because the supply is limited, the cost of collecting them is rising and the quality varies considerably. Natural resins have a limited use in the paint industry.

***Synthetic resins*** (Table 2.5) are manufactured under strict quality control to specifications designed for a precise purpose. They may be either (a) organic, obtained from carbonaceous materials, the chief

**Table 2.1** Properties of pigments

| | Origin | | | | | | Resistance to | | | | |
|---|---|---|---|---|---|---|---|---|---|---|---|
| *Pigment* | ***Earth*** | ***Organic*** | ***Inorganic*** | ***Specific gravity*** | ***Opacity*** | ***Tinting strength*** | ***Alkali*** | ***Acid*** | ***Light*** | ***Other characteristics*** | ***Principal use*** |
| Whites | | | | | | | | | | | |
| Titanium dioxide (rutile) | | | / | 4.2 | 1 | 1 | 1 | 1 | 1 | Completely stable | In most paints |
| Zinc oxide | | | / | 5.6 | 3 | 3 | 1 | 4 | 1 | Becomes brittle | Fungicidal paints/stain block primers |
| Lithopone | | | / | 4.3 | 2 | 2 | 1 | 4 | 2 | Chalks on exposure to weather | Undercoats |
| White lead | | | / | 6.7 | 3 | 3 | 1 | 5 | 1 | Discoloured by sulphides, toxic | Exterior wood primers |
| Antimony oxide | | | / | 5.7 | 2 | 3 | 1 | 4 | 1 | Discoloured by hydrogen sulphide | Fire-retardant paints |
| Calcium plumbate | | | / | 5.7 | 2 | 3 | 1 | 4 | 1 | Rust-inhibitive, toxic | Primers for wood and galvanised iron |
| Zinc phosphate | | | / | 3.3 | 3 | | 3 | 3 | 1 | Harmful to environment in larger quantities, rust-inhibitive | Metal and some types of universal primer |
| Yellows | | | | | | | | | | | |
| Yellow lead chromes | | | / | 5.6 | 3 | 3 | 5 | 4 | 3–5 | Unsuitable for water paints, toxic | Tinter for oil paints |
| Zinc chromate | | | / | 3.4 | 4 | 4 | 3 | 3 | 1 | Rust-inhibitive | Metal primers |
| Ochre | / | | | 2.9 | 2 | 3 | 1 | 2 | 1 | Tends to retard drying | Tinter for all paints |
| Iron oxide yellow | | | / | 5.1 | 2 | 3 | 1 | 2 | 1 | Excellent fastness on weathering | Ideal for finishing paints |
| Raw sienna | / | | | 4.0 | 4 | 5 | 1 | 1 | 2 | Transparent in oil | Scumbles and glazes |
| Hansa yellow IOG | | / | | 1.6 | 3 | 3 | 2 | 2 | 1 | Excellent fastness on exposure | Decorative paints only |
| Reds | | | | | | | | | | | |
| Orange and scarlet chromes | | | / | 6.1 | 3 | 3 | 4 | 4 | 3–5 | Unsuitable for water paints | Tinter for undercoat and gloss |
| Toluidine red | | / | | 1.4 | 2 | 2 | 1 | 3 | 2 | Affected by heat | Decorative finishes |
| Red iron oxide | / | | | 5.1 | 3 | 3 | 1 | 3 | 1 | Stable | Finishing paints |
| Red lead | | | / | 9.0 | 2 | 3 | 3 | 5 | 4 | Rust-inhibitive, toxic | Metal and timber primers |
| Para red | | / | | 1.5 | 2 | 2 | 2 | 2 | 3 | Reduced strengths fade in water | Decorative finishes only |
| Bordeaux | | / | | 1.3 | 2 | 2 | 1 | 1 | 2 | Unsuitable for cellulose paints | Tinter for most paints |

**Table 2.1** Properties of pigments (cont'd)

| | Origin | | | | | | Resistance to | | | | |
|---|---|---|---|---|---|---|---|---|---|---|---|
| | *Earth* | *Organic* | *Inorganic* | *Specific gravity* | *Opacity* | *Tinting strength* | *Alkali* | *Acid* | *Light* | *Other characteristics* | *Principal use* |
| Dioxazine violet | | / | | 1.4 | 2 | 1 | 1 | 1 | 1 | Expensive | Tinter for all finishes |
| Blues | | | | | | | | | | | |
| Ultramarine blue | | | / | 2.3 | 4 | 3 | 3 | 4 | 3 | Loses gloss on weathering | Tinter for all paints |
| Prussian blue | | | / | 1.8 | 5 | 1 | 5 | 3 | 3 | Fades in absence of light | Tinter for oil/alkyd paints |
| Phthalocyanine blue | | | / | 1.6 | 1 | 1 | 1 | 1 | 1 | Completely stable | All types of paint |
| Greens | | | | | | | | | | | |
| Phthalocyanine green | | | / | 2.2 | 1 | 1 | 1 | 1 | 1 | Completely stable | All types of paint |
| Mid chrome green | | | / | 4.2 | 3 | 3 | 5 | 4 | 4 | Turns blue on exposure | All types of oil/alkyd finishes |
| Chromium oxide green | | | / | 5.1 | 4 | 4 | 1 | 2 | 1 | Heat-resistant | Chemical/heat-resistant paints |
| Browns | | | | | | | | | | | |
| Raw umber | / | | | 4.0 | 2 | 3 | 1 | 2 | 1 | Semi-transparent | Scumbles, glazes and tinters |
| Burnt umber | / | | | 3.5 | 2 | 3 | 1 | 2 | 1 | Completely stable | Tinter for all paints |
| Vandyke brown | / | | / | 1.5 | 5 | 4 | 1 | 1 | 1 | Brittle and transparent in oil | Water-colour graining |
| Burnt sienna | / | | | 3.5 | 3 | 4 | 1 | 1 | 1 | Completely stable | Tinter for all paints |
| Blacks | | | | | | | | | | | |
| Carbon black | | | / | 1.8 | | 2 | 1 | 1 | 1 | Brown undertone | Black gloss finishes |
| Lamp black | | | / | 1.8 | | 3 | 1 | 1 | 1 | Blue undertone | Tinter for all paints |
| Graphite | | | / | 2.2 | | 5 | 1 | 1 | 1 | Leafing properties | Anti-corrosive paints |
| Metallics | | | | | | | | | | | |
| Micaceous iron oxide | / | | | | 1 | | 1 | 1 | 1 | Leafing properties | Paints for primed steel-work |
| Aluminium powder | | | / | 2.6 | | | | | 1 | Leafing or non-leafing types | Sealers and heat-resisting paints |
| Bronze powder | | | / | 8.9 | | | | 4 | 1 | Tarnishes on exposure | Decorative metallic paints |
| Zinc dust | | | / | 7.0 | 1 | 4 | | | 1 | Protects iron and steel | Zinc-rich paints for metal |
| Metallic lead | | | / | 11.4 | | | | 1 | | Rust-inhibitive, settles on storage | Metal primers |

**Table 2.2** Properties of extenders

| Extender | Origin: Earth | Organic | Inorganic | Specific gravity | Principal use |
|---|---|---|---|---|---|
| Barytes/barium sulphates (also blanc fixe) | / | | | 4.5 | Pigment in water paints<br>Extender mainly in undercoats, contributes to opacity and keeps pigments suspended |
| Calcium carbonates (chalks/calcites) | / | | | | Cheap, all-round extender for filling properties and sheen control, mostly for interior coatings |
| Terra alba | / | | | 2.3 | Water paints<br>Primers<br>Sealers<br>Fillers |
| Witherite | / | | | 4.3 | Improves durability and colour permanence |
| Paris white or whiting | / | | | 2.7 | Pigment in cheap water paints<br>Gives body to undercoats |
| Silica | / | | | 1.9 | Very good flattening agent<br>Improves inter-coat adhesion<br>Road line paints |
| China clay | / | | | 2.6 | Aluminium silicates/kaolin, contributes to opacity, structure<br>Control agent, aids suspension of pigments |
| Fuller's earth | / | | | 2.5 | Anti-settling agent |
| Talc | / | | | 2.8 | Magnesium silicate, leafy<br>Flattening agent, anti-settling and sheen/abrasion control properties |
| Asbestine | / | | | 2.9 | Suspending and flattening agent<br>Used in wood fillers |
| Mica | / | | | 2.8 | Resistant to shrinkage, chalking and cracking<br>Good moisture-barrier<br>Transparent |
| Wollastonite | / | | | 2.9 | Extender for decorative paints |
| Opacifying beads | | / | | | Organic polymers, forming hollow spheres, contribute to opacity |

**Table 2.3** Properties of drying oils

| Oil | Properties | Use |
|---|---|---|
| Linseed oil (obtained from the seed of the flax plant) | Pale yellow in colour<br>Dries within 7 days under good drying conditions<br>Saponified by the action of strong alkalis<br>Darkens with age<br>Limited water resistance | Mixed with red or white lead for metal and timber primers<br>Oil-modified resin mediums<br>Chemically combined with alkyd, epoxy and urethane resins |

**Table 2.3** Properties of drying oils (cont'd)

| *Oil* | *Properties* | *Use* |
|---|---|---|
| Boiled oil (linseed oil heated with driers added) | Dark brown in colour<br>Dries within 12 hours<br>Improved water resistance, compared with linseed oil<br>Good adhesion<br>Saponified by alkalis<br>Darkens and becomes brittle with age | Oil-modified resin mediums |
| Stand oil (partially polymerised oil) | Pale yellow in colour<br>High viscosity<br>Excellent flow<br>Good flexibility<br>Not alkali-resistant | Oil-modified resin mediums<br>Some wood and metal primers<br>Manufacture of alkyd resins |
| Soya bean oil | Pale yellow<br>Very slow-drying<br>Does not darken with age<br>Not alkali- or water-resistant | Combined with resins, particularly in pale and non-yellowing paints |
| Tung or China wood oil | Milky white in colour<br>Drying time similar to linseed oil<br>Resistant to alkali and water | Combined with resins for use in alkali-resisting and water-resisting coatings |
| Tobacco seed oil | Slow-drying | Chemically combined with alkyd resins |
| Oiticica oil | Similar properties to tung oil | Sometimes used as an alternative to tung oil |
| Dehydrated castor oil | Can dry in under 5 hours<br>Pale colour, non-yellowing | Chemically combined with alkyd and epoxy resins |

**Table 2.4** Properties of natural resins

| *Name* | *Properties* | *Use* |
|---|---|---|
| Copal<br>Fossil resins obtained mainly from the Congo | Very hard<br>Pale colour | Oil varnishes |
| Manila copal<br>Fossil resins obtained from Manila | Bleed-resistant | Road-marking paints |
| Kauri<br>Fossil resin obtained from New Zealand | Pale colour<br>Combines well with oil | Oil varnishes |
| Damar<br>Recent resin tapped from living trees in Malaysia | Very flexible | Cellulose lacquers<br>Shellac varnishes |
| Lac | Soluble in alcohol | Knotting |
| Recent resin obtained from excreta of tree-feeding insects in India | Bleed-resistant | Button polish<br>French polish |

**Table 2.5** Properties of synthetic resins

| *Name* | *Properties* | *Use* |
|---|---|---|
| Oil-modified alkyd resins<br>Containing over 60% of a drying oil | Good flow<br>High gloss<br>Good flexibility<br>Weather-resistant<br>Pale colour<br>Not alkali-resistant | Some air-drying primers<br>Most undercoat, gloss and eggshell finishes<br>Varnishes |
| Modified to a gel or thixotropic structure | Good brushability<br>High gel strength<br>Fast gel recovery rate<br>Heavy film build<br>Non-drip | Thixotropic paints<br>One-coat paints |
| Containing non-drying oils | Very slow/non-drying<br>Pale colour<br>Soft | Plasticiser in cellulose paints |
| Epoxy or epoxide resins<br>2-pack or 'cold cure' type | Excellent adhesion<br>Excellent water resistance<br>Excellent chemical resistance<br>Excellent abrasion resistance<br>Extremely hard film | Chemical-resistant coatings<br>Abrasion-resistant coatings<br>Water-resistant coatings |
| 1-pack epoxy esters (modified with drying oils) | Chemical-resistant, but much less than 2-pack types | Maintenance paints for factories |
| Polyurethane resins<br>2-pack or 'cold cure' type | Extremely hard film<br>Excellent abrasion resistance<br>Excellent weather resistance<br>Excellent chemical resistance but less alkali-resistant than epoxy | Chemical-resistant finishes<br>Abrasion-resistant coatings |
| 1-pack based on drying oils (urethane oil) | Films are harder, tougher and more water-resistant than most alkyd resins | Interior and exterior gloss finishes<br>Clear wood finishes |
| Polyvinyl acetate (PVA) emulsions<br>Copolymer | Good colour, non-yellowing<br>Good adhesion<br>Good water resistance<br>Alkali-resistant<br>Good washability | Adhesives<br>Emulsion paints |
| Acrylic emulsions<br>Copolymer | Excellent adhesion<br>Water-white colour<br>Non-yellowing<br>Very good washability<br>Good alkali resistance<br>Very good external durability | Emulsion paints<br>Timber primers<br>Quick-drying undercoats<br>Adhesives<br>Masonry paints |
| Maleic resin | Very pale in colour<br>Non-yellowing<br>Good gloss<br>Poor resistance to chemicals | Combined with alkyd resins for non-yellowing paints<br>Pale varnishes |

**Table 2.5** Properties of synthetic resins (cont'd)

| Name | Properties | Use |
|---|---|---|
| Urea/formaldehyde resin (2-pack acid catalysed) | Very pale colour<br>Non-yellowing<br>Hard glossy films<br>Heat-resistant | Clear lacquers for bar counters and furniture |
| Silicone resin | Water-resistant<br>Heat-resistant up to 475°C | Clear waterproofing solutions<br>Heat-resisting paints |
| Polyvinyl butyryl | Excellent adhesion | Etch primers |

**Table 2.6** Properties of oil-modified resins

| | | |
|---|---|---|
| Oil-modified (natural) resins (natural resin and linseed or other drying oil) | High gloss<br>Flexible film<br>Saponified by alkalis<br>Limited water resistance<br>Yellows on ageing | Some primers<br>Some undercoats<br>Varnish<br>Gold size |
| Oil-modified (synthetic) resins (synthetic resin and tung or other oils) | Water-resistant<br>Alkali-resistant | Acid- and alkali-resistant paints and varnishes (limited resistance)<br>Boat varnish |

source being crude petroleum oils; or (b) inorganic: only one resin commonly used in paint manufacture, silicone resin, is obtained from material such as sand and quartz.

Most surface coatings contain a synthetic resin-based film-former. Most decorative solvent-based paints are based on oil-modified resins; these are not mixtures of oils and resins, but the result of a chemical combination of the two during the manufacture of the resin. Most decorative water-based paints are based on emulsion copolymers.

### Oil-modified resins (Table 2.6)

Combinations of oil and resin are usually carried out during the manufacture of the resin, the resultant oil-modified resin being chemically and physically stable. A few mixtures of oil and resin are still used. These are prepared by melting the resin and stirring in the oil.

### Other film-forming substances (Table 2.7)

A few organic substances other than oils and resins are used in some paints and varnishes.

## 2.1.5 Thinners and driers

### Thinners (solvents)

Thinners are colourless and are selected according to the type of paint in which they are employed. They are incorporated into the liquid part of the paint but evaporate from the film during the drying process.

***Properties***

(i) Act only on the medium of the paint.
(ii) Remain long enough in the wet paint to allow easy application.
(iii) Are free from residue which may affect the film.
(iv) Are colourless and do not cause discoloration.
(v) Are not strong enough to soften underlying coatings.

***Use*** When used in decorative paints, their main use is to reduce the viscosity of primers, fillers and undercoats, making them easier to apply and ensuring penetration into absorbent surfaces. Thinning should only be carried out to manufacturers' instructions, as over-thinning can lead to poor opacity, loss of gloss and other film weaknesses.

***Types of solvent*** Seven main classes of solvent are used in paint:

| | |
|---|---|
| Hydrocarbons | Ketones |
| Alcohols | Chlorinated hydrocarbons |
| Esters | Water |
| Ethers | |

**Table 2.7** Properties of other film-forming substances

| Medium | Properties | Use |
|---|---|---|
| Bitumens (asphalt, bitumen and pitch)<br>Water-based versions available | Very dark colour<br>Excellent water resistance<br>Resistant to acids<br>Resistant to alkalis<br>Low cost<br>Colour bleeds through oil-resin paints<br>Softens on heating | Damp-proof compounds<br>Bituminous paints |
| Rubber (natural and synthetic, treated to form chlorinated rubber, isomerised rubber, cyclised rubber, neoprene, hypolon) | Excellent water resistance<br>Chemical-resistant<br>Very flexible<br>Quick-drying<br>Good flow | Chemical-resistant paints<br>Water-resistant paints |
| Cellulose (nitrocellulose: cellulose treated with nitric acid) | Rapid-drying<br>Hard film<br>Chemical-resistant<br>Almost impossible to brush<br>Tends to be brittle | Cellulose paints and lacquers<br>Cellulose stoppers and fillers |

***Hydrocarbons***

(a) Those obtained during the distillation of crude oil: petrol, white spirit and paraffin.
(b) Aromatic hydrocarbons, obtained from the distillation of coal tar: solvent naphtha, benzene, toluene and xylol. These are used in chlorinated rubber and nitrocellulose paints.

***Alcohols***

These used to be produced by distilling fermented potatoes and starch. Industrial alcohols are now produced by a chemical process.

Propyl alcohol, butyl alcohol, amyl alcohol, benzyl alcohol, isopropanol alcohol, ethyl alcohol and methylated spirits are commonly used for dissolving resins and as thinners for spirit paints and varnishes.

***Esters***

These are derivatives of acids and include methyl acetate, ethyl acetate, butyl acetate and amyl acetate. They are used as solvents for nitrocellulose finishes.

***Ethers***

Prepared by chemical action on certain other solvents. Glycol ether and methyl glycol are used as solvents for cellulose resins and many other synthetic finishes.

***Ketones***

Manufacture is similar to that of alcohols. They include acetone, methyl ketone and methyl ethyl ketone, and are used extensively in nitrocellulose finishes.

***Chlorinated hydrocarbons***

Produced by a reaction of chlorine on various materials such as methane. Their main disadvantage is toxicity, but they have the advantage of being non-flammable. Two main types are in use: (a) methylene chloride, a very strong solvent used as the main constituent in paint removers; (b) trichlorethylene, used as a degreasing agent.

Both types give off poisonous gases when burnt.

***Water***

All emulsion paints are thinned with water. Water is cheap, readily available, non-flammable and non-toxic.

***Flash point***

Flash point is the lowest temperature at which a substance will give off flammable vapour. With the exception of chlorinated hydrocarbons and water, all the above solvents are flammable, having different flash points. (See Section 6.6.)

### Driers

Driers are additives to paint, used to speed up the drying process. They increase the capacity of the film-former to absorb oxygen from the atmosphere. An oil paint which takes approximately 4 days to dry without driers will dry overnight with the addition of 1 per cent drier.

***Properties***

(i) Speeds up the drying process.
(ii) Mixes easily with drying oil.

***Use*** Normally incorporated by the manufacturer into the drying oil. Can be used by the decorator in oil colours for signwriting or scumble and glazes where driers are not normally added.

Where driers are used, they should only be added very sparingly as overuse will lead to cracking or shrivelling. Solvent-based paints typically have drier already added by the paint manufacturer.

***Composition*** Driers are soluble compounds of metals, most commonly manganese or calcium mixed with other ingredients. The two main types are naphthenates and octanoates.

The metals have their own characteristics and are often used in combination to give the desired properties.

## 2.1.6 Properties of paint

By the careful selection and mixing of the film-formers, pigments and thinners available, the paint manufacturer can produce materials that under most known conditions will satisfy the requirements of four reasons for painting:

(a) to protect the surface (*preservation*)
(b) to make the surface washable (*sanitation*)
(c) to decorate the surface (*decoration*)
(d) to identify the surface (*identification*)

In addition to these properties, paint must have certain basic qualities to ensure that it (i) is appliable; (ii) will dry in a reasonable length of time; (iii) can adapt to the physical changes of the surface; (iv) will maintain its function for an acceptable period.

### Consistency

This can also be referred to as the fluidity/viscosity of paint. Certain trade terms are used to describe it, such as 'round' for thick paint, 'sharp' for thin paint and 'viscous' for syrupy or sticky paint.

Consistency depends on:

(a) *Type of film-former:* gloss paints and varnishes feel 'sticky' when being applied, compared with the smooth feel of emulsion paint. The consistency of paint changes as the thinner evaporates or the set begins to take place.
(b) *Type of pigment or extender:* coarse pigments drag on the brush, and therefore influence the paint's fluidity; for example, a masonry paint which contains granules of stone is much more difficult to brush than one containing no coarse aggregate.
(c) *Temperature:* an ideal temperature for painting is 15°C. Below this temperature, the resins and oils become more viscous and 'drag' on the brush or roller. Paints become very difficult to apply at temperatures below 5°C, and water-based paints can be damaged irreversibly. Higher temperatures can also affect brushing and rolling consistency as the thinner evaporates quickly, making the paint less fluid.
(d) *Thixotropy:* many paints are deliberately made in a very 'round' condition and are called thixotropic or *gel* coatings. These include a component that gives the paint the qualities of thickness in the container but fluidity when brushed or rolled. Thixotropic paints allow thicker films to be applied.
(e) *Surface condition:* absorbent surfaces readily absorb the liquid part of the paint, so that brushing and rolling may become difficult.

### Opacity (hiding power)

Opacity is the property of a paint to obliterate the surface to which it is applied. Although the pigment is the principal component responsible for opacity, a careful blending of appropriate film-formers and pigments is also important. Many strong-coloured paints have poor opacity, mainly because they contain no white base pigments, and the coloured pigments that produce the hue have poor opacity. Many yellow, orange and red paints are almost transparent.

### Spreading capacity/spreading rate

This is defined as the area covered by a given quantity of paint. It is influenced by both the film-former and the pigment and is closely related to paint consistency; for example, a smaller area will be covered by a litre of heavy-bodied or 'sticky' paint than of a thin or smooth paint. Rough, textured or absorbent areas will require a greater quantity of paint to cover them than will a smooth surface.

### Adhesion

The resin in the paint provides its ability to stick to a surface. An alkyd gloss paint will adhere to a smooth plastic surface better than a spirit varnish.

The condition of the surface is vitally important to the maximum adhesion of the paint film.

### Elasticity (flexibility)

Timber and metal surfaces naturally expand and contract throughout their life. The degree of movement depends upon temperature and humidity, and varies considerably. Paints applied to such surfaces must be able to stretch and shrink to the same degree. Most oil and oil/resin film-formers have this ability. Some surface coatings, such as shellac knotting, are brittle and readily crack when applied to flexible substrates.

### Density

Pigments with high specific gravity tend to settle out of paint to form a solid mass at the bottom of the container. They also affect the spreading property of the paint as they tend to 'drag'. The most obvious effect of dense pigment is on the weight of the paint: a litre of lead primer will be considerably heavier than a litre of leadless primer.

### Drying speed, drying time

Paints are manufactured to 'stay open' long enough for easy and even application. The hardening process is much longer and varies according to the type of film-former used. Oil paints set-up in less than $\frac{1}{2}$ hour, and touch-dry in 4–6 hours, can be overcoated after 12 hours but do not attain maximum hardness for many weeks. Emulsion paints have a similar set-up time, but are touch-dry after 1–2 hours and can be recoated in 2–4 hours.

Another important aspect of drying speed is a paint's 'wet-edge' time. This is similar to its set-up time, and is the length of time that an applied coat takes to become so tacky that the next section cannot be brushed into it without showing a join. It is a characteristic of paint which must be accurately judged when covering large areas.

### Flow

Flow is the quality of paint that enables the painter to apply coatings so that brush or roller marks are not visible. Some paints, such as gloss paints, have good flowing properties and help the painter to produce a texture-free finish. Other paints, such as undercoats and flat finishes, have little flow and require different application techniques to produce an acceptable finish.

### Durability

Correctly specified paint systems, applied skilfully to sound, properly prepared surfaces will provide the required protective, washable or decorative coating. The effective life of the coating depends principally on (a) the destructive effect of the environment, i.e. air pollution, dampness, salt spray, ozone, ultraviolet light, industrial chemicals, heat and cold; (b) the natural breakdown of the film-former as it ages and becomes more and more brittle. A paint which may last for 10 years in a rural area may break down in 2 years in a heavily polluted industrial area. An average life may be stated as 5 years.

# 2.2 Primers, undercoats and stoppers

## 2.2.1 Acrylic primer/undercoat

A water-thinnable paint made to fulfil the combined functions of primer and undercoat. Available mainly in white, although some manufacturers supply a limited range of colour shades.

***Composition***

***Pigment*** Titanium dioxide, extenders and coloured pigments.

***Film-former*** Acrylic latex.

***Thinner*** Water.

***Properties***

Quick drying enables same-day overcoating.

Water-based: cleaning and thinning can be carried out quickly and cheaply.

Flexible film with good adhesion.

Non-toxic.

Non-reversible coating.

***Use*** Can be applied direct to interior and exterior woodwork, fibre hardboard, fibre insulating board (except fire-retardant treated types), chipboard, paper and dry plaster. Not to be used over bare metal, e.g. steel fixings, which should be 'spot primed' with a metal primer.

***Application***

***Drying method*** By coalescence.

***Time*** 30 minutes normally, but may be considerably longer in cold, damp or very humid conditions.

***Overcoating*** 2 hours or less.

***Cleaning solvent*** Clean water.

***Spreading capacity*** 60–80 $m^2$ per 5 litres, depending upon the porosity of the surface.

*Note:* Whilst this product has good crack-filling properties, it can have poor flow.

## 2.2.2 Alkali-resisting primer

Primer specially designed for surfaces which are alkaline in nature and require a solvent-based finish.

***Composition*** Based on resins and pigments which are resistant to alkali.

***Pigment*** Titanium dioxide and/or alkali-resistant pigments.

***Film-former*** Oil-modified resin containing tung oil.

***Thinner*** Follow manufacturer's recommendation 2010.

***Properties***

Alkali-resistant.

Resistant to limited water pressure from inside the substrate. Non-reversible coating.

***Use*** (i) Primer on all new and old building materials of an alkaline nature (see 3.4.5).

(ii) Primer for surfaces which have been coated or impregnated with fire-retardant chemicals, sealing general stains and sealing of fire-damaged surfaces.

(iii) Binding down powdery substrates.

***Application***

Brush recommended for maximum adhesion. On smooth surfaces, one full coat is sufficient, but on very porous or textured surfaces, two coats may be required.

***Drying method*** Autoxidation.

***Time*** Under normal conditions, 8–12 hours.

***Overcoating*** 16–24 hours. Can be overcoated with any decorative system, but if water-based materials are used, some cissing may occur due to the oily nature of the primer. It is advisable not to rub down the surface before overcoating, to avoid breaking the surface.
***Cleaning solvent*** White spirit.
***Spreading capacity*** Depends entirely on the porosity and texture of the surface, but the following will give a general guide.

| | $m^2$ per 5 litres |
|---|---|
| Smooth, hard plaster surfaces | 45–55 |
| Cement rendering slightly textured | 35–45 |
| Stucco, rough cast | 15–30 |

## 2.2.3 Aluminium wood primer

A dull, metallic grey oil-based primer.
***Composition***
***Pigment*** Non-leafing aluminium paste. Non-leafing aluminium particles are scattered throughout the film at various angles and provide good build and adhesion for subsequent coats (Fig. 2.5).
***Film-former*** Tung oil/phenolic resin.
Tung oil/coumarone resin.
***Thinner*** Follow manufacturer's recommendation 2010.
***Properties***
Good sealing properties over surfaces which are likely to bleed.
Good opacity.
Very good spreading capacity.
Self-knotting.
Good moisture-barrier.

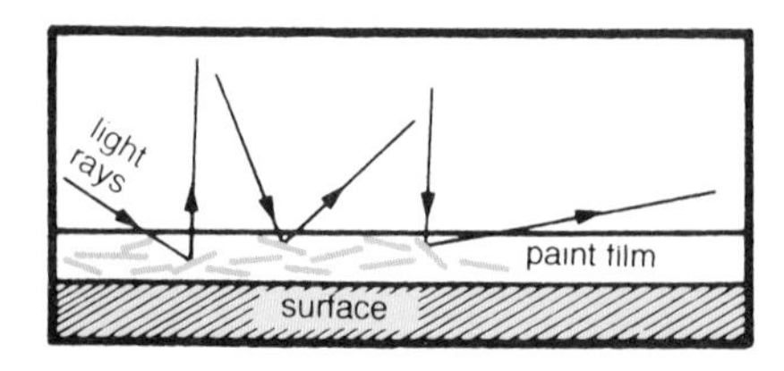

**Fig. 2.5** Structure of aluminum wood primer

Heat-reflecting.
Not all types are lead-free.
Non-reversible coating.
***Use*** (i) Wood primer specially suitable for resinous timber such as Columbian pine, Oregon pine, afrormosia, gurgion and iroko.
(ii) Sealer over surfaces previously treated with coal-tar wood preservatives.
(iii) To seal old bitumen-coated surfaces, but where thick layers of bitumen are present, cracking may occur.
***Application***

***Drying method*** Oxidation.
***Time*** 4–6 hours. Because of the laminar structure of the pigment, the lower parts of the film may remain soft after the surface has dried, due to trapped solvents and lack of oxygen.
***Overcoating*** Allow 24 hours before applying a decorative system.
***Cleaning solvent*** White spirit.
***Spreading capacity*** 80 $m^2$ per 5 litres.

*Note:* Because of its colour, an extra undercoat may be required if a light-coloured finishing coat is used.

## 2.2.4 Quick-drying acrylic metal primer

A protective water-based primer for use on non-ferrous metal surfaces.
***Composition***
***Pigment*** Zinc phosphate, titanium dioxide, mineral extenders and colouring pigments.
***Film-former*** Acrylic co-polymer emulsion.
***Thinner*** Water.
***Properties***
Excellent adhesion.
Non-toxic.
Non-reversible.

***Use*** Primer for use on new, bright and weathered galvanised metal and other non-porous surfaces, both inside and outside. Not suitable for hand-prepared ferrous surfaces.

***Application***

***Drying method*** Coalescence.

***Time*** Overcoatable in 4–6 hours, longer in cold, damp conditions.

***Cleaning solvent*** Water.

***Spreading capacity*** 60–90 m$^2$ per 5 litres depending on profile.

## 2.2.5 All-purpose primer

Also called multi-surface primer or universal primer.

Suitable for the priming of most interior and exterior surfaces.

***Composition***

***Pigment*** Zinc phosphate, titanium dioxide, mineral extender and colouring pigments.

***Film-former*** Alkyd resin.

***Thinner*** Follow manufacturer's recommendation 2010.

***Properties***

Good rust inhibition.

Non-toxic.

Non-reversible coating.

***Use*** Suitable for interior and exterior use on wood, metal and plaster surfaces, giving a smooth, sound surface and provides good adhesion for subsequent coats of paint. Two coats required over badly rusted metal. For highly resinous wood, galvanised steel and new plaster, use specialist primers.

***Application***

***Drying method*** Autoxidation.

***Time*** Overcoatable in 6 hours, longer in cold, damp conditions.

***Spreading capacity*** Dependent on the type and porosity of surface being coated, guide as follows:

35–55 m$^2$ per 5 litres.

Metal 40–60 m$^2$ per 5 litres.

Plaster 25–45 m$^2$ per 5 litres.

## 2.2.6 Calcium plumbate primer

Good all-purpose primer, off-white or grey in colour. Designed for use as a first coat on galvanised iron.

***Composition***

***Pigment*** Should contain not less than 33 per cent lead in the form of calcium plumbate mixed with small quantities of titanium, asbestine or barytes.

***Film-former*** Linseed oil.

***Thinner*** Follow manufacturer's recommendation 2010.

***Properties***

Can be used on exterior or interior work.

Excellent rust inhibition.

Prevents the spread of corrosion on ferrous metals where paintwork is scratched or broken.

Excellent adhesion.

Good opacity.

Good flow and levelling properties.

Non-reversible coating.

***Use***
(i) Specially suited for priming galvanised iron without the need for weathering.
(ii) Primer for iron and steel where light colours are used.
(iii) Priming hardwoods such as oak. Composite units, e.g. wood and metal window frames, can be primed with the same primer.

***Application***

***Drying method*** Oxidation.

***Time*** Touch-dry in 6–8 hours.

***Overcoating*** The primed surface should be left for 24 hours before overpainting.
***Cleaning solvent*** White spirit.
***Spreading capacity***
***Woodwork*** 40–60 $m^2$ per 5 litres.
***Metalwork*** 50–60 $m^2$ per 5 litres.
*Note:* Not recommended over aluminium; not suitable for immersion in water for long periods; toxic.

Extensively used in the past as a good all-purpose primer, but now scarcely used because of its toxic lead content. Replaced mainly by metal primer zinc phosphate. These primers contained a minimum of 33 per cent lead in the form of calcium plumbate along with small quantities of other pigments.

## 2.2.7 Etch primer

Designed for pre-treatment of non-ferrous metals to ensure maximum adhesion of the paint system. Available in one- or two-pack materials, the latter offering greater stability and adhesion. Sometimes called wash primers.
***Composition*** Both types have the same basic components in differing proportions. The main difference is that in the two-pack paint, the acid is supplied separately and added just before use. The main components are:
***Pigment*** Zinc phosphate.
***Film-former*** Polyvinyl butyral resin.
***Thinner*** Butanol.
***Activator*** Phosphoric acid.

Plasticisers and extenders are also added.

***Properties***

Etch primers work on a principle of interaction between pigment, binder and substrate, resulting in excellent adhesion.
Can be used on composite units where different types of metal are used.
Matt finish, giving good key.
Corrosion-resistant.
Lead-free.
Non-reversible coating.

***Use*** For pre-treatment of non-ferrous metals, including metal spray and galvanised steel, where the surface has not been pre-treated by the manufacturer. It may also be used as an alternative to mordant solution in a closed-workshop situation.
*Note:* This material is moisture-sensitive and must be protected from condensation and must be overcoated within the recommended time.
***Application***

They should be applied in dry conditions as they are sensitive to water.
***Drying method*** Polymerisation.
***Time*** Touch-dry in 1–4 hours.
***Overcoating*** Overcoat within 12–16 hours. Follow manufacturer's recommendations for finishing systems.
***Cleaning solvent*** Only the manufacturer's recommended cleaning solvent should be used.
***Spreading capacity*** 80–120 $m^2$ per 5 litres. On abrasive-cleaned or cast metals, spreading capacity will be reduced owing to the greater surface area.
*Note:* Limited shelf life; once mixed, two-pack etch primers have a pot life of 8–24 hours; they give minimal protection if left unpainted, providing only a very thin, dry film (typically 5 mm).

## 2.2.8 Wood primer

A general-purpose solvent-based wood primer for interior and exterior use. Mostly supplied in white but some manufacturers offer in pink.
***Composition***
***Pigment*** Titanium dioxide and mineral extender.
***Film-former*** Oil-modified alkyd. The composition of the film-former can be adjusted to suit the requirements of the film; i.e. for outside use, a long oil resin would give a more flexible film than a short oil resin, which would dry quicker and harder, fulfilling the requirements of an interior primer.

***Thinner*** White spirit.

***Properties***

Satisfies the porosity of the wood.

Provides good adhesion for succeeding coats.

Non-toxic.

Non-reversible coating.

***Use*** For all types of soft and hard woods, inside and outside. Thin up to 10 per cent for use on hardwoods. Wood that is highly resinous should be primed with aluminium wood primer.

***Application***

Brushing ensures better penetration.

***Drying method*** Autoxidation.

***Time*** Touch-dry in 4–6 hours.

***Overcoating*** Any decorative paint system can be used. Overnight drying should be allowed.

***Cleaning solvent*** White spirit.

***Spreading capacity*** Depends upon porosity of wood. Guide 50–70 $m^2$ per 5 litres.

## 2.2.9 Preservative primer

Solvent-based penetrating primer for sealing and protection of exterior woodwork.

***Composition***

***Pigment*** Generally absent.

***Film-former*** Alkyd resin.

***Thinner*** White spirit.

*Note:* These products contain fungicides, and contact with the skin, plant and animal life should be avoided.

***Properties***

Low-viscosity product.

Penetrates deep into the wood to guard against rot and seals the grain to help keep out water.

Non-reversible coating.

***Use*** As part of specialist exterior paint systems that offer prolonged protection.

***Application***

***Drying method*** Autoxidation.

***Time*** Generally touch-dry within 2 hours.

***Overcoating*** Either within 2 hours or after fully dry, generally 16–24 hours.

***Cleaning solvent*** White spirit.

***Spreading capacity*** 75–125 $m^2$ per 5 litres, depending on surface porosity.

## 2.2.10 Knotting solution

Used to prevent staining and discoloration from knots and resinous streaks in softwood.

***Composition***

***Pigment*** Generally unpigmented but may contain low levels of titanium dioxide or extender pigment.

***Film-former*** Shellac.

***Thinner*** Methylated spirit and methoxy propanol.

***Properties***

Very quick-drying.

Interior and exterior use.

Highly flammable.

***Use*** Sealing knots and resinous streaks. Use a white/bleached shellac knotting solution when using water-based paints.

***Application***

***Drying method*** Evaporation.

***Time*** 5–10 minutes.

***Overcoating*** 10–15 minutes.

***Cleaning solvent*** Methylated spirit, immediately after use.

***Spreading capacity*** Not applicable as only used in small quantities.

## 2.2.11 Flexible filler

Also called flexible stopper. A water-based, lightweight filler for wood and masonry.

***Composition***

***Pigment*** Titanium dioxide and inert inorganic fillers.

***Film-fomer*** Acrylic latex.

***Thinner*** Water.

***Properties***

Excellent resistance to shrinkage and cracking.

Good flexibility and adhesion.

Easy to rub down.

***Use*** Use both inside and outside for making good minor imperfections in wood, plaster and brickwork.

***Application*** Putty knife or other suitable tool.

***Drying method*** Coalescence.

***Time*** Drying time is variable depending on depth of the repair. Allow at least overnight drying before rubbing down, longer in cold, damp conditions.

***Overcoating*** After rubbing down.

***Cleaning solvent*** Clean water.

***Spreading capacity*** Not applicable.

## 2.2.12 Spirit-based aluminium sealer

Designed for use over surfaces which have a tendency to soften or bleed through subsequent coats of paint.

***Composition*** The ingredients of these paints should be of a type that will not disturb or dissolve the surface or part of the surface to which they are applied.

***Pigment*** Aluminium flake non-leafing.

***Film-former*** Shellac or manila gum.

***Thinner*** Industrial methylated spirit.

***Properties***

Forms barrier film to insulate bleeding surfaces.

Quick-drying.

Good opacity.

Resistant to mild chemical fumes.

Seals water-stained surfaces.

Reversible coating.

***Use*** (i) Sealer for bituminous surfaces, dry, creosoted surfaces and nicotine stains.

(ii) Sealer for bleeding colours or inks in previously painted or papered surfaces.

***Application***

***Drying method*** Evaporation.

***Time*** 15–30 minutes.

***Overcoating*** Allow 1 hour before applying finishing system or papering.

***Cleaning solvent*** Methylated spirit.

***Spreading capacity*** Approximately 30–35 $m^2$ per 5 litres.

## 2.2.13 Primer sealer

A solvent-based sealer for highly porous interior surfaces.

***Composition***

***Pigment*** Titanium dioxide, zinc oxide and inert extenders.

***Film-former*** Oil-modified resin.

***Thinner*** White spirit.

***Properties***

Excellent sealing and adhesion properties.

Effective against a wide variety of water-soluble stains.

Some alkali resistance.

Non-reversible coating.

***Use*** Suitable for sealing highly porous, dry, friable interior surfaces prior to the application of an appropriate finishing system.

***Application***

***Drying method*** Autoxidation.

***Time*** 8–12 hours.

***Overcoating*** 16–24 hours.

***Cleaning solvent*** White spirit.

***Spreading capacity*** 25–45 $m^2$ per 5 litres, according to porosity.

## 2.2.14 Stabilising primer (also called stabilising solution)

A solvent-based primer to seal unstable areas, which remain powdery or chalky after the surface has been thoroughly prepared. May be clear or unpigmented, the clear having better penetrating properties.

***Composition***

***Pigment*** Sometimes added to give opacity, but generally absent.

***Film-former*** Oil-modified resin.

***Thinner*** White spirit.

***Properties***

Binds powdery surfaces.

Seals porous surfaces.

Will take any form of decoration.

For inside and outside use.

Non-reversible coating.

***Use*** To stabilise powdery, highly porous, lime-washed, distempered or cement-based ceilings and walls. This product is not intended to take the place of adequate preparation.

***Application***

***Drying method*** Autoxidation.

***Time*** 4–8 hours.

***Overcoating*** 16–24 hours, longer in cold, damp conditions.

***Cleaning solvent*** White spirit.

***Spreading capacity*** 40–60 $m^2$ per 5 litres, according to porosity.

## 2.2.15 Plaster sealer

A water-based primer formulated for use on dry, porous interior surfaces.

***Composition***

***Pigment*** Generally pigment-free.

***Film-former*** Alkyd emulsion polymer.

***Thinner*** Water.

***Properties***

Seals dry porous interior surfaces.

Not for use as a stain sealer.

Non-reversible coating.

***Use*** As a sealer on hardboard and papered surfaces including plasterboard.

***Application*** Brush; a short pile roller may also be used on larger areas that have a smooth surface.

***Drying method*** Autoxidation.

***Time*** Dependent upon temperature and humidity.

***Overcoating*** 18–24 hours.

***Cleaning solvent*** Clean water.

***Spreading capacity*** 60–90 $m^2$ per 5 litres, according to porosity.

## 2.2.16 Red lead primer

A rust-inhibitive primer designed to protect iron and steel from corrosion. Bright red in colour. Although still available for special specifications requiring lead paint, it is almost obsolete due to the toxic lead content. Being replaced with lead-free primer such as zinc phosphate metal primer.

***Composition*** Red lead pigment in a linseed oil medium, thinned with white spirit.

***Properties***

Excellent rust inhibition.

High film build.

Can be used effectively over hand-cleaned surfaces without significant loss of performance.

Non-reversible coating.

***Use*** (i) Heavy structural steelwork such as industrial plants, installations, gas holders and storage tanks.

(ii) Light structural steel such as railings, rain-water pipes and gutters.

***Application***

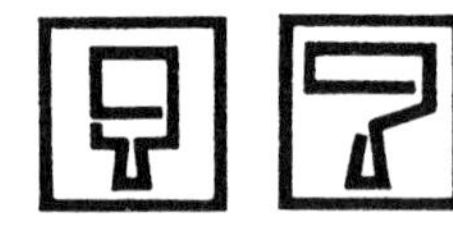

***Drying method*** Oxidation.
***Time*** 2–4 hours, may take longer in damp or cold conditions.
***Overcoating*** The primer should be left to dry for at least 24 hours before overcoating. Excellent results are obtained when overcoated with micaceous iron oxide paint. Most oil-based decorative systems can be used over red lead primer.
***Cleaning solvent*** White spirit.
***Spreading capacity*** 60–85 m$^2$ per 5 litres.
*Note:* Has poor flow and levelling properties; toxic; use controlled by Control of Lead at Work Regulations (see 6.4.1).

Intensively used in the past to protect iron and steel from corrosion, but now withdrawn due to the toxic lead content. Replaced mainly by metal primer zinc phosphate.

## 2.2.17 Zinc chromate metal primer

These are not now used commercially. A highly toxic yellow metal primer, sometimes referred to as a universal metal primer.

***Composition***
***Pigment*** Zinc chromate (red oxide is sometimes added).
***Film-former*** Alkyd or oil-modified resin.
***Thinner*** Follow manufacturer's recommendation 2010.
***Properties***
Rust-inhibitive.
Good flow and levelling properties.
Good resistance to strong solvents which may be present in subsequent coats.
Resistant to marine atmospheres.
Can be sprayed.
Non-toxic.
Suitable for interior or exterior.
Non-reversible coating.
***Use*** (i) Clean or etched aluminium and light alloys, sprayed zinc or sherardised surfaces.
(ii) Ferrous metals.
(iii) Composite wood and metal components.
***Application***

***Drying method*** Oxidation.
***Time 16–24 hours.***
***Overcoating*** Should be carried out as soon as possible after the primer is dry: approximately 24 hours. Most types of decorative and industrial paints can be used, including chlorinated rubber.
***Cleaning solvent*** White spirit or manufacturer's recommended cleaning solvent.
***Spreading capacity*** 55 m$^2$ per 5 litres.
*Note:* Widely used in the past, but now withdrawn due to concerns over chromate pigments.

## 2.2.18 Zinc phosphate metal primer

A special rust-inhibitive primer developed as an alternative to red lead.

***Composition***
***Pigment*** Zinc phosphate titanium dioxide, iron oxide and mineral extenders.
***Film-former*** Alkyd resin.
***Thinner*** Follow manufacturer's recommendation 2010.
***Properties***
Good/excellent rust inhibition.
High film build (dry film 40–50 microns).
Non-reversible coating.
***Use*** Suitable on iron and steel, and non-ferrous metals including weathered or pre-treated galvanised steel. For use on new galvanised surfaces, a pretreatment with mordant solution is necessary. Not suitable for permanent immersion in water.

***Application***

***Drying method*** Autoxidation.
***Time*** Recoatable in 6–16 hours – longer in cold, damp conditions.
***Cleaning solvent*** White spirit, sometimes xylene or specific brush cleaner.
***Spreading capacity*** 40–50 $m^2$ per 5 litres.

## 2.2.19 Zinc-rich primer and zinc-rich epoxy primer

A specialist primer based on pure zinc dust for the protection of blast-cleaned steel at works or after fabrication. Sometimes referred to as cold galvanising.
***Composition*** 85–95 per cent (by weight) pure zinc dust dispersed in a medium such as chlorinated rubber or epoxy resin.
***Properties***
The film is 'sacrificial', i.e. the zinc corrodes instead of the steel (cathodic protection).
The zinc particles, when in contact with each other and the bare steel, will gradually cold-weld themselves together.
Excellent water resistance.
Will protect scratched or broken films against further corrosion.
Not essential to overcoat immediately as it can protect a surface for many years.
Film thickness 50–75 microns per coat (average thickness of one coat of paint is 25 microns).
Non-reversible coating.
Readily attacked by acids.
***Use*** (i) Ideal (especially epoxy type) where speed of drying is important, e.g. between tides on marine installations.
(ii) For factory-made units which may be exposed during transport and storage on site.

***Application*** Preparation of the metal must be by blast cleaning. Must not be applied over old paint and not recommended for 'spot priming'. Small areas can be brush-painted.

***Drying method*** Chlorinated rubber, by evaporation; epoxy resin, by polymerisation.
***Time*** Epoxy-based types can be handled after only 5–10 minutes; other types may take 30–60 minutes.
***Overcoating*** 4 hours minimum.
*Note:* Do not overcoat with paints containing saponifiable resins such as oil-modified resin or alkyd-based paint, unless a non-saponifiable resin-based barrier coat has been applied first.
***Cleaning solvent*** The manufacturer's recommended cleaning solvent, immediately after use.
***Spreading capacity*** 50–60 $m^2$ per 5 litres.
*Note:* Very expensive.

## 2.2.20 Alkyd undercoat

A heavily pigmented paint which dries to a semi-matt finish. Designed for use under gloss paints.
***Composition***
***Pigment*** Titanium dioxide with extenders and coloured pigments as required.
***Film-former*** Alkyd resin.
***Thinner*** Follow manufacturer's recommendation 2010.
***Properties*** Fills small imperfections, easy to sand.
Good adhesion to the primer.
Good opacity and build.
Good flow and levelling.
High film build.
Of lighter tone than the finish.
Non-reversible.
***Use*** To provide a foundation and to provide build for a top coat over primed interior and exterior surfaces.

***Application***

***Drying method*** Autoxidation.
***Time*** 2–6 hours.

***Overcoating*** Overnight or 6–16 hours. Paints containing strong solvents may disturb and lift the dried film and should not be used over oil undercoats.
***Cleaning solvent*** White spirit.
***Spreading capacity*** 80–90 $m^2$ per 5 litres.

# 2.3 Finishing paints

## 2.3.1 Interior and exterior metallic paints (formerly bronze/aluminium paint)

Available in a wide range of metallic colours from Silver to Gold, e.g. Aluminium Paint.

***Composition*** Metallic paints can be made from mixtures of real metal, e.g. aluminium, zinc, copper, or alternatively by using pearlescent pigments that can be used in various proportions to give a number of different colours. Metallic paints can be either water-borne (quick-drying) for use on interior surfaces or solvent-borne for exterior use.

Aluminium dust should not be dispersed directly into water, as the metal will chemically react to produce hydrogen gas. Aluminium dust can be dispersed but into the following media:

(i) Cellulose lacquers
(ii) Alkyd resins (oil-based)

Non-reactive metals and pearlescent pigments can also be dispersed as above but also in

(iii) PVA or acrylic emulsions

***Properties***

Designed to provide protection for interior and exterior wood and metal surfaces. Aluminium paint can withstand temperatures up to 260°C.

Interior metallic paints can be used for the enrichment and gilding of ornamental plasterwork in theatres and in the home, picture-frame decoration, theatre props, furniture and radiators, whereas exterior metallic paints are ideal for wrought-iron railings, fleur-de-lis, coats of arms, decorative memorabilia and stencils, etc.

***Drying***

Cellulose lacquers dry by evaporation, and this process may take several hours depending upon temperature and humidity.

Alkyd resin or oil-based paints dry by a process called 'autoxidation' (in the presence of metal drier, e.g. cobalt, which is soon to be banned and replaced by manganese) and can take 16–24 hours to fully dry.

PVA or acrylic copolymer emulsions dry by coalescence and will take 1–2 hours depending upon temperature and humidity.

For exterior solvent-based metallic paints allow a minimum of 24 hours before applying additional coats. Quick-drying interior metallic paints can normally be overcoated within 4 hours.

***Cleaning solvent***

The choice of cleaning solvent will depend upon the type of metallic paint.

Cellulose lacquers will require cellulose thinner.
Oil-based will require white spirit.
PVA or acrylic will require water.

Brushes and rollers should be well cleaned in solvent, then detergent, and finally rinsed in clean water to ensure all particles of metal are removed.

***Spreading rate*** Typically 18 $m^2$ per litre on most surfaces.

## 2.3.2 Eggshell or satin finish

An interior decorative finish which can be either water-based (quick-drying), or solvent-based which dries to a mid-sheen finish.

***Composition***

***Pigment*** Titanium dioxide (white), mineral extenders and a wide range of coloured pigments.

***Film-former*** Alkyd resin or oil-based binders (some types reinforced with polyurethane for extra hardness). Acrylic copolymer emulsion.

***Properties***

Hard-wearing.
Thixotropic.
Good flow.
Washability.
Non-toxic.
Easily cleaned with detergent.
Film tougher and more washable than emulsion paints.
Requires no undercoat.
Available in wide range of colours.
Excellent opacity (except some yellows and oranges).
Resists normal domestic steam and heat.
Non-reversible.

***Use*** Following 2010, *NOT* recommended on broad areas, only trim.

***Application***

***Quick-drying eggshell***

***Solvent-based eggshell***

***Drying***

Quick-drying eggshell (acrylic emulsion) dries by coalescence and will take 4–6 hours depending upon temperature and humidity. Additional coats can be applied after 6 hours.

Solvent-based eggshell is an alkyd resin or oil-type paint which dries by autoxidation and can take 16–24 hours to fully dry. Additional coats should only be applied after 24 hours.

***Cleaning solvent***

The choice of cleaning solvent will depend upon the type of eggshell paint. Quick-drying eggshell can be cleaned with water whereas solvent-based eggshell can be cleaned with white spirit.

***Spreading capacity*** 80–85 $m^2$ per 5 litres.

## 2.3.3 Matt emulsion paint

An interior water-based decorative emulsion for use on walls and ceilings which dries to flat sheen or finish, e.g. vinyl matt.

***Composition***

***Pigment*** Titanium dioxide (white), mineral extenders, opaque polymer and a wide range of coloured pigments.

***Film-former*** There are many different types depending upon the desired film properties and these include:
Polyvinyl acetate or PVA copolymer emulsion.
Ethylene-vinyl acetate or EVA copolymer emulsion.
Acrylic copolymer emulsion.
Styrene-acrylic copolymer emulsion.

***Properties***

Matt emulsion paints are traditionally highly pigmented (formulated above chlorinated polyvinyl chloride – CPVC) which can impact upon the final film performance, such as poor washability and crack resistance. However, matt emulsion paints formulated below CPVC show marked improvements in the film.

Typical properties of a matt emulsion paint formulated above CPVC:

Permeable – allows moisture vapour to pass through the film.
Flexible.
Excellent opacity or hiding power.
Washable.
Alkali-resistant.
Good sealing properties on porous materials, e.g. new plaster.
Quick-drying.
Low odour.

Typical properties of a matt emulsion paint formulated below CPVC:

Permeable – allows moisture vapour to pass in and out of the film.
Flexible.

Excellent opacity or hiding power.
Scrub-resistant.
Stain-resistant.
Tougher than conventional matt and silk emulsion paints.
Burnish-resistant.
Washable.
Alkali-resistant.
Good sealing properties on porous materials, e.g. new plaster.
Quick-drying.
Low odour.

***Use*** Extensively used for walls and ceilings. Suitable for use over plasterboard, fibre insulating board, wood, hardboard, plaster, brick, cement rendering, stucco, asbestos sheeting and fabrications, foamed polystyrene and wallpaper.

***Application***

***Drying method*** Coalescence.
***Time*** 2–4 hours depending upon temperature and humidity.
***Overcoating*** Allow 4 hours. Emulsion painted surface can be papered or painted over after cleaning and denibbing.
***Cleaning solvent*** Water. Cleaning should be carried out immediately after use.
***Spreading capacity*** 65–80 m$^2$ per 5 litres on smooth, non-porous surfaces.
*Note:* The dry film can be softened by methylated spirits.

## 2.3.4 Vinyl silk emulsion

Similar to matt emulsion paint but with poorer opacity and drying to an eggshell sheen.

***Composition***
***Pigment*** Titanium dioxide (white), mineral extender, opaque polymer and a wide range of coloured pigments.
***Film-former*** Polyvinyl acetate or PVA copolymer emulsion. Ethylene-vinyl acetate or EVA copolymer emulsion.
***Thinner*** Water.

***Properties***
Silk emulsion paint is for use on interior walls and ceilings. It enhances low-relief wall coverings and textured plaster.
Washable.
Scrub-resistant.
Durable.
Alkali-resistant.
High sheen.
Low odour.
Non-yellowing.

***Use*** Where the properties of emulsion paints are required but with a higher sheen, e.g. in kitchens, bathrooms, hospitals, schools, food-preparation plants.

***Application***

***Drying method*** Coalescence.
***Time*** 1–2 hours.
***Overcoating*** 4 hours.
***Cleaning solvent*** Warm water.
***Spreading capacity*** 65–80 m$^2$ per 5 litres.

## 2.3.5 Gloss finish

Interior and exterior decorative paint having a full gloss finish. Used as the main protective coating in the decorating industry.

***Composition***
***Pigment*** Titanium dioxide (white) and a wide range of pigments to produce the full colour range.
***Film-former*** Alkyd resins modified with drying oils. Some types contain a small proportion of other resins such as silicone and polyurethane resins to improve the film performance.
***Thinner*** Most formulations will be compliant post-2010 so thinning will not be recommended.

***Properties***

Traditional gloss paints are solvent-based for use on interior and exterior wood, metal and masonry surfaces. They can be used on heated surfaces such as hot water pipes and radiators. However, gloss paints do have a tendency to yellow with time, and this chemical change takes place in the dark.

Typical film properties for traditional gloss paint are as follows:

High gloss.
Tough and durable.
Washable.
Good gloss retention.
Excellent flow and levelling.
Good adhesion.
Good flexibility.
Whites tend to yellow.
Good weather resistance.
Heat resistance up to 93°C.
Non-reversible coating.

***Use*** General interior and exterior finishes. Decorative finish over suitably primed and undercoated timber, metals, plaster, concrete, brickwork and building boards.

***Application***

***Drying method*** Autoxidation.
***Time*** Tack-free in 4–6 hours.
***Overcoating*** Leave overnight to dry.
***Cleaning solvent*** White spirit.
***Spreading capacity*** 75–85 m$^2$ per 5 litres.

## 2.3.6 Water-based gloss

An alternative to traditional solvent-based systems, it offers reduced odour and also allows for overcoating on the same day.

***Composition***

***Pigment*** Titanium dioxide and coloured pigments to produce the full colour range.
***Film-former*** Acrylic polymer.
***Thinner*** Water.
***Properties*** Suitable for painting timber and appropriately primed metal surfaces.

***Application***

***Drying method*** Coalescence.
***Time*** 2 hours.
***Overcoating*** 6 hours.
***Cleaning solvent*** Water.
***Spreading capacity*** 16 m$^2$ per litre on smooth surfaces.

## 2.3.7 Multi-colour finish

An interior decorative finish which has 2, 3 or 4 separate colours in the form of spots, flecks or streaks applied in one application. Each fleck of paint exists as an individual both in the tin and during application. Then, as the film dries, it joins up with its neighbours to form a multi-colour finish. The paints contain two incompatible materials, which allow several colours to be used in one container without becoming mixed together. Three types are available:

(a) Cellulose/water medium (spray applied).
(b) Vinyl/resin medium (spray applied).
(c) A brush-applied clear coating containing coloured flecks applied over an opaque coloured ground.

### (a) Cellulose multi-coloured finish (limited availability)

***Composition***

***Pigment*** Titanium white and range of coloured pigments.
***Film-former*** Pigments ground in cellulose medium dispersed in stabilised water.

***Properties***

Anti-static, does not attract dust.
Non-toxic.
Washable.
Does not chip.

Excellent adhesion.
Camouflages surface irregularities and joints in boards.
Non-reversible coatings.
Hard-wearing and abrasion-resistant.
Liable to lift oil paint substrates.
Shelf life of 6–9 months.
Strong odour and fumes which can be unacceptable in hospitals, food factories, occupied buildings.

***Use*** All types of interior decoration where a decorative and hard-wearing surface is important, e.g. in toilets; corridors in schools, hospitals and public buildings. Suitable for application over glazed tiles.

***Application***

Medium or large set-ups give best results. Pressure should be about 20–30 psi (1.5–2 bar).

***Drying method*** Evaporation.

***Time*** Touch-dry in 2 hours; hard overnight.

***Cleaning solvent*** Water, and the manufacturer's cleaning solvent.

***Spreading capacity*** 16–20 $m^2$ per 5 litres.

### (b) Vinyl multi-coloured finish

***Composition***

***Pigment*** Titanium white and range of coloured pigments.

***Film-former*** Pigments ground in low-odour spirit-carried vinyl resin dispersed in stabilised water.

***Properties***

Anti-static.
Non-toxic.
Washable.
Does not chip.
Excellent adhesion.
Camouflages surface irregularities and joints in boards.
Non-reversible coatings.
Hard-wearing and abrasion-resistant (although inferior to cellulose multi-colour).
Can be applied over oil paint substrates.
Shelf life of 6–9 months.
Low odour.

***Use*** All types of interior decoration where a decorative and hard-wearing surface is important, e.g. corridors in schools, hospitals and public buildings; restaurants; clubs; food factories. Particularly suitable for use in occupied buildings because of its low odour.

***Application***

Medium or large set-ups give best results. Pressure should be about 20–30 psi (1.5–2 bar).

***Drying method*** Coalescence.

***Time*** Touch-dry in 4 hours; hard overnight.

***Cleaning solvent*** Water or manufacturer's recommended cleanser.

***Spreading capacity*** 14–20 $m^2$ per 5 litres.

### (c) Brush applied, vinyl medium

Consists of two parts: (i) an undercoat which provides the ground colour, and (ii) the finishing coat which carries the coloured flakes.

***Composition – undercoat***

***Pigment*** Titanium dioxide with extenders and coloured pigments.

***Film-former*** Copolymer emulsion.

***Composition – finishing coat***

***Pigment*** No pigment.

***Film-former*** Coloured flakes are suspended in clear copolymer emulsion.

***Properties***

Non-toxic.
No objectionable odour.
Good adhesion.
Washable.
Abrasion-resistant.
Good flexibility.
Camouflages poor substrates.

***Use*** For interior and exterior surfaces, particularly in public buildings where resistance to abrasion is of prime importance.

***Application*** Undercoat is brushed or rolled. Finishing coat is brush applied. It may be necessary to roll the coating after to remove excess resin.

***Drying method*** Coalescence.

***Time*** Undercoat, 2–4 hours. Finishing coat, touch-dry in 1 hour.
***Cleaning solvent*** Water.
***Spreading capacity*** Undercoat – up to 70 $m^2$ per 5 litres. Finishing coat – up to 50 $m^2$ per 5 litres.

## 2.3.8 Masonry paint

### (a) Emulsion-based

A water-based, quick-drying, smooth or textured finish for exterior walls (not common or Fletton bricks).

***Composition***

***Pigment*** Titanium dioxide (white), mineral extender, opaque polymer, sand and a wide range of coloured pigments. Mica, sand or aggregate is used for reinforcement and to provide texture to the film.
***Film-former*** Acrylic copolymer emulsion. Styrene-acrylic copolymer emulsion.
***Thinner*** Water.

***Properties***

Masonry emulsion paints are for exterior walls and brickwork excluding timber. The smooth or textured finish combined with a fungicide and algaecide can help the surface stay cleaner for longer.

Typical film properties for masonry emulsion paints are as follows:

Tough and durable.
Flexible.
Good opacity.
UV-resistant.
Shower-resistant within 30 minutes.
Waterproof.
Mould- and algae-resistant.
Covers hairline cracks.
Alkali-resistant.
Non-reversible coating.

***Use*** For protection and decoration of new and old cement rendering, concrete, brickwork, asbestos sheeting, pebbledash, stucco and other types of masonry.

***Application***

***Drying method*** Coalescence.
***Time*** 1–4 hours depending upon temperature and humidity.
***Overcoating*** Can be overcoated as soon as the film is dry. Any type of decorative finish can be used over masonry paints as long as the surface is sound.
***Cleaning solvent*** Water. It is very difficult to remove all aggregate material from brushes and rollers, and they should not subsequently be used in normal finishing systems.
***Spreading capacity*** 60–70 $m^2$ per 5 litres on smooth, non-porous surfaces. Will vary according to the porosity and texture of the surface.

### (b) Oil-based

A durable finish for outside use on all suitably primed or sealed surfaces.

***Composition***

***Pigment*** Titanium dioxide (white) with mineral extenders and coloured pigments.
***Film-former*** Alkyd resins modified with drying oils.
***Thinner*** Follow manufacturer's recommendation 2010.
***Additives*** Fungicides.

***Properties***

Masonry gloss paints are solvent-based for use on exterior wood, metal and masonry surfaces. Typical film properties for traditional gloss paint are as follows:

High gloss.
Tough.
Durable.
Washable.
Good gloss retention.
Excellent flow and levelling.
Good adhesion.
Good flexibility.
Good weather resistance.
Mould-resistant.

***Use*** Interior, exterior protection and decoration of dry, primed plaster, brick, stucco, concrete, asbestos, sheeting, cement rendering, wood and metal.

*Application*

***Drying method*** Oxidation.
***Time*** 16–24 hours.
***Overcoating*** Overnight drying should be allowed. Can be overpainted with any kind of decorative paint.
***Cleaning solvent*** White spirit.
***Spreading capacity*** 85 $m^2$ per 5 litres on smooth, non-porous surfaces. Will vary according to the porosity and texture of the surface.
*Note:* Other masonry paints are based on synthetic rubber, pliolite resins.

## 2.3.9 Cellulose coating

Quick-drying, highly flammable material, based on nitrocellulose. Available as primer, sealer, filler and pigmented, clear or metallic finish.

***Composition***
***Pigment*** Titanium dioxide (white) and a wide range of coloured pigments, extenders, slate powders and metallic powders.
***Film-former*** Basically nitrocellulose, but many types are modified with synthetic resins for greater flexibility and improved resistance.
***Thinner*** Follow manufacturer's recommendation 2010.

***Properties***
Very fast-drying
Very hard and abrasion-resistant.
Water-resistant.
Reversible coating.
***Use*** (i) Industrial and car refinishing.
(ii) Aluminium and bronze paints.

***Application***

***Drying method*** Evaporation (some polymerisation with modified types).
***Time*** Touch-dry in 10–15 minutes.
***Overcoating*** 30–60 minutes depending on type, temperature and humidity.
***Cleaning solvent*** Manufacturer's special thinner.
***Spreading capacity*** For finishing coats, these materials are sometimes thinned 50/50 with solvents. Therefore, the spreading capacity covers a wide range, according to use.

*Note:* Cellulose finishes provide very thin films; they are often brittle and can sometimes crack, especially on timber surfaces.

Being reversible coatings, they are almost impossible to apply by brush except on very small areas.

The solvent odour from cellulose materials is strong, and good ventilation and the wearing of an appropriate vapour mask are essential when spraying these materials.

These materials are highly flammable (see Part 6).

## 2.3.10 Moisture-vapour permeable (MVP) or microporous coatings

A paint system which allows moisture vapour to permeate through the dried film, but will not allow liquid water through. Such coatings allow the moisture content of timber to escape without causing the paint to blister or flake. Two types are available:

*(a)* Spirit-thinned
*(b)* Water-thinned

### (a) Spirit-thinned MVP coatings

***Composition***
***Pigment*** Titanium dioxide (white) and a range of lightfast pigments to give a full colour range.
***Film-former*** Oil-modified alkyd resin.
***Thinner*** Thinning not recommended 2010.

***Properties***
Mid-sheen to full gloss.
Good flow.
Excellent flexibility.
Excellent durability.
Non-reversible coating.
Low maintenance.

***Use*** All interior and exterior finishes can be applied over any bare timber substrate, or directly on to similar coatings in sound condition.

***Application***

***Drying method*** Oxidation.

***Time*** Touch-dry in 3–5 hours.

***Overcoating*** Leave overnight to dry.

***Cleaning solvent*** White spirit.

***Spreading capacity*** 12–15 $m^2$ per litre.

***Maintenance*** Wash, spot prime, make good, apply two coats MVP coating every 4–6 years according to exposure.

### (b) Water-thinned MVP coatings

***Composition***

***Pigment*** Titanium dioxide (white) and a range of lightfast pigments to give a full colour range.

***Film-former*** Acrylic resin.

***Thinner*** Water.

***Properties***

Semi-gloss finish.

Good flow.

Excellent flexibility.

Excellent durability.

Non-reversible coating.

Low maintenance.

***Use*** All interior and exterior finishes. Can be applied over any bare timber substrate, or directly on to similar coatings in sound condition.

***Application***

***Drying method*** Coalescence.

***Time*** Touch-dry in 2–4 hours.

***Overcoating*** 4 hours in good drying conditions.

***Cleaning solvent*** Water.

***Spreading capacity*** 12–15 $m^2$ per litre.

***Maintenance*** Wash, spot prime, make good, apply two coats MVP coating every 4–6 years according to exposure.

# 2.4 Special paints

## 2.4.1 Anti-condensation paint

An oil-based paint which dries with a textured finish and helps to reduce condensation. Similar paints based on emulsion resins are also available.

***Composition*** Oil-based or emulsion products act as the 'carrier' for additives such as cork flour and various suitable beads, which can provide a low thermal conductivity. These beads or granules act as an insulator between the warm, moist atmosphere and the cold surface. Water-based product formulations are becoming more important carriers (Fig. 2.6).

***Properties***

Texture finish.

Reduces condensation.

Retains its properties when overpainted.

Does not absorb moisture.

Non-reversible coating.

***Use*** For interior ceilings and walls which are subjected to the effects of condensation. These paints are suitable for all surfaces, but it should be noted that when applying the emulsion type, all bare metal should be well primed and undercoated with an oil paint before applying anti-condensation paint.

***Application*** The paint can be sprayed or applied liberally by brush and finally hair-stippled to ensure an even texture. Two coats are normally required for full effectiveness.

***Drying method*** Oil-based, by autoxidation; emulsion-based, by coalescence.

***Time*** 2–24 hours according to thickness of coating, type of medium and conditions during application.

***Cleaning solvent*** For oil-based types: white spirit. For emulsion type: water.

***Spreading capacity*** 25–30 m$^2$ per 5 litres. Maximum 20 m$^2$ per 5 litres for second coat.

## 2.4.2 Anti-graffiti paint

Specialist paint systems used as a defence against graffiti, suitable for areas such as subways, car parks, shopping centres and communal areas. These can be single-pack, easily strippable (sacrificial) water-based coatings, but the majority tend to be two-pack. Both solvent-based and water-based two-pack epoxy or polyurethane systems are used, either as solid colours or as a clear finish.

A sealer may be required to stabilise unsound/powdery surfaces, although a primer will usually suffice to provide a sound surface for overcoating with a finish. For maximum protection, two (or more) coats of finish may be required. Anti-graffiti coatings are resistant to repeated washing, steam cleaning and solvent cleaning.

***Composition***

***Pigment*** Titanium dioxide, extenders and coloured pigments.

***Film-former*** Epoxy resin or polyurethane resin.

***Activator*** Polyamine or polyisocyanate.

***Thinner*** Water or manufacturer's recommended thinner if solvent-based.

***Properties***

High-performance coatings.

Excellent adhesion.

Solvent-resistant non-reversible coatings.

***Use*** Can be applied to interior and exterior brickwork and plasterwork and primed metal surfaces. Do not

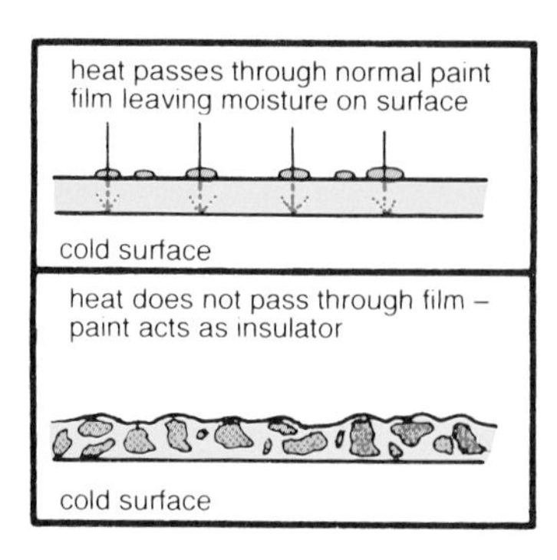


**Fig. 2.6** Anti-condensation paint

apply over non-cross-linked flexible coatings, e.g. acrylated rubber or chlorinated rubber or underbound coatings, e.g. distemper.

***Application***

*Note:* Clear coat – *NOT* to be spray applied.

Roller and brush for polyurethanes.

Spraying for polyisocyanate-cured paints should be avoided due to the risk of respiratory sensitisation from inhaling isocyanates; if unavoidable, air-fed respiratory equipment should be worn.

***Drying method*** Polymerisation.

***Time*** 4–6 hours; application at temperatures below 10°C should be avoided.

***Overcoating*** 16 hours minimum. Maximum 7 days.

***Cleaning solvent*** Water or manufacturer's recommended thinner if solvent-based.

***Spreading capacity*** Primer, sealer, finish 40–60 m$^2$ per 5 litres.

## 2.4.3 Anti-fungal paint

Contains a biocide (fungicide), which will kill fungi when they start to feed and grow on the paint film.

***Composition*** Most common are traditional paints, which have fungicides added by the manufacturers. The type of fungicide and the environmental conditions will determine how long the paint remains anti-fungal.

Some paints also exist with carefully chosen ingredients, which resist or discourage fungal growth. These remain effective throughout the life of the film.

***Properties*** In addition to the properties of the base paint, it will also be a mould-inhibiting and non-reversible coating.

***Use*** Where conditions are favourable for the growth of fungi, i.e. damp and humid conditions as found in bakeries, breweries, dairies and greenhouses. It is important that surfaces should be thoroughly prepared and sterilised before painting. Most leading manufacturers supply an antiseptic solution for this purpose. Extreme care should be taken to avoid contamination when using anti-fungal paints in places where food is being prepared. Their use should be completely avoided in children's rooms.

***Application***

The addition of anti-fungal compounds makes the paint semi-toxic, therefore spray application is not recommended. Where there is no alternative to spray application, adequate ventilation should be provided and masks worn over mouth and nose, preferably with a separate air supply.

***Overcoating*** Further applications must be of an anti-fungal type if such properties are required to remain.

***Cleaning solvent*** Manufacturer's recommended solvent.

*Note:* Tools, etc., used in preparation, should be thoroughly sterilised to avoid spreading spores to other areas.

## 2.4.4 Bituminous paint

A black paint used widely for the protection of steelwork and concrete.

***Composition***

***Film-former*** Bitumens, which are available in different forms, blended together to produce the required type of finish. Heat radiation from the sun causes softening of the film and for this reason aluminium pigments may be added to give reflective properties; it is then known as aluminium bituminous paint.

***Thinner*** Follow manufacturer's recommendation 2010.

***Properties***

Chemical-resistant.
Moisture-resistant.
Relatively low cost.
Gloss finish which is reduced by exposure to sunlight.
Reversible coating.

***Use*** To provide a weather-resistant film on iron and steelwork, roof sheeting, storage tanks, concrete fabrications, bridges, piers, soil pipes and gutters.

***Application***

***Drying method*** Evaporation.

***Time*** 30–120 minutes.

***Overcoating*** These paints are soluble in oil paint solvents and should be overpainted only with further bitumen coatings or be sealed to prevent bleeding through subsequent coats (see 3.4.7).

***Cleaning solvent*** White spirit, paraffin or solvent naphtha. Avoid using the same brushes for other types of paint to avoid contamination.

***Spreading capacity*** Varies with temperature and nature of surface at the time of application; average of 30–40 m$^2$ per 5 litres.

*Note:* A limited range of coloured bitumen paints is available, but these are more expensive and less resistant to moisture and chemicals.

## 2.4.5 Cement paint

Supplied in dry powder form and sold by weight rather than volume. The powder is mixed with water before use.

***Composition*** White Portland cement with extenders, waterproofing agents and accelerators. Alkali-resisting pigments are used to produce pastel shades.

***Properties***

Alkali-resistant.
Good filling properties.
Rough-textured matt surface.
Low cost.
Non-reversible coatings.

***Use*** For the protection and decoration of brick, concrete, cement rendering, pebbledash, asbestos sheeting. Can also be used on similar interior surfaces in outbuildings, cellars, garages, farm buildings. Specially suitable for surfaces where damp and humid conditions would present problems for ordinary paints.

*Note:* Not suitable for use on gypsum plasters.

***Application***

Synthetic brushes should be used, as alkali destroys pure bristle. Must not be applied over other types of decorative paint. Porous surfaces should be damped down to control suction and the paint applied while the surface is still damp.

***Drying method*** By setting of the cement.

***Time*** 2–4 hours according to the conditions.

***Overcoating*** Cement paints should be sealed with alkali-resistant primer/sealer before overcoating with ordinary decorative paints.

***Cleaning solvent*** Water.

***Spreading capacity*** Fair-faced brickwork and smooth surfaces, 8–15 m$^2$ per 5 kg; roughcast, pebbledash and Tyrolean, 6–8 m$^2$ per 5 kg.

*Note:* Available in limited colour-range; poor flow and brushing properties.

## 2.4.6 Chlorinated rubber paint

Chemical and water-resistant paint used for special purposes. Special primers are available for most types of surface. Thixotropic materials are available for heavy film build. A glossy finish offering enhanced water resistance is also available.

***Composition***

***Pigment*** Titanium dioxide and a limited range of chemical-resistant pigments.

***Film-former*** Chlorinated rubber.

***Thinner*** Manufacturer's recommended thinner.

***Properties***

Very good resistance to diluted alkalis.

Excellent water resistance even when immersed.

Produces very flexible film.

Quick-drying.

Excellent resistance to acids and acid fumes.

Gives off chlorine gas when heated (limited fire-retardant properties).

Reversible coating.

***Use*** (i) Applied to clean rust-free metal and wood, concrete, asbestos, brickwork and insulation boards. Suitable for areas where special water or chemical resistance is required, such as factories, laboratories, swimming pools, water storage tanks, dairies and floors, etc.

(ii) Road marking or line paint. When used for this purpose minute glass spheres, known as Ballotini spheres, are sometimes mixed with the paint to increase its light reflectance. They may be added to the paint before brush application, but when spraying they may need to be dusted on to the wet paint after application. Sand can also be added to these paints to improve skid resistance.

***Application***

***Drying method*** Evaporation.

***Time*** Touch-dry in 30 minutes–2 hours.

***Overcoating*** 8–16 hours by spray; 24 hours by brush. (Being a reversible coating, application by brush may sometimes be difficult, as the wet film softens previous coatings.)

***Cleaning solvent*** Manufacturer's recommended solvent.

***Spreading capacity*** 45–55 $m^2$ per 5 litres; 15–25 $m^2$ per 5 litres for thixotropic types.

*Note:* These paints are not suitable for temperatures above 65°C, hot water above 43°C, surfaces in contact with drinking water or for immersion in solvents (the film will soften, swell and/or dissolve).

## 2.4.7 Acrylated rubber paints

Chemical and water-resistant paints used when a chlorine-free product is required. These share most of the attributes and limitations of chlorinated rubber paints.

***Composition***

***Pigment*** Titanium dioxide, mineral pigments and lightfast pigments.

***Film-former*** Acrylated rubber.

***Thinner*** Manufacturer's recommended thinner.

***Properties***

Resistant to splashing with alcohols, concentrated acids, acid fumes and alkalis.

Excellent resistance to high levels of moisture, heavy condensation.

Reversible coating.

***Use*** Apply to clean interior and exterior ferrous and non-ferrous metal and to masonry surfaces. Suitable for the exterior of water-storage tanks. Not for use on swimming pools (acrylated rubber does not have good resistance to body fats/oils).

***Drying method*** Evaporation.

***Application***

***Time*** Touch-dry in 1–2 hours.

***Overcoating*** 16–24 hours.

**Cleaning solvent** Manufacturer's recommended solvent.

**Spreading capacity** 40 $m^2$ per 5 litres, typically 20–25 $m^2$ per 5 litres for thixotropic type.

*Note:* These paints are not recommended for heated surfaces above 80°C, surfaces in contact with drinking water or for immersion in solvents.

## 2.4.8 Road line paint

Also called Road Marker or Line Paint. A high-build coating for marking car parks, demarcation zones, etc.

**Composition**

**Pigment** Titanium dioxide, mineral extenders and glass spheres.

**Film-former** Chlorinated rubber or acrylated rubber resin.

**Thinner** Manufacturer's recommended thinner.

**Properties**

Abrasion-resistant.

Fast-drying

Water-resistant.

Reversible coating.

**Use** For exterior use, marking car parks, demarcation zones, crossings, etc. Not intended for general road marking where a high-build hot melt plastic coating is commonly used. Not recommended where spillage of petrol or solvents is likely or on heated surfaces.

**Application**

**Drying method** Evaporation.

**Time** 5–10 minutes.

**Overcoating** 1 hour. Resistant to vehicular traffic after 2–4 hours dependent upon number of coats applied.

**Cleaning solvent** Manufacturer's recommended solvent.

**Spreading capacity** 10–15 $m^2$ per 5 litres.

## 2.4.9 Machinery enamel

Also called Protective Enamel or Polyurethane Enamel. A solvent-based glossy finish for machinery, plant, farm equipment, etc.

**Composition**

**Pigment** Titanium dioxide, coloured pigments.

**Film-former** Polyurethane modified alkyd.

**Thinner** Follow manufacturer's recommendation 2010.

**Properties**

Good resistance to mineral oils, fresh and salt water.

Limited resistance to battery acid and cutting oils.

Surfaces subject to continuous splashing or immersion may require a specialist two-pack coating.

High gloss.

Non-reversible coating.

**Use** Interior and exterior use. For light machinery and plant, farm equipment and metal surfaces in general.

**Application**

**Drying method** Autoxidation.

**Time** 2–4 hours.

**Overcoating** 4–6 hours, longer in cold, damp conditions.

**Cleaning solvent** White spirit.

**Spreading capacity** 75–85 $m^2$ per 5 litres.

## 2.4.10 Floor paint (single pack)

A solvent-based finish for interior floors.

**Composition**

**Pigment** Titanium dioxide, mineral extender and coloured pigments.

**Film-former** Alkyd or modified alkyd resin.

**Thinner** Follow manufacturer's recommendation 2010.

***Properties***

Resistant to pedestrian traffic and rubber-wheeled vehicular traffic.

Resistant to heat up to 90°C.

Resistant to repeated and regular washing with water and detergents, but not suitable for use on floors that are likely to be continuously wet.

Available in gloss and semi-gloss finishes.

An aggregate may be included to improve slip resistance.

***Use*** As a decorative coating for concrete, wood and steel floors. Suitable for warehouses, factories, workshops and utility areas, etc. Refer to manufacturer's instructions for guidance on nature of floor to be painted. Not suitable for freshly laid concrete floors.

***Application***

***Drying method*** Autoxidation.

***Time*** 4–6 hours.

***Overcoating*** 16–24 hours.

*Note:* It is essential to allow the coated floor to dry for a minimum of 24 hours under well-ventilated conditions before use.

***Cleaning solvent*** White spirit.

***Spreading capacity*** 60–80 $m^2$ per 5 litres.

## 2.4.11 Damp-proofing compound

A waterproof, dark-coloured flexible coating, used as a barrier against dampness.

***Composition*** Prepared from bituminous materials, emulsified with an agent such as an alkaline soap. Sometimes pigmented with lime-resistant pigments and dyes.

***Properties***

Water-resistant.

Unaffected by humidity or temperature change.

Alkali-resistant.

Non-reversible coating.

***Use*** (i) To hold back dampness on brick, plaster, cement, asbestos and most building materials. These compounds are not designed to cure damp but will help to check it for a limited time.

(ii) Exterior stucco and concrete, particularly below ground level.

(iii) Sealing porous roof felt.

***Application***

***Drying method*** Evaporation.

***Time*** Within 1 hour.

***Overcoating*** Water-based acrylic and emulsion paints can be applied over them. A bitumen sealer should be applied before using oil-based paints. If the surface is to be papered, two coats of emulsion paint should be applied and allowed to dry thoroughly.

***Cleaning solvent*** Water and detergent.

*Note:* Brushes used for this material should not be used for any other type of paint.

## 2.4.12 Epoxy resin paint

Covers a wide range of two-pack solvent-based and water-based products. Chemical-resistant primers, undercoats and finishes, generally pigmented but also available clear. Can be used to provide anti-graffiti paints and primers for difficult surfaces.

***Composition***

***Pigment*** Titanium dioxide (white), metallic pigments, micaceous iron oxide, zinc phosphate, extender pigments and a wide range of coloured pigments.

***Film-former*** Epoxy resins.

***Thinner*** Manufacturer's recommended thinner. Activator: polyamine or polyamide.

***Properties***

Limited resistance to acids, alkalis, solvents and water.

Excellent adhesion to metals and other smooth surfaces.

Abrasion-resistant.

Non-reversible coating.

***Use*** (i) Where tough chemical-resistant surfaces are required, e.g. in chemical plants, dairies, laundries, laboratories.

(ii) Floor paints and other surfaces subject to abrasion.

(iii) Swimming pools or water-storage tanks.

***Application***

Can cause health problems when sprayed; avoid or use air-fed mask. Has been banned in some countries.

***Safety*** Two-pack epoxies can cause health hazards if splashed in the eyes, on the skin, or if spray mist is inhaled. Always read and observe the manufacturer's safety instructions.

***Drying method*** Polymerisation.

***Time/overcoating*** 2–4 hours; maximum hardness and chemical resistance after 2–10 days; overcoating should be done within this time to ensure good inter-coat adhesion; pot-life: day of mixing only.

***Cleaning solvent*** Manufacturer's recommended solvent.

***Spreading capacity*** Varies greatly dependent on product. Guide 30–60 $m^2$ per 5 litres.

*Note:* A single-pack epoxy paint based on epoxy esters is also available. It is thinned with white spirit and its chemical, abrasion and water resistance are considerably inferior to that of the two-pack type.

## 2.4.13 Flame-retardant paint

Paint designed to reduce flammability and surface spread of flame on combustible surfaces. Previously undecorated surfaces such as plasterboard are non-combustible, but when a wall has been repeatedly decorated, the build-up of paint can be flammable and cause rapid flame spread in a fire. There are two main types:

(a) *Intumescent* – where the paint swells or froths when heated and produces a hard crust which acts as a barrier to prevent combustion.

(b) *Non-intumescent* – those containing substances which fuse together upon heating, to prevent combustion. They also give off a fire-retarding gas when heated.

***Composition***

(a) Chemical compounds which, when heated, expand rapidly to form a cellular skin approximately 30 mm thick over the surface. This acts both as a barrier to the flame and as a poor conductor of heat which helps to keep the substrate relatively cool.

(b) Paint with antimony oxide, chlorinated or brominated materials, or minerals as active pigments.

***Properties***

Greatly reduces the surface spread of flame on combustible substrates (see 3.4.1).

Available in a range of colours, or with a range of coloured topcoats.

Thick films and low spreading rates are often required, up to 4 mm dry film thickness.

Often contains hazardous materials – check safety data sheet before use.

Non-reversible coating.

***Use*** To upgrade the classification of flammable building substrates in order to satisfy fire regulations in public buildings, exhibitions, film and television sets.

Intumescent paints are very useful for the protection of exposed steel girders and pillars in factories and warehouses where a fire would severely distort the steel supports, rendering the shell of the building unsafe.

***Application***

***Drying method*** Emulsion-type, by coalescence; oil-type, by oxidation.

***Time*** Emulsion-type, 2 hours; oil-type, 4–6 hours; barrier-type, 16 hours–3 days.

***Overcoating*** After overnight drying. Overcoating with orthodox paints greatly reduces the flame-retardant properties.
***Cleaning solvent*** Water or white spirit, depending on type.
***Spreading capacity*** Very variable, depending upon type. 1–35 $m^2$ per 5 litres.

## 2.4.14 Micaceous iron oxide paint

A high- or medium-build material with outstanding protective properties for use on primed steelwork in exposed environments.

***Composition***
***Pigment*** Micaceous iron oxide, mineral extenders.
***Film-former*** Alkyd resin or oil-modified resin.
***Thinner*** White spirit.
***Properties***
High-build.
Good protection of sharp angles and edges.
Corrosion resistance.
Non-slip properties due to slight texture.
Non-toxic.
Chemical-resistant and resistant to atmospheric pollution.
Low gloss.
Tough film.
Water-resistant.
Non-reversible.
***Use*** For protection of steel structures typically in marine and industrial environments. For severe attack or direct contact with chemicals, an epoxy or polyurethane product may be required. Widely used on railways.
***Application***

***Drying method*** Autoxidation.
***Time*** 2–6 hours.
***Overcoating*** After 24 hours.
***Cleaning solvent*** White spirit.
***Spreading capacity*** 50–60 $m^2$ per 5 litres.

## 2.4.15 Fluorescent and luminescent paints (fluorescent and phosphorescent types)

The difference between fluorescent and phosphorescent materials is the duration of the afterglow. 'Normal' pigments selectively reflect specific wavelengths of light (what we see) and dissipate the other wavelengths. In a fluorescent colour, the segments of light are not dissipated but emitted as light of the same colour, and we see this as an exceptionally bright/dazzling colour, i.e. neon colours. They have no visible afterglow in the dark.

Phosphorescent materials absorb and store the energy from light when exposed to either daylight or artificial light. They have a long afterglow.

Zinc sulphide normally carried in a clear medium will appear almost colourless in daylight but glow with a yellow-green light for a considerable time. However, an even brighter, longer (and more expensive) afterglow can be achieved using strontium and calcium complexes.

Careful formulation is required for fluorescent and phosphorescent paints. Products are supplied in water- or solvent-based media as suits the material being carried. Care has to be taken by the formulator to ensure no 'quenching' of the effect occurs, such as would be caused by the addition of pigments. If quenching occurs, the desired fluorescent or phosphorescent effect will be significantly reduced or destroyed. Therefore, these effects are generally carried in clear formulations.

***Use*** As effect paints or particularly for safety applications, e.g. to mark emergency exit routes.

## 2.4.16 Dry-fall paint/dry-fog paint

A quick-drying paint for spray application, designed for high walls or ceilings. Overspray dries to a dust as it falls to the floor, where it can be swept up with a dry cloth or brush.

***Composition*** Emulsion-type paint with specially formulated drying properties.

***Properties***

Quick-drying.

Non-toxic.

Easy and quick to apply.

***Use*** Useful for large interior areas that must be painted with minimal disruption of operations, such as shops, factories, warehouses and industrial buildings. Contents of rooms should be sheeted to ease clean-up.

***Application***

***Drying method*** Dries by evaporation. Spray-tip size, pressure, fall distance, temperature and humidity all affect the dry-fall characteristics. Good results should be achieved at approximately 4 metres fall distance, 25°C and 50 per cent relative humidity as atmospheric conditions.

***Overcoating*** 2–4 hours.

***Cleaning solvent*** Water.

***Spreading capacity*** Up to 45 $m^2$ per 5 litres, depending on product and surface porosity.

## 2.4.17 Blockfiller

A water-based material suitable for interior use on concrete. It is designed to fill surface imperfections in concrete block and poured concrete and leaves a relatively smooth matt finish.

***Compostion***

***Pigment*** Titanium dioxide and extenders.

***Film-former*** Acrylic latex.

***Thinner*** Water.

***Properties***

Quick-drying to enable same-day overcoating.

Water-based – cleaning can be carried out quickly and cheaply.

Good block-filling ability.

Non-toxic.

Non-reversible coating.

***Use*** It is designed to fill surface imperfections in concrete block and poured concrete and upgrades the appearance of the substrate. Can act as a finish or as a primer for subsequent finishes. This product is designed only for the use of filling voids and pores within the surface of concrete.

***Application***

Recommended – airless spray followed by laying off with a medium-pile roller (can be applied by roller, brush or conventional spray).

***Thinning*** Thinning is not normally recommended.

***Drying method*** Dries by evaporation.

***Time*** Touch-dry within 1 hour.

***Overcoat*** 4 hours or less.

***Cleaning solvent*** Water.

***Spreading capacity*** Up to 5 $m^2$ per litre depending on the porosity of the substrate.

## 2.4.18 Hygienic coatings

The term ‘hygienic coating’ is not defined by any standard and is left to the manufacturer’s definition. Depending on the product, a hygienic coating can be paint creating an ‘easy to clean’ surface, which also resists disinfectants or steam cleaning. By this definition, for example, anti-graffiti paints would act as such a coating. Some manufacturers also include in their product ranges anti-fungal paints for hygienic purposes – for example, mould-free walls. Recent developments include paints with claimed ‘anti-bacterial/anti-microbial’ properties, basically inhibiting the growth of the bugs or at least halting their growth.

Otherwise, properties of hygienic coatings depend on the base paint (analogue anti-fungal paint).

*Note:* Not recommended for operating theatres, as these need to be cross-link technology.

## 2.4.19 Polyurethane paint

Chemical-, water- and abrasion-resistant pigmented and clear systems available as a two-pack material. Generally used as a protective topcoat and can be used as an anti-graffiti coating. Solvent-based and water-based types are available.

***Composition***

***Pigment*** Titanium dioxide (white), mineral extenders and coloured pigments.

***Film-former*** Polyurethane or acrylic urethane resin.

***Thinner*** Follow manufacturer's recommendation 2010.

***Properties***

Chemical-resistant.

Water-resistant.

Abrasion-resistant.

Good adhesion.

Solvent-resistant, graffiti-cleanable.

Non-reversible coating.

***Use***
(i) Where tough chemical-resistant surfaces are required, e.g. in chemical plants, dairies, laundries and laboratories.
(ii) Floors and other surfaces subject to scratching and abrasion.
(iii) Where an anti-graffiti finish, particularly clear, is required.

***Application***

***Safety*** Two-pack polyurethanes can cause health hazards if splashed in the eyes, on the skin or if spray mist is inhaled. Always read and observe manufacturer's safety instructions.

***Drying method*** Polymerisation.

***Time*** Touch-dry in 3–5 hours. Maximum hardness and chemical resistance after 2–10 days. Pot-life of these products may be very limited.

***Overcoating*** Allow 24 hours between coats. Products may need to be overcoated within 48 hours to ensure good inter-coat adhesion. Refer to manufacturer's literature.

***Cleaning solvent*** Manufacturer's recommended solvent.

***Spreading capacity*** 40–65 $m^2$ per 5 litres.

*Note:* A single-pack polyurethane paint based on urethane oils is also available. It is thinned with white spirit and its chemical, abrasion and water-resistance are considerably inferior to the two-pack type, although slightly better than alkyds.

## 2.4.20 Texture-effect paint

An emulsion-type paint, gaining texture upon drying due to coarse aggregates incorporated within the paint.

***Composition***

***Pigment blend*** Typically titanium dioxide plus extenders comprising clays, calcium carbonates, talcs, etc., added to create desired properties. Other options include using a clear formulation (no pigmentation) as the 'carrier' for the selected aggregate and then tinting it as desired.

***Binder*** Typically acrylic polymer and adhesion-promoted versions.

***Thinner*** Generally ready for use, but can be thinned with water.

***Aggregates*** Formulator can select from a large range of materials, such as sand, glass beads and polymer beads.

Aggregate can be selected for particle shape, size, colour and textural feel/look. Many of the bead-type materials can also offer other paint-film qualities such as hardness, sheen and durability depending on size, shape and volume used.

***Properties*** As per standard emulsion paint. Versions can be made for interior or exterior use.

*Application*

*Note:* Spray may be possible with some products if filters are removed.

***Drying method*** Dries by evaporation.

***Time*** Recoat time 2–4 hours depending upon prevailing conditions.

***Spreading capacity*** Depending on formulation 25–55 $m^2$ per 5 litres.

## 2.4.21 Relief-texture paint (interior)

Sometimes referred to as 'plastic paint', it forms an extra-thick surface coating which can be manipulated to give a range of relief textures. There are two main types available:

### Type A

Supplied in powder form.

***Composition***

***Pigment/aggregate*** China clay or whiting, exfoliated mica, mica and hemi-hydrate plaster.

***Film-former*** Gum arabic.

***Thinner*** Water.

***Properties***

Excellent adhesion.

Crack-resistant.

Flame-retardant (class 0).

Does not encourage mould growth.

Reversible coating.

***Use*** (i) Crack-resistant systems for plasterboard ceilings.

(ii) Texture finish to other smooth surfaces, e.g. plaster, cement rendering, concrete.

***Application***

Applied by brush to the surface and manipulated while still wet. Various tools are used to produce the required texture, e.g. stipplers, combs, sponges and palette knives. Some textures are laced with a plastic spatula to remove sharp edges and smooth out the high relief (see 1.1.12). Other effects can be achieved by direct application with spray or roller.

***Drying method*** Evaporation.

***Time*** In good conditions, the surface will dry in 12 hours, but initial setting takes only $\frac{1}{2}$ hour, after which time the coating cannot be reworked.

***Overcoating*** To make the surface washable and also to improve its appearance, it may be coated with emulsion paint but following 2010 not with oil-based paints. A further treatment can be scumbling and wiping.

***Cleaning solvent*** Water.

***Spreading capacity*** For average texture approximately 0.5–0.8 kg/$m^2$.

### Type B

Supplied as a paste ready for use.

***Composition***

***Pigment/aggregate*** Titanium white, earth colours and coarse aggregates such as powdered rock.

***Film-former*** Emulsions, sometimes blended with latex.

***Thinner*** Water.

***Properties***

Excellent adhesion.

Flexible film – resists cracking.

Self-coloured.

Flame-retardant (class 0).

Does not encourage mould growth.

Alkali-resistant.

Non-reversible coating.

***Use*** (i) Heavy coatings applied to disguise most building surfaces including bricks and blocks.

(ii) Provides a wide range of decorative textures on virtually all common building surfaces.

(iii) Some types can be used on external surfaces.

***Application*** Roller. Special rollers available for heavy applications. It can be manipulated while still

wet to form various textures using stipplers, sponges, combs and plastic spatulas (see 1.1.12).
***Drying method*** Coalescence.
***Time*** 20–30 minutes to set. Touch-dry in 4 hours. Dry in 24 hours.
***Cleaning solvent*** Water.
***Spreading capacity*** 5.5–10 $m^2$ per 25 kg.

## 2.4.22 Self-cleaning paint

Formerly an exterior paint designed to chalk such that the surface of the paint film (and attached dirt) is continually eroded and washed away by rain to maintain a clean surface. These paints are no longer common, although solvent-borne exterior paints may claim mild chalking as a self-cleaning benefit. Overcoating of such surfaces requires removal of the chalk layer and usually a primer/sealer coat if a water-borne paint is to be used.

Various modern paints with stay-clean claims are becoming available. These have either a very hydrophobic (water-repellent) or a very hydrophilic (water-wettable) surface that encourages dirt to be washed away by rain. The effect is often a reduced rate of dirt pick-up, rather than no dirt pick-up at all.

## 2.4.23 Heat-resistant paint

Remains protective and decorative when used on surfaces subject to heat.
***Composition*** The components are carefully selected according to the temperature of the surface. White finishes will discolour at temperatures above 65°C and coloured pigmentation becomes impracticable after about 200°C. Aluminium powder is used for most finishes in the higher temperature range. Commonly in use are the following (see Fig. 2.7):

up to 90°C

***Alkyd gloss finish*** Most makes will withstand temperatures of around 90°C, but white and very pale tints will discolour at 70°C and above.

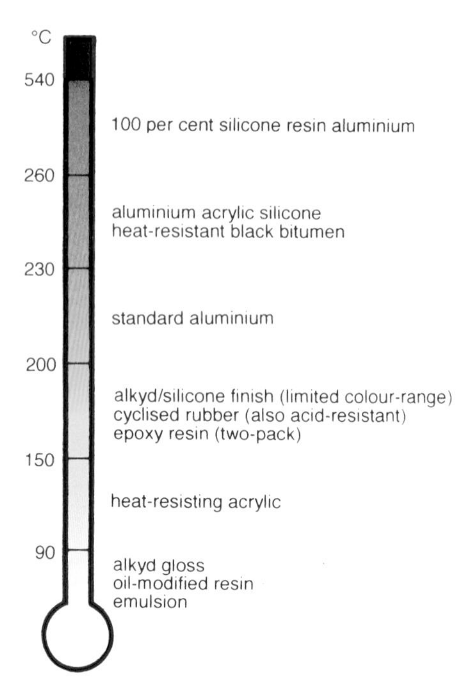


**Fig. 2.7** Heat-resistance of paints

up to 150°C

***Heat-resisting acrylic paints*** Normally applied as a two-coat system on clean bare metal, omitting primer. May lift previous coatings. Surfaces should be cold on application and brought up to 100°C between coats and after finishing in order to effect proper curing.

up to 230°C

***Aluminium paints*** Based on modified alkyd resin and aluminium powder. Suitable for inside or outside use. Two coats are normally applied to clean bare metal.

up to 260°C

***Aluminium acrylic silicones*** Blends of acrylic and silicone resins with aluminium powder. Manufacturers' solvents are required for cleaning.

up to 260°C

***Heat-resistant black*** Based on high-grade bitumens and used on boiler casings, furnaces and exhaust stacks. Primers are not normally used. These paints are cheap compared with others in this range.

up to 540°C

***Heat-resistant silicone aluminium*** Very expensive paints used on furnace fronts, heat exchangers and exhaust systems inside or outside. Silicone resins decompose at about 350°C but the aluminium becomes sintered to the metal and gives protection.

up to 500°C

***Zinc dust/alkali silicates*** A two-pack material designed as a primer for steelwork and offering a tough, durable, heat-resistant base for systems which are subject to severe exposure and high temperatures (500°C).

***Use*** Radiators and pipes on a normal domestic central heating system reach temperatures of about 65°C and most decorative paints can be used. Special heat-resistant paints are used for furnaces, exhausts, heat exchangers, boilers, steam pipes, chimney stacks.

***Choice of finish*** Some paints will discolour but remain protective and this may prove hazardous when the paint is used as an identification colour on pipelines. On a furnace or metal chimney stack, the main consideration would be protection rather than colour retention.

# 2.5 Clear finishes and wood treatments

## 2.5.1 Wood preservatives

The Health and Safety Executive registers all wood preservatives used in the UK, and the registration covers areas of use, application methods and safety precautions. The use of timber preservatives by professionals and 'DIYers' has been subjected to changes brought about by legislation; there are new rules under The Biocidal Products Directive Regulations 2009/10.

There are three main types of preservatives, which give protection against decay and insect attack.

***Types and composition***

(i) *Coal tar oils* These are principally creosote which is obtained from coal tar distillates. Since July 2003, the sale and use of creosote has been restricted to professional users re-treating structures previously treated with creosote and to industrial pre-treatments for certain applications.

(ii) *Organic solvent preservatives* Made by dissolving the preservative in solvents such as white spirit or naphtha. After treatment, the solvent evaporates, leaving the preservative in the wood. Chemicals used include zinc or copper metal carboxylates, permethrin and propiconazole, often applied by low-pressure impregnation.

(iii) *Water-borne preservatives* Chemicals such as salts of copper dissolved in water. Application is by pressure impregnation only. May have restrictions on the chemicals used and areas of use in the future.

***Properties***

*Type (i)* Low cost, strong odour, good protection against fungal and insect attack, poor drying, light or dark brown, non-corrosive, harmful to vegetation, difficult to paint over.

*Type (ii)* Good penetration, good protection against fungal and insect attack, non-corrosive, most types can be painted over when the solvent has evaporated.

*Type (iii)* Usually odourless, can be used to treat damp timber, often takes weeks to dry. Can be decorated over, toxic to fungal spores.

***Use*** Type (i) Industrial pre-treatment of timbers for railways, electricity poles, fencing, harbours and waterways. Treated timber cannot be used indoors, come into contact with plants or be used for play equipment.

Type (ii) Interior and exterior joinery such as windows and doors and fencing. But not structural timbers.

Type (iii) Principally structural and landscape timbers such as timber frames, fencing, bridges, electricity posts and railway sleepers.

***Application*** The method of application determines the amount of preservative in the timber. The more severe the conditions that the treated wood will face, the greater the loading required.

Pressure impregnation: the most effective, performed by specialist factory applicators.

Immersion/steeping: soaking of timber for a few minutes up to several days or weeks.

Dipping: submerging the ends of timber for 10–60 minutes to give extra protection.

Brush or spray: easiest, most common method for site work, generally least effective, and for timber in less hazardous areas.

***Drying method*** *Type (i)* does not fully dry. *Type (ii)* by solvent evaporation. *Type (iii)* chemicals 'fix' in the wood as the water evaporates.

***Time*** From a few days to several weeks, depending on the level of penetration, type and drying conditions.
***Overcoating*** *Type (i)* Coal tar preservatives can be overpainted with similar-based materials. If conventional paints are used, the surface should be sealed with two coats of an aluminium primer. *Types (ii)* and *(iii)* Most can be finished with paints, wood stains or varnishes when dry.
***Cleaning solvent*** *Type (i)* and *(ii)* white spirit.
***Precautions*** Guidance on the use of personal protective equipment, and on the safe use of the preservatives is given on the pack as part of the registration requirements.

## 2.5.2 Wood stains (translucent)

Durable finishes for wood in a range of translucent colours, allowing the grain of the wood to be seen through the finish. Can be solvent- or water-based. Apply to woodstained or bare wood.
***Composition***
***Pigment*** Natural and synthetic iron oxide pigments, other synthetic pigments.
***Film-formers*** Solvent-based – oil-modified alkyds. Water-based – acrylic polymer emulsions.
***Thinners and clean-up*** Follow manufacturer's recommendation 2010.
***Properties***
Enhances timber grain.
Good flexibility and adhesion.
Low to mid sheen.
Resists cracking and peeling.
Easy maintenance.
Good application characteristics.
***Application***

***Drying method*** Solvent-based – oxidation. Water-based – coalescence.
***Overcoating*** Solvent-based – 16–24 hours. Water-based – 2–4 hours.
***Spreading rate*** 15–25 $m^2$ per litre on smooth, planed wood but can be as low as 10 $m^2$.
***Maintenance*** Sand lightly and wash with detergent and water, apply 1–2 coats.

## 2.5.3 Wood stains (opaque)

Offer attractive alternative to translucent wood stains in both colour and finish. Offer a paint-like solid colour with the flexibility of a wood stain. Ideal for covering blemishes in poor-quality timber.
***Composition*** Alkyd/acrylic resin blend dispersed in water.
***Thinners*** Not required.
***Application***

***Drying method*** Coalescence.
***Time*** Touch-dry in 1 hour. Recoatable after 16 hours.
***Spreading rate*** 8–10 $m^2$ on smooth, planed timber.

## 2.5.4 Oil resin varnishes

'Traditional' clear coatings, which dry to form a continuous protective film. Largely superseded by more modern systems, particularly indoors (see section 2.5.6). Are classified as 'long oil' or 'short oil', depending on the type of oil present.
***Composition***
***Film-former*** Oil-modified resins such as tung or phenolic, or oil-modified alkyds.
***Properties***
Sheen level from high gloss to matt.
Chemical resistance.
Long oil varnishes are flexible, hard-wearing and water- and weather-resistant.
Short oil varnishes are brittle, less flexible and for interior use.

Coloured varnishes suitable for interior and exterior use are available.

***Use*** Floor and furniture seals, protecting bare and stained wood, yacht varnish and a protective finish for graining, marbling and lettering.

***Application***

***Drying method*** Oxidation.

***Time*** Touch-dry in 6–12 hours.

Overcoat 16–24 hours.

***Cleaning solvent*** White spirit.

***Spreading rate*** 60–80 m² per 5 litres.

## 2.5.5 Interior varnishes

Clear surface coatings in a range of sheen levels from high gloss to matt. Not suitable for exterior use. May have pigments added to stain the finish in one operation.

***Type and composition***

*Type (i)* Two-pack based on polyurethane resin and single-pack based on urethane oils or urethane-modified alkyds. Solvent-based.

*Type (ii)* Single-pack water-based acrylic, styrene acrylic or urethane acrylic polymer emulsions.

***Properties***

*Type (i)* Harder-wearing and more abrasion-resistant than oil resin varnishes. Excellent water, heat and chemical resistance along with good adhesion. Two-pack more expensive.

*Type (ii)* Quick-drying, less yellow colour to film than solvent-based, low odour, good physical and chemical-resistant properties.

***Use*** Floors and timbers in hard-wearing areas, furniture, and all bare and stained interior timber surfaces. Two-packs for chemical-resistant surfaces, e.g. factories, bars and dairies.

***Application***

***Drying method*** *Type (i)* Two-pack by polymerisation, single-pack by oxidation. *Type (ii)* Coalescence.

***Time*** *Type (i)* 3–6 hours, overcoat 12–24 hours. Delaying overcoating beyond recommended time can lead to adhesion problems. *Type (ii)* 2–4 hours.

***Cleaning solvent*** *Type (i)* Manufacturer's recommended solvent for two-packs, white spirit. *Type (ii)* Water.

***Spreading rate*** 10–20m² per litre.

## 2.5.6 Water-repellent solution

Colourless solution, which penetrates into porous surfaces, to impart water repellence, non-film-forming.

***Composition*** Silicone resin, resins or aluminium soaps dissolved in hydrocarbon solvent.

***Thinner*** Follow manufacturer's recommendation 2010.

***Properties***

Penetrating colourless treatment for exterior use.

Can be applied to damp but not wet surfaces.

Helps prevent weather damage such as spalling due to frost.

Treated surface stays breathable allowing moisture vapour to escape.

Not suitable for horizontal surfaces, below ground or to treat the effects of condensation or rising damp.

***Use*** Porous exterior surfaces such as brick, stone, cement rendering; some can be used to impart water repellence to bare timber.

***Application*** Brush, non-atomising spray. Flood solution on to surface working from top downwards.

***Drying method*** Solvent evaporation.

***Time*** Variable depending on drying conditions and coverage achieved. Water repellence builds up a few days after application.

***Overcoating*** Will accept most masonry-type paints once dry, but sometimes not if of the silicon-wax type. Weather for 6 months before applying cement-based paints.

## 2.5.7 Spirit varnish

Quick-drying varnish for interior use only.

***Composition*** Principally shellac. May have castor oil or balsam as a plasticiser.

***Thinner*** Follow manufacturer's recommendation 2010.

***Properties***

Hard, glossy, brittle film.

Good sealing properties over resinous wood.

Poor resistance to water, heat and chemicals.

Reversible coating.

***Use*** Available as

(i) Spirit varnish, French polish, button polish, white polish – for varnishing or polishing furniture and wooden articles. Spirit-soluble dyes can be added to the varnish to stain and finish the wood in one operation.

(ii) Knotting (genuine shellac or patent) to seal knots and resinous areas in bare wood to prevent bleeding into paintwork leading to discoloration. Patent knotting, which also contains manila resin, seals bituminous surfaces and non-fixed colours or stains caused by water and nicotine.

***Application*** By brush to small areas such as knots. Difficult to apply by brush to large areas due to rapid drying, preventing an 'open edge' being maintained. Polishes usually applied by a 'rubber' made by folding a wad of cotton wool into a lint-free cloth, adding the polish to the cotton wool and then applying by squeezing through the cloth whilst 'rubbing' the wood surface.

***Drying method*** Evaporation.

***Time*** 15–30 minutes.

***Overcoating*** Other spirit varnishes will soften the coating leading to possible application problems. Not usually recommended for overcoating.

***Cleaning solvent*** Industrial methylated spirit.

## 2.5.8 Graining and marbling materials

Graining and marbling may be carried out in either oil or water media. Both are often used together, allowing suitable drying times between each, to create the desired patterns.

Although ready-made graining colours and ground colours are available, they are usually made up from a range of materials to suit the colours and patterns to be created and the personal requirements of the painter. Oil glaze or scumble is used in many oil-based preparations.

### Oil glaze or scumble

Transparent material applied over a previously painted ground coat to give a special two-tone or broken-colour effect.

***Composition***

***Extender*** China clay or aluminium stearate.

***Film-former*** Resin-modified oil varnish and beeswax.

***Thinner*** Follow manufacturer's recommendation 2010.

***Properties***

Non-flowing.

Easy brushing.

Easily tinted or coloured.

Non-reversible coating.

***Use*** (i) Decoration of relief texture by stippling and wiping the highlights.

(ii) Broken-colour effects such as stippling, rag rolling and dragging.

(iii) Added to graining and marbling colour to prevent flowing.

***Application***

***Drying method*** Oxidation.

***Time*** Touch-dry in 4–6 hours, but 24 hours should be allowed before further treatments are applied.

***Overcoating*** To protect the glaze from abrasion and to enhance the depth of colour and texture, varnish is applied to the dried glaze.

***Cleaning solvent*** White spirit.

***Spreading capacity*** When used considerably thinned, which is usual, a rate of approximately 50 $m^2$ per litre would be expected.

## Acrylic scumble glaze

A clear emulsion glaze designed to have an extended wet-edge time. It can be tinted, and some formulations have the ability to carry many special optical-effect pigments, such as pearl pigments, and multi-colour flip materials and texturing materials if required.

The product allows all of the traditional effects to be achieved, although the methodology for obtaining those effects has to be modified due to the nature of an emulsion glaze. As a rule, achieving a job with emulsion glaze is far quicker than with oil glaze.

***Binder*** Acrylic copolymer often as an adhesion-promoted version.

***Thinner*** Water.

***Properties***

Low odour.

Non-yellowing.

Open time – depends on exact product, depending on conditions and effect being achieved, typically 5–15 minutes.

***Time*** Depends greatly on conditions. Touch-dry in around 4 hours.

***Overcoating*** Can be from 6–24 hours, depending on prevailing conditions and effect being achieved. Products can be overcoated with a water-based clear coat for protection, either a satin or gloss working best, especially if optically active materials (pearls, etc.) are in the formulation. A solvent-based varnish (gloss) can also be used to give the job a greater sense of depth and fullness. This works particularly well on doors and trim.

## Oil-graining colour

***Ready-made*** Semi-prepared oil glaze obtainable in a range of colours suitable to imitate most common woods. They can be intermixed, or further coloured with oil stainer. They only require stirring and thinning with white spirit before use. Raw linseed oil can be added to slow the drying time required for large-scale work. They dry to a matt finish which is ideal to receive overgraining in water-graining colour without undue cissing.

***Hand-made*** Oil stainers thinned with white spirit. A little oil scumble glaze is usually added to make the material 'hold-up', or prevent it flowing, and liquid driers are also added to ensure overnight drying.

## Oil-marbling colour

For general use, the medium is a mixture of two parts of white spirit and one part of raw linseed oil, with a little liquid drier to ensure overnight drying. Into the medium are mixed the oil stainers and sometimes white undercoat to obtain the required colours.

Unlike graining colour, the main property of marbling colour is that it flows. This allows the various colours to blend into each other and create the soft marble effects.

There will be special cases where the addition of a pale varnish, or the use of a medium of pale varnish and white spirit is preferred, and even the addition of a small amount of scumble glaze if required by the painter.

## Water-graining/marbling colour

Powder colours soaked overnight in water to produce a creamy paste, or powder colours ready prepared in a water medium. The colour is further thinned for use with water containing a binding agent of vinegar, stale beer, or fuller's earth.

After application and drying, water colours are often fixed or 'strapped-down' with thinned varnish to prevent them being softened and damaged during further operations.

# 2.6 Paint defects

## 2.6.1 Defects of surface coatings

**(Table 2.8; see also Tables 3.11 and 3.12)**

**Table 2.8** Defects of surface coatings

| *Defect* | *Cause* | *Prevention* |
|---|---|---|
| Bittiness (Plate 1)<br>Pieces of dust, grit or skin on the surface of a paint film | Using paint which has not been strained | Remove skin and strain the paint into a clean kettle |
| | Applying paint to dust-laden areas | Remove all dirt and dust with tack-rag or vacuum cleaner before painting |
| | Using dirty equipment and utensils | Clean all brushes, rollers and utensils before use |
| Blooming<br>A whitish appearance on the surface of varnish or enamel, sometimes accompanied by loss of gloss | Application of paint in cold, damp conditions | Avoid application during such conditions |
| | Application of paint in steamy or humid conditions | Apply paint in dry conditions |
| | Water from abrading or washing absorbed into the coating | Ensure that the surface is thoroughly dry before overpainting |
| | Prolonged exposure to pollutants contained in the atmosphere | Remove crystalline deposits occasionally and wipe over surface with a mixture of equal parts of linseed oil and acetic acid (vinegar) |
| Discoloration<br>The pigment colour changes or becomes paler | Paint film exposed to strong sunlight | Use lightfast colours |
| | Paint film exposed to chemical atmospheres | Use chemical-resistant finish |
| | Paint applied to chemically active substrates | Use correct primer/sealor |
| Dry spray<br>Sprayed paint film dries to a rough, gritty finish | High air pressure causes the paint to rebound and settle elsewhere on the partly dried film | Use pressure at a level which allows the paint to adhere in a cohesive film |
| | The gun is moved too fast across the surface: the paint may be too thin to form a continuous film | Move the gun at a speed to produce a wet film |
| | The gun is used too far away from the surface: the solvent evaporates before the coating reaches the surface | Use the gun closer to the surface |

**Table 2.8** Defects of surface coatings (cont'd)

| *Defect* | *Cause* | *Prevention* |
|---|---|---|
| Fat edges<br>Heavy ridge of paint along edges of a painted surface | When both faces of a right-angled surface are being painted, e.g. down the edge of a door, the edge normally receives a double coat | Careful brushwork at edges and corners |
| Flashing or sheariness<br>Patches of uneven sheen on flat or eggshell finishes | Edge sets before overlapping of joints, mainly on ceilings or large wall areas | Increase the number of painters or use larger brushes in order to cover the work more quickly |
| | High room temperatures cause rapid setting | Reduce room temperature |
| | Swift air movement accelerates drying process | Close windows and doors during application to cut down air movement |
| | Application over too porous a surface causes uneven sinkage | Apply over well-sealed and undercoated surfaces only |
| Grinning<br>The previous surface colour shows through the finishing coat | Attempting too wide a colour change at one application, e.g. white over black | Apply additional undercoats until obliteration is achieved |
| | Overbrushing or uneven application | Apply more liberally and evenly |
| | Paint too thin | Thin only to manufacturer's instructions |
| | Applying finish over wrong colour undercoat | Undercoat should be slightly paler than finish |
| Lifting or picking-up<br>Newly applied paint softens or disturbs the previous coating, making it difficult to apply | The previous coat of paint is not fully dried | Allow adequate drying time, taking into account any abnormal conditions such as damp, cold |
| | Previous coating is not resistant to the solvent in the new paint, e.g. chlorinated rubber over a new oil-based system | Ensure that the new paint is compatible with previous coatings. If in doubt, a small test patch should be prepared |
| Misses or holidays<br>Patches where the paint has not been applied | Careless application | Apply paint carefully and methodically |
| | Working under artificial or inadequate lighting conditions | Provide adequate lighting |
| | On textured surfaces, often caused by using wrong method or type of application | Use most suitable method of application, e.g. on roughcast, pebbledash, use a long pile roller or spray |
| | Colour of undercoat too close to colour finish | Use suitable coloured undercoat under the finish |

**Table 2.8** Defects of surface coatings (cont'd)

| *Defect* | *Cause* | *Prevention* |
|---|---|---|
| Orange peel (Plate 2) Sprayed coating that dries to a textured finish resembling the skin of an orange | Paint too viscous, preventing it from flowing out | Thin the paint according to manufacturer's instructions |
| | Poor paint atomisation because of low air pressure | Increase air pressure |
| | Holding the gun too close to the work causes the air pressure to disrupt the paint as it strikes the surface | Adjust the position of the gun to about 150–200 mm from the surface |
| Rain-spotting or cratering (Plate 3) Craters in the surface of a dry paint film | Rain spots falling on the surface of the wet paint | Avoid painting exposed work when it is likely to rain |
| | Droplets of condensation on the wet paint film | Avoid painting in damp or humid atmospheres |
| | Heavy dew forming on the wet paint film | Paints should be applied so that they can become touch-dry before dew starts to fall |
| Retarded drying The paint remains tacky after a prolonged drying period | Grease or wax polish not removed from surface | Remove all traces of grease and wax |
| | Insufficient air and light, e.g. inside cupboards, door and window frames, basements | Allow adequate ventilation during drying |
| | Use of an excess of universal tinters to produce strong colours from a pale base | When colour matching, the base colour should be as near as possible to the required shade Avoid using excess of universal tinters |
| | Use of incorrect solvent such as paraffin | Use correct solvent only |
| Ropiness (Plate 4) Heavy brush marks in the surface of the film. This defect causes reduced gloss and dirt retention | Poor application | Apply correctly with careful laying-off |
| | Applying finishing paints to poorly applied undercoats | Wet-abrade to remove all brush marks |
| | Applying paint over a soft film | Ensure previous coat is thoroughly dry |
| Runs, sagging or curtaining (Plate 5) Uneven paint film flows down the vertical surface, along which wrinkling normally follows | Uneven application | Even application |
| | Paint runs from the corners of mouldings and ornamental work or projections | Avoid heavy coats on such areas |
| | A wet edge starts to set and when joined or lapped the paint runs | Plan the work so that the wet edge is kept open |
| Saponification Soft, sticky paint film, sometimes exuding a brown, soapy liquid | Application of oil paint over new surfaces containing alkali | Sealer should be of an alkali-resisting type |
| | Application of paint to surfaces likely to be splashed by alkali solutions | Use chemical-resistant paints, e.g. chlorinated rubber |

**Table 2.8** Defects of surface coatings (cont'd)

| *Defect* | *Cause* | *Prevention* |
|---|---|---|
| Sulphiding<br>Paint films becoming dark or black | The action of sulphurous fumes on lead-based finishing paints | Use lead-free systems in industrial atmospheres |
| Wrinkling, shrivelling or rivelling (Plate 6)<br>Varnish or gloss film becomes rough and contoured like the shell of a walnut | Paints applied too thickly may skin over, leaving the underneath of the film soft. This prevents the film drying evenly throughout its thickness | Apply paint evenly with a normal film thickness |
| | The paint is subjected to rapid drying conditions | Avoid painting in hot, sunny conditions |
| Yellowing (Plate 7)<br>White paint gradually yellows | Linseed oil or phenolic resin-based paints which receive little or no light, e.g. inside cupboards or behind pictures | Use non-yellowing whites |

## 2.6.2 Defects in the tin (Table 2.9)

**Table 2.9** Defects of paint in the tin

| *Defect* | *Cause* | *Prevention* |
|---|---|---|
| Fattening, feeding or livering<br>The thickening of paint to an unusable viscosity which cannot be reduced by thinning | Mixing incompatible materials | Never mix different types of coating |
| | Materials stored longer than their shelf life | Use materials within the stipulated shelf life |
| | Two-pack materials mixed and left longer than the pot-life | Discard materials after the recommended pot-life |
| | Use of wrong thinner | Use recommended thinner |
| Settling<br>The separation of pigments from the binder during storage or use | Heavily pigmented paints stored for long periods | Invert tins at regular intervals |
| | Metallic powder content, e.g. aluminium | Stir regularly during use |
| | Tinted scumble glaze thinned for use | Stir regularly during use |
| Skinning<br>This defect may take the form of a continuous skin which is easily removed or a very thin discontinuous film which breaks up and contaminates the paint | Unused paint returned to cans | Store paint in airtight cans where possible |
| | Exposure to air of paste and knifing fillers | Cover surface with water or white spirit to exclude air (some fillers are supplied with an artificial skin) |

**Plate 1** Bittiness

**Plate 2** 'Orange peel'

**Plate 3** Rain-spotting or cratering

**Plate 4** Ropiness

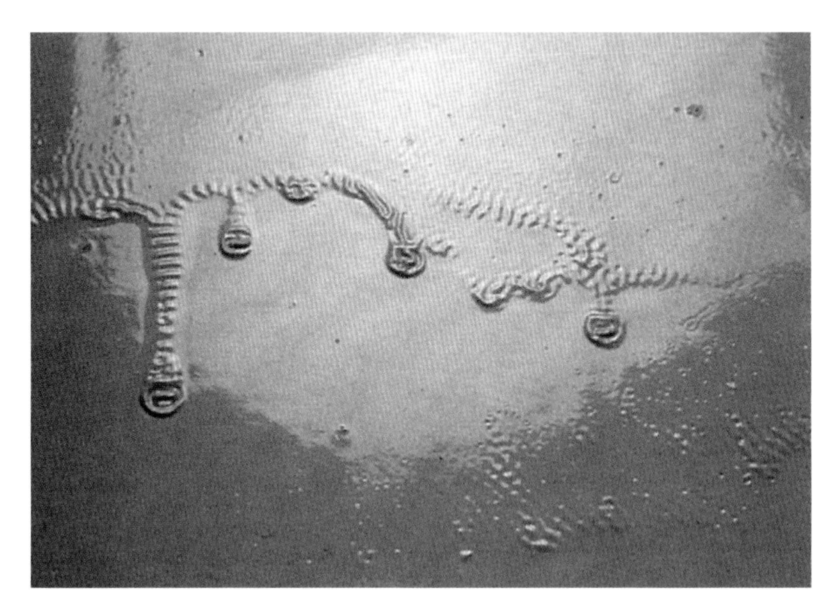

**Plate 5** Runs, sagging or curtaining

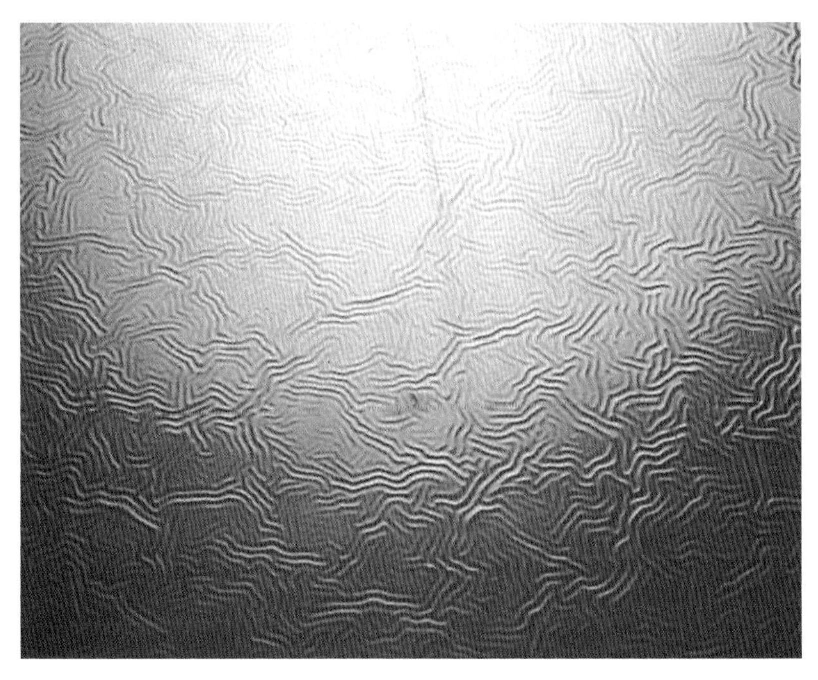

**Plate 6** Wrinkling, shrivelling or rivelling

**Plate 7** Yellowing

# 2.7 Gilding and bronzing materials and defects

## 2.7.1 Gilding

The application of metallic leaf to a previously applied gold size. Gilding is used to decorate both:

(i) *Exterior surfaces*, i.e. building finials, decorative ironwork, usually as a highlight rather than full coverage, signwriting and heraldic painting.

(ii) *Interior surfaces*, i.e. decorative plasterwork on cornices, ceilings and wall panels, painted or stained joinery details such as mouldings and pateras, signwriting and heraldic painting.

### Metallic leaves

Base metals or alloys are beaten into very fine films. The films are cut into easy-to-use sized leaves ranging in size from 80 mm × 80 mm to 150 mm × 150 mm. The leaves are supplied in books, one leaf between each of the 25 pages. They are available either:

***Loose*** Each leaf free between the pages.

***Transfer*** Each leaf lightly adhered to a fine backing paper for easy handling. Some suppliers offer either a light or firm transfer depending on the type of work being carried out, or to the gilder's preference.

Metallic leaves are made from gold, silver, aluminium and alloys of zinc and copper.

### Gold leaf (Table 2.10)

The most widely used leaf which has been used as decoration for many thousands of years. Types available:

(i) The most expensive is beaten from pure gold (24 carat). Rarely used by the decorator.

(ii) The most commonly used leaf is beaten from 23 carat to 23.5 carat gold which is gold mixed with a small amount of silver to make it slightly harder and easier to work with. Suitable for all exterior surfaces.

(iii) Both above golds are available as:

*Fine or regular* which is a standard beaten thickness.

*Double fine, extra thick or triple* which is beaten to a less fine thickness for use on very exposed surfaces.

(iv) Cheaper forms are mixed with silver and/or copper and vary in quality and colour from 22 carat (yellow) to 16 carat (green). Suitable for interior use.

(v) White gold is predominantly silver. It has a good silver colour and is suitable for interior use only.

***Properties***

(i) Cannot be handled because of the fineness to which it has been beaten. Loose gold can be manipulated only with a gilder's tip.

(ii) A very malleable metal and when applied can be burnished by polishing to a high sheen. The purer golds can be burnished until the leaf-edge marks are removed.

(iii) Does not tarnish; 23 carat and 24 carat golds will keep their colour and burnish without any protective coating on both exterior and interior surfaces. Less pure golds may require coating.

***Protection***

23 carat and 24 carat gold will not tarnish and do not need a protective coating. On surfaces which are subject to constant abrasion, all golds should be coated with a rabbit skin size, or gelatine solution followed by a gloss varnish. Varnish applied directly to gold will destroy the natural metallic burnish and eventually darken. This process of sizing and varnishing should be used for all the less pure golds if they are likely to be exposed to polluted air.

### Other metallic leaves (Table 2.11)

Gold leaf is the most widely used metallic leaf. Other metallic leaves are widely available but used only rarely by the painter and decorator.

**Table 2.10** Types of gold leaf

| Grades | Uses | Resistance to tarnish |
|---|---|---|
| 23, 23.25, 23.5ct | All exposed and exterior surfaces | Complete resistance |
| 22ct (yellow), 18ct (lemon), 16ct (green) | Interior surfaces, i.e. decorative plasterwork, embellishments, signwriting, heraldic painting | Resistant in most interior situations without a protective coating |
| White gold | All interior surfaces where a silver effect is required | Will tarnish rapidly if not protected by a clear coating |

**Table 2.11** Other metallic leaves

| Type | Base | Sizes available | Forms available |
|---|---|---|---|
| Silver leaf | Silver | 80 mm × 80 mm | Loose and transfer in books of 25 leaves |
| Schlag metal Dutch metal | Zinc and copper | Varies from 130 mm × 130 mm to 150 mm × 150 mm | Loose and transfer in books of 25 leaves or packets of 500 or 1000 leaves, gold, silver and copper |
| Aluminium leaf | Aluminium | Varies from 95 mm × 95 mm to 140 mm × 140 mm | Loose and transfer in books of 25 leaves or packets of 250 leaves |

All of these leaves are considerably cheaper than gold leaf and although their resistance to tarnishing varies, none of them are completely resistant and must be protected with a varnish or lacquer immediately after application. Because of this they are not recommended for exterior use.

## 2.7.2 Bronzing

The application of metallic powders, either to a previously applied gold size, or mixed with a varnish medium and used in the form of a paint.

### Metallic powders

Alloys of copper and zinc and aluminium, which are atomised, then broken into very fine, small flakes by grinding in a ball mill. The flakes are polished to various degrees of finish.

### Bronze powders

Alloys of copper and zinc, plus a small quantity of aluminium.

***Types***

***Standard*** Cheapest and most coarse. Suitable for mixing into paints.

***Lining*** A fine powder with a high sheen.

***Burnishing*** A very fine powder of high lustre and very expensive.

***Tarnish-resistant*** Coated during manufacture with silica to resist discoloration. Extremely expensive. All types available in a wide range of colours from: Pale (most like gold), Rich or Copper (deeper than gold) to Penny Bronze (very dark).

***Properties***

(i) Wide range of gold-based colours.
(ii) All tend to tarnish on exposure and should be coated with varnish or lacquer to maintain their burnish.
(iii) Wide range of qualities to suit all jobs.

***Use*** (i) Embellishing interior decorative plasterwork.
(ii) Decorating temporary structures such as exhibition stands, stage and television sets.
(iii) In association with gold leaf to reduce costs.

### Aluminium powder

***Types*** Usually one standard grade available.

***Properties***

(i) Good silver colour.
(ii) Dulls quickly on exposure. Needs lacquering to maintain brilliance.

***Use*** Similar uses to bronze powders when a silver effect is required.

### Gold sizes

The materials used to adhere metallic leaves and powders to a prepared surface. They are available in two forms as mordants or mediums depending on the method employed in gilding or bronzing.

### Mordants

Applied to the surface and allowed to set to a tacky film before metallic leaf is applied to it, or before powders are dusted on to it.

Three principal types are:

(i) *Japan gold size* A quick-drying, short oil varnish sometimes called signwriter's size, or quick size. Available in a variety of grades depending on required gilding time, e.g. 1 hour; 2–4 hours; 4–6 hours. These times represent the approximate time the size takes to reach the ideal gilding tack.

***Use*** Small areas, generally for signwriting.

***Properties*** Provides minimum burnish.

(ii) *Oil gold sizes* A range of sizes made from a blend of drying oils, resins and drying agents. They are available in many grades, e.g. 3-hour size; 12-hour size, and 24-hour size. Sometimes called French size or old oil size.

***Use*** Because of their slow setting time, they are most suitable for large areas of gilding, e.g. church finials, or extensive cornice and other relief-moulding embellishment.

***Properties*** Allow more even film and produce a good burnish.

(iii) *Gloss paint*

***Use*** Equally useful on small or large areas of gilding, allowing a more leisurely application of the leaf.

***Properties*** Available in a wide range of colours; thick film can be applied which allows a good burnish; sets in 2–4 hours.

### Mediums

Mixed with metallic powders to enable them to be applied as paints.

***Japan gold size*** Quick-setting, suitable for small areas only.

***Bronzing oil medium*** Specially prepared varnishes which do not react with the acidic nature of the powders. Suitable for all areas of bronzing.

***Bronzing cellulose medium*** Specially prepared cellulose lacquers for spraying on to properly prepared surfaces.

## 2.7.3 Gilding defects (Table 2.12)

**Table 2.12** Gilding defects

| *Defect* | *Cause* | *Prevention* |
|---|---|---|
| Misses<br>Patches where the leaf or powder has not adhered | Careless application | Apply size carefully. Tint size so that coating can be seen clearly as it is applied |
| Show as dark or lustreless areas | | Apply size evenly to avoid thin areas drying faster than thicker areas<br>Ensure surface is of even absorbency |
| Wrinkling<br>Shrivelling of film so that it becomes contoured like the shell of a walnut | Size applied too thickly | Apply size evenly in normal thickness film<br>Apply size in a thinner film |
| | Leaf applied too soon after application of the size. The leaf cuts off oxygen supply and the size remains soft underneath | Allow size to reach correct level of tackiness before applying leaf |

**Table 2.12** Gilding defects (cont'd)

| *Defect* | *Cause* | *Prevention* |
|---|---|---|
| | Pressure of thumb or cotton wool when applying leaf may wrinkle soft areas of size | Apply minimum of pressure when laying leaf |
| Lack of burnish<br>The metallic finish looks dull | Use of too thin size | Apply size in thicker film<br>Use an oil gold size or gloss paint |
| | Bittiness of size caused by dust setting on size while setting-up | Screen-sized areas with greased paper, aluminium foil or plastic |
| | Lack of polishing | Burnish well with cotton wool pad |
| Visible joins of leaves<br>Edges of all leaves clearly seen in finish | Insufficient burnishing | Polish well with a cotton wool pad |
| | Use of hard metals which do not lose their leaf edges | Only use top-quality gold leaf |
| Patchiness<br>Uneven sheen or colour of a bronzing finish | Careless application | Apply size in even film<br>Apply mixed paint evenly<br>Stir and strain mixed paint thoroughly to ensure even distribution of metallic powder |
| Ragged edge<br>Edges of gilded work appear ragged | Excess chalk on surface during sizing | Dust off chalk with a soft brush |
| Lack of lustre | Application of the gold to the size, prematurely | Wait for the right degree of tack which can sometimes last for days as opposed to hours |

# 2.8 Stencil and masking materials

## 2.8.1 Stencil-plate materials

Materials from which stencil plates or templates are made are usually obtained ready prepared. A limited range can be made as required from commonly available materials. However, each should possess the following properties:

(i) Easily cut with a sharp knife to produce a clean, smooth edge.
(ii) Non-absorbent to paint and cleaning solvents used in the stencilling process.
(iii) Resist damage during use and when cleaning.
(iv) Remain flat during use and during storage.

### Types

***Self-prepared stencil paper*** Good-quality drawing paper coated both sides with a mixture of linseed oil and liquid driers. After being suspended for 2–3 days to allow thorough drying, it is ready for use. If a quicker drying coating is required, shellac knotting can be used to coat the paper.

***Properties***

Easy to cut.
Inexpensive.
Easily damaged, especially whilst cleaning.

***Use*** (i) Small stencil designs.
(ii) Projects with few repeated applications and cleaning processes.

***Ready-prepared stencil paper*** Good-quality card of varying thickness, impregnated with a drying oil. Available in A2-sized sheets, and either brown or yellow in colour.

***Properties***

Better resistance to damage.
Thicker varieties are difficult to cut.
More expensive than self-prepared stencil paper.

***Use*** (i) Small and larger stencil designs.
(ii) Multi-plate stencils.
(iii) Projects with a medium number of repeated applications and cleaning processes.

***Plastic stencil sheet*** Transparent acetate sheet containing a plasticiser to prevent splitting at the intersections of cuts.

***Properties***

Excellent resistance to damage.
Easy to cut.
Transparency allows easy and accurate alignment.
Expensive.

***Use*** (i) Small and larger stencil designs.
(ii) Small sections of designs requiring accurate registration.
(iii) Multi-plate designs.
(iv) Projects with many repeated applications and cleaning processes.

***Metal stencils*** Ready-prepared designs cut from thin metal sheeting. Also available are individual letters and numerals which can be linked together to produce simple signs.

***Properties***

Resists unlimited applications and washing processes.
Easy to store without damage.
Expensive.
Limited range of designs and sizes.
Usually single-plate designs only.

***Use*** (i) Small repeating patterns.
(ii) Furniture decoration.
(iii) Friezes and borders.
(iv) Simple signs.

### Liquid masking or strippable coating

A water-based film-former with limited surface adhesion when dry. After brush or spray application, and thorough drying, the stencil design is cut directly into the dry film. Sections of the film are easily removed, and paint applied. When the job is completed, the remaining film is removed, leaving the design painted on the surface.

***Properties***

Does not rely on tie bars to hold stencil together.

Easy to hand-cut.

Can be cut with knives set in compasses and trammels.

Will not lift from the surface whilst the stencilling paint is applied.

Produces very clean edges to the finished work.

***Use*** (i) One-off designs requiring clean-edged definition.

(ii) Complex multi-colour projects.

(iii) Work requiring a high level of accuracy.

(iv) Geometric designs.

## 2.8.2 Other stencil materials

### Masking tape

Rolls of self-adhesive tape available in various widths from 6 mm to 300 mm wide. Made from paper or plastic film and possessing either standard-tack or low-tack adhesive qualities.

***Use*** (i) In association with paper to mask areas from overspray when spray painting.

(ii) Dividing one painted area from another in a clean line.

(iii) Adhering drawings, pounces and stencil plates to the prepared surface.

(iv) Low-tack tapes used for similar purposes on surfaces which may be damaged by the standard-tack variety.

(v) Interior and exterior types available.

(vi) Different types available for short and long duration use.

### Lining tape

Rolls of low-tack plastic adhesive tape, which after application allow a central section to be removed, exposing the surface in a fine parallel line. Paint can be applied, and when tacky, the remaining tape removed to leave a fine painted line on the surface.

***Use*** Painting fine lines on sign boards, panels and vehicles.

### Liquid adhesive

Liquid adhesive in an aerosol spray container.

***Use*** Sprayed on the back of stencil plates and templates to hold them in position whilst paint is applied by brush or spray.

# PART 3

# Preparation of Surfaces for Paint and for Clear Finishes

# 3.1 Abrasives

Abrasives may be classified broadly into two groups: coated abrasives and abrasive powders.

## 3.1.1 Coated abrasives (Table 3.1)

Natural or synthetic abrasive materials crushed, graded and stuck to paper or cloth with either a water-soluble or a waterproof adhesive. Used dry or with water, depending on type.

### Paper

Weights of paper range from grades A to D. Grade A heavy-weight: lightweight, flexible. Used to support finer grades of abrasive particles, for hand abrading. Grade D: heavyweight, rigid. Used to support coarser grades of abrasive particles, for hand and mechanical abrading.

*Note:* Aluminium oxide abrasive papers can be obtained wrapped around and adhered to a rectangular block of synthetic sponge. They are used with water and available in fine, medium and coarse grades.

### Size

(i) Standard sheets of 280 mm × 230 mm.
(ii) Ready-cut rectangles or discs suitable for orbital or disc sanders.
(iii) Rolls suitable for belt sanders.

### Grades of abrasive (Table 3.2)

Coated abrasives are graded by a number which corresponds to the size of abrasive particle used. For

**Table 3.1** Properties of coated abrasives

| *Abrasive* | *Manufacture* | *Grades* | *Use/Properties* |
|---|---|---|---|
| Emery cloth or paper | Natural emery affixed to stout paper or cloth with a water-soluble adhesive | Coarse<br>Medium<br>Fine | Used dry or with a spirit lubricant for etching and degreasing metals before painting<br>Very hard cutting edge<br>Expensive |
| Garnet cabinet paper | Natural garnet 'open-coated' to grade D paper with a water-soluble adhesive | 40<br>60<br>80<br>100 | Used dry for the hand or mechanical abrading of wood before painting or varnishing<br>Very sharp cutting edge |
| Garnet finishing paper | Natural garnet 'open-coated' to grade A paper with a water-soluble adhesive | 150<br>180<br>220<br>240 | Used dry for the 'flatting-down' between coats of varnish or french polish<br>Largely superseded by aluminium oxide papers |
| Production paper | Synthetic aluminium oxide 'open-coated' to grade D paper with a water-soluble adhesive | 40<br>60<br>80<br>100<br>120 | Used dry for heavy abrading operations by hand or by mechanical sanding tools<br>Tough, very sharp cutting edge |

**Table 3.1** Properties of coated abrasives (cont'd)

| *Abrasive* | *Manufacture* | *Grades* | *Use/Properties* |
|---|---|---|---|
| Aluminium oxide paper | Synthetic aluminium oxide 'open-coated' to grade A paper with a water-soluble adhesive | 150<br>180<br>220<br>240 | Due to its long life and controlled cut, it is superseding the use of garnet finishing papers |
| Glasspaper | Glass, natural flint or quartzite mineral affixed to grade D paper with a water-soluble adhesive | Strong (S)<br>Coarse (C)<br>Medium (M)<br>Fine (F)<br>Also 1, 0, 00 and Flour | Used dry for the hand or mechanical abrading of plaster, wood, boarding before painting, and denibbing between coats of paint<br>Clogs readily, therefore has a short life<br>Tends to scratch the surface |
| Silicon carbide paper (self-lubricating) | Synthetic silicon carbide 'open-coated' to grade A paper with a water-soluble adhesive. Paper additionally coated with zinc stearate which acts as a self-lubricant | 220<br>240<br>320<br>400<br>500 | Mainly used dry in the vehicle-refinishing industry for abrading filling materials, 'flatting-down' between coats of paint and before painting<br>Non-clogging |
| Silicon carbide paper (wet or dry) | Synthetic silicon carbide affixed to a waterproofed grade A paper with a waterproof adhesive | 120<br>240<br>320<br>400<br>600 | Used with water for abrading *all* types of previously painted surfaces before painting<br>Superseded the use of pumice stone and other abrasive powders<br>Tough and long-lasting if kept unclogged during use |
| Tungsten carbide | Synthetic mineral usually affixed to thin metal sheet | Fine<br>Medium<br>Coarse | Used for dry abrading of coarse painted surfaces, or new wood before priming<br>Tough, long-lasting<br>Extremely open-coated, so preventing clogging |

**Table 3.2** Comparison of grades of abrasive

| | *Fine* | | | *Medium* | | | *Coarse* | | |
|---|---|---|---|---|---|---|---|---|---|
| Glasspaper | 00 | 0 | 1 | $1\frac{1}{2}$ | F2 | M2 | S2 | $1\frac{1}{2}$ | 3 |
| Aluminium oxide | 180 | | 150 | 120 | 100 | 50 | 60 | 50 | 40 |
| Silicon carbide | 600 | | 400 | 280 | 240 | 180 | 150 | 120 | 100 |

**Table 3.3** Properties of abrasive powders and other abrasives

| *Abrasive* | *Origin or manufacture* | *Use/properties* |
|---|---|---|
| Pumice or soda block | Pumice powder and soda or detergent pressed into a block | Used with water for the initial cutting down of very coarse paintwork before painting or paperhanging<br>Liable to scratch severely |
| Wire or steel wool | Fine, tangled strands of steel or stainless steel wire | Coarse, medium and fine grades, for<br>(i) degreasing and etching metal and plastics<br>(ii) scouring timber after stripping with liquid paint remover<br>(iii) flatting varnished or lacquered surfaces to produce a matt finish (advisable to wear protective gloves) |
| Shot or grit | Abrasive particles obtained from many sources (see 1.2.3) | High-pressure abrasive cleaning (see 1.2.3) |
| Rubbing and polishing compound | Mild abrasive powders mixed to a paste ready for use (water-soluble) | Used for rubbing and polishing synthetic and nitrocellulose finishes |

example, a coated abrasive graded 240 contains particles which will pass through a sieve containing 240 holes per square inch. Common grades range from 120 (coarse) to 600 (extremely fine).

### Open-coated abrasives

Many coated abrasives tend to 'clog up' when used dry. Manufacturers overcome this problem by supplying a paper sparsely coated with abrasive particles which is ideal for mechanical sanding.

## 3.1.2 Abrasive powders and other abrasives (Table 3.3)

Usually naturally occuring materials, crushed, graded and either (a) used with water and a felt block; or (b) pressed into a block and used with water; or (c) made into a paste ready for use.

# 3.2 Making good

The materials used in the process of repairing and surfacing before and during painting may be classified as stoppers and fillers.

## 3.2.1 Stopper (Table 3.4)

A stiff material, which dries with the minimum of shrinkage, used for making good large holes and cracks.

## 3.2.2 Filler (Table 3.5)

A smooth paste used for filling slight surface imperfections.

### Caulking

Caulking is a process of covering joints in dry-lining. *Joints* (Figs. 3.1 and 3.2). After the joints between boards or at angles have been filled, paper tape is

**Table 3.4** Properties of stoppers

| ***Stopper*** | ***Use/properties*** |
|---|---|
| Linseed oil putty | Commonly used on wood for stopping holes and fixing window glass<br>Less expensive than hard stopper<br>Becomes brittle with age |
| Plaster | Stopper for gypsum and lime plasters<br>Quick setting and drying<br>Rake out and undercut for good adhesion<br>For deep holes, sand is added to the plaster to reduce shrinkage and cracking |
| Sand and cement mix | 4 parts sand to 1 part of cement<br>Used for<br>(i) deep holes in plaster<br>(ii) making good external rendering and brickwork |
| Mastic | Very flexible material based on oil, latex or resin<br>Mainly used around external door and window frames<br>Usually applied with a caulking gun |
| Cellulose stopper | Used extensively in the vehicle-refinishing industry under nitrocellulose finish<br>Quick drying |
| Plastic wood | Mixture of wood, flour and resin<br>Used under clear wood finishes, ready-mixed or two-pack<br>Quick setting<br>Expensive |
| Metal | Mixture of aluminium or iron powders and resin |
| Stopper | Used for stopping and rebuilding surfaces and arrises<br>Usually two-pack<br>Quick setting<br>Extremely hard<br>Expensive |

**Table 3.4** Properties of stoppers (cont'd)

| *Stopper* | *Use/properties* |
|---|---|
| Tinted | Water-borne |
| Stopper | Used under clear wood finishes<br>Available in a variety of colours to match stained wood |
| Waterproof sealant | Very flexible; based on latex or silicone resin<br>Waterproof<br>Tinted<br>Filling cracks around baths and basins |
| Two-pack stopper | Very hard<br>Takes screws and fixings<br>Previous paint coatings should be removed<br>Minimum shrinkage<br>Rapid setting (15 minutes)<br>Can be used on all clean, sound substrates |

**Table 3.5** Properties of fillers

| *Filler* | *Use/properties* |
|---|---|
| Plaster-based ('alabastine' type) | Used on plaster and for filling grain in timber<br>Very absorbent when dry |
| Plaster/cellulose-based ('polyfilla' type) | Used on plaster and for filling grain in timber<br>Good adhesion, slight shrinkage on drying<br>Absorbent when dry |
| Plaster/vinyl-based ('instant polyfilla' type) | For use on internal surfaces<br>Supplied ready-mixed<br>Very smooth paste<br>Shrinks on drying |
| Cement/vinyl-based ('exterior polyfilla' type) | Used for external surfaces<br>Excellent adhesion and water resistance<br>Brittle and difficult to abrade |
| Synthetic emulsion-based grain fillers ('fine-surface polyfilla' type) | For internal and external surfaces<br>Supplied ready-mixed<br>Very smooth paste<br>Can be worked to a fine feather edge<br>Quick setting |
| Oil-based paste filler (spachtel type) | For internal or external painted surfaces<br>Supplied ready-mixed<br>Low absorption when wet<br>Slow drying<br>Shrinks on drying |
| Brush or spray filler | (i) Ready-prepared with oil, synthetic or cellulose medium<br>(ii) Oil-based, thinned with white spirit<br>Several coats brushed or sprayed to flat or curved internal or external surfaces<br>Low absorption |

**Table 3.5** Properties of fillers (cont'd)

| *Filler* | *Use/properties* |
|---|---|
| Oil-based rubbing or grain filler | For internal and external use<br>Supplied ready-mixed in natural and a range of colours<br>Slow drying<br>Applied by coarse cloth |
| All-purpose filler | Available in powder form or ready-mixed<br>Suitable for interior and exterior surfaces<br>Waterproof<br>Flexible<br>Can be used on most materials<br>Good adhesion to metal, wood, masonry<br>Very hard when set |
| Lightweight filler ('one-time' type) | No sanding, sagging or shrinking<br>Ready-mixed and lightweight<br>For use on plaster, wallboard, wood and stucco<br>Not suitable for plastic or metal |
| Flexible acrylic filler ('decorators' caulk' type) | Overpaintable in 1 hour<br>Mould-resistant<br>Low shrinkage<br>Internal/external use |

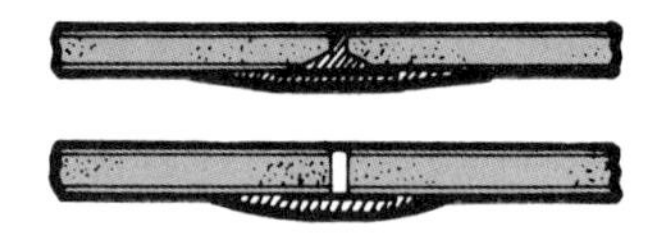

**Fig. 3.1** Filling joints

**Fig. 3.2** Caulking process

sandwiched between a thin layer of filler. This work is carried out with a wide spatula or caulking tool.

# 3.3 Preparation of previously painted surfaces

## 3.3.1 Unsound surfaces

Surfaces which are flaking, peeling, blistering, cracking, coated with size-bound distemper or otherwise unsound should be stripped back to bare surface by one of the following methods, and then treated as new surfaces.

### Washing off

Surfaces, especially ceilings in older buildings, may be coated with size-bound distemper which must be regarded as an unsound surface.

(i) Remove distemper by washing with water.
(ii) Bind any residue of distemper with thinned alkali-resisting primer or proprietary binder/sealer (stabilising solution).

### Burning-off

The fastest method of removing old, defective paint and varnish coatings: they are softened by intense heat and scraped off with a stripping knife or shavehook.

***Use*** Mainly on timber surfaces.

***Limitations*** See Table 3.6.

***Equipment*** See 1.1.4.

***Safety precautions*** (see also 6.2).

(i) Do not overfill petrol and paraffin lamps.
(ii) Allow ample warming-up time before pumping paraffin lamps.
(iii) Supply adequate ventilation for inside work.
(iv) Locate and have ready at hand suitable fire-fighting equipment before starting work (Fig. 3.3).
(v) Protect floor from hot droppings with thin metal or non-flammable sheet.

**Table 3.6** Comparison of effects of liquid paint remover and heat stripping

| *Material treated* | *Liquid paint remover* | *Heat stripping* |
|---|---|---|
| Coatings of large, flat surfaces | Slow<br>Messy<br>Expensive<br>Evaporation before the film is softened | Satisfactory |
| Coatings on intricately carved detail | Satisfactory | Disfiguration of detail by chipping with shavehook |
| Coatings adjacent to or on glass | Satisfactory | Cracking of the glass by heat |
| Coatings adjacent to or on flammable surfaces | Satisfactory | Ignition or melting |
| Coating to be replaced by clear finish | Satisfactory | Disfiguration of the surface by scorching |
| Coatings on plaster and cement renderings | Satisfactory | Cracking and 'blowing' of the plaster from its backing |
| Coatings on thin metal sheet | Satisfactory | Buckling of the metal by heat (or melting of lead) |

**Fig. 3.3** Fire-fighting equipment

(vi) Avoid playing flame on any adjacent flammable surfaces.
(vii) Extinguish flaming droppings *immediately*.
(viii) Must not be used for removing old lead paint coatings.

### Use of liquid paint remover

Paint removers soften old coatings of paint, varnish and textured coatings, allowing them to be removed with stripping knife or shavehook. They are based on either solvent or alkali.

***Use*** Where removal by burning-off is not practical or safe (see 6.4.2).

***Limitations*** See Table 3.6.

A. *Solvents*
  (a) Highly flammable solvent, such as acetone plus wax to slow down evaporation and thicken the material.
  (b) Non-flammable solvent, such as a chlorinated hydrocarbon plus methyl cellulose to slow down evaporation and thicken the material.

***Properties***

(i) Softens the paint film and penetrates through to the substrate. The solvent evaporates, helping to lift the film.
(ii) Does not damage timber, building boards, plaster, metal or stone surfaces.
(iii) Does not destroy bristle or hair.
(iv) Will remove most types of air-drying paint.

***Safety precautions***

(i) Avoid causing sparks or naked flames (see 6.5).
(ii) Provide adequate ventilation (see 6.6).
(iii) Protect linoleum and some plastics which may be damaged.
(iv) Protect hands with gloves.

B. *Alkalis*
  (a) Proprietary brands based on caustic soda.
  (b) Mixed on site by adding 0.5 kg caustic soda to 4 litres starch paste.

***Properties***

(i) Saponifies the drying oils in the paint film.
(ii) Destroys bristle and hair brushes.
(iii) Difficult to remove all traces of alkali after use, especially on absorbent surfaces.
(iv) Tends to lift the grain of timber and may lift veneers.
(v) Cheap and non-flammable.
(vi) Particularly effective used as a dip.
(vii) Corrosive on non-ferrous metals, especially aluminium.

***Safety precautions***

(i) Wear rubber apron, gloves and goggles to prevent damage by accidental splashing.
(ii) Protect surrounding areas, particularly linoleum and fabrics.

### Mechanical abrading (see 1.2.3)

Most types of coating can be removed by the use of disc sanders or belt sanders.

***Safety*** Respirators should be worn to prevent dust being inhaled.

## 3.3.2 Sound surfaces

Some surfaces may be firm and sound but contain a few areas where the coating is flaking as a result of corrosion or dampness. The procedure then is to scrape off all the loose paint film to a hard edge and to expose the surface if possible. The bare areas should be cleaned, abraded and primed (spot-priming). When the primer is dry, the surface should receive the following treatment:

(i) Wash and rinse to remove all traces of grease and grime which, if left, would reduce the adhesion of the undercoat. A variety of washing materials are available (see Table 3.7) which, when mixed with water and are ready for use, are often referred to as pickle. These are usually applied and scrubbed on with an old brush.

**Table 3.7** Types of washing material

| *Material* | *Advantages* | *Disadvantages* |
|---|---|---|
| Sugar soap | Inexpensive<br>Etches the surface | Softens paint film if mixed and used too strong<br>Residue left on surface may slow down or prevent drying of paint applied over |
| Washing soda | Inexpensive<br>Does not froth up | Softens paint if mixed and used too strong (1 handful to 1 litre boiling water is usual strength for normal washing)<br>Difficult to rinse off all traces<br>Residue may affect drying of subsequent coats |
| Detergent powder | Removes grease very efficiently | Softens paint if mixed and used too strong<br>Expensive<br>Difficult to remove because of its frothing action |
| Soap powder | Milder than detergent, does not soften paints | Expensive<br>Difficult to remove froth and 'greasy' film<br>Residue may affect drying of subsequent coatings |
| Emulsifying liquids | Excellent degreasing properties, particularly in removing thick grease and oils from absorbent surfaces | Expensive |

(ii) Abrade to provide a smooth surface and a 'key' for the undercoat, again to increase adhesion. Use the abrasive (Table 3.1) on a felt, rubber or wooden block to maintain even pressure and a flat surface. Wet-abrading may be combined with the washing process.

*Note:* For health reasons, do not dry-abrade painted surfaces which may contain lead (see 6.4).

(iii) 'Make good' or stop and fill all holes, cracks and other irregularities in the surface.

# 3.4 Preparation of new, stripped or untreated surfaces

## 3.4.1 Properties of building materials (Table 3.8)

All building materials can be classified as either (a) *porous materials* which have air voids in their composition, formed by liquid, usually water, evaporating or chemically combining with the material during its manufacture or formation (Fig. 3.4); or (b) *non-porous materials* which have no air voids in their composition, being manufactured or formed under conditions of extreme heat or pressure and allowed to cool; this drives off all water content, leaving no air voids (Fig. 3.5).

Materials which are commonly painted can, for the purpose of surface preparation, be further classified into six groups according to their physical and chemical properties.

(a) *Very absorbent* materials have air voids linked by capillary tubes, readily allowing the passage of liquids (Fig. 3.6).
(b) *Absorbent* materials have air voids partially separated by impervious material such as resin. They are reluctant to allow the passage of liquids (Fig. 3.7).
(c) *Non-absorbent* materials do not allow liquids to enter or pass through them, because either they are non-porous, or they are porous but have air voids completely surrounded by impervious material, so that the movement of water is prevented (Fig. 3.8) (also known as impervious materials).
(d) *Chemically active* materials are alkaline and promote a chemical reaction with some drying oils in paints. The most common alkali found in building materials is lime (Fig. 3.9).
(e) *Corrodible* metals will rust under certain conditions (Fig. 3.10).
(f) *Non-drying or bleeding* materials (i) contain or are contaminated with compounds which retard or prevent the drying of paints applied to them; (ii) will 'bleed' through and discolour paints applied to them (Fig. 3.11).

### Flame-spread classification

All the above materials can be classified according to their flame-spread properties.

The Building Regulations specify minimum requirements in terms of flame-spread classifications for building materials. BS 476: Part 8: 1997 lays down the method of assessing materials so that they can be given a 'class' of flame spread. The method is dependent on the distance that a flame will travel across a sample of material of given size. The classes of flame spread are:

Class 0 Non-combustible surfaces
Class 1 Surfaces of *very low* flame spread
Class 2 Surfaces of *low* flame spread
Class 3 Surfaces of *medium* flame spread
Class 4 Surfaces of *rapid* flame spread

## 3.4.2 Very absorbent surfaces

### Insulating board

Wood fibres and resin lightly compressed to form a rigid, porous, lightweight building board. Used extensively for the internal cladding of ceilings and walls. Excellent sound- and heat-insulating properties. Available:

(a) untreated
(b) ready-primed
(c) fireproofed (see 3.4.5)
(d) bitumen-impregnated (see 3.4.7)

**Table 3.8** Properties of building materials

| *Materials* | *Porous* | *Non-porous* | *Very absorbent* | *Absorbent* | *Non-absorbent* | *Chemically active* | *Corrodible* | *Non-drying or bleeding* | *Flame-spread classification* |
|---|---|---|---|---|---|---|---|---|---|
| Bitumen and coal-tar coated surfaces | | | | | | | | X | 1–4 |
| Blockboard and plywood | X | | | X | | | | | 3 |
| Brickwork | X | | X | | | X | | | 0 |
| Building blocks | X | | X | | | X | | | 0 |
| Cement rendering | X | | X | | | X | | | 0 |
| Chipboard | X | | | X | | | | | 3 |
| Concrete | X | | X | | | X | | | 0 |
| Expanded polystyrene | X | | | | X | | | | 4 |
| Fabrics | X | | X | | | | | | 4 |
| Ferrous metals | | X | | | | | X | | 0 |
| Fireproofed boards | X | | X | X | | X | | | 1 |
| Glass | | X | | | X | | | | 0 |
| Glass-reinforced cement board | X | | X | | | X | | | 0 |
| Glass-reinforced plastics | | X | | | X | | | | 3 |
| Glazed tiles | | X | | | X | | | | 0 |
| Gypsum plaster | X | | X | | | X | | | 0 |
| Hardboards and MDF | X | | | X | | | | | 3 |
| Hardwoods | X | | | X | | | | | 2 |
| Insulating board | X | | X | | | | | | 4 |
| Lime plaster | X | | X | | | X | | | 0 |
| Non-ferrous metals | | X | | | X | | X | | 0 |
| Oil-contaminated surfaces | | | | | | | | X | 1–4 |
| Paper | X | | X | | | | | | 4 |
| Plasterboard | X | | X | | | | | | 1 |
| Plastics | | X | | | X | | | | 4 |
| Resinous woods | X | | | X | | | | X | 3 |
| Softwoods | X | | | X | | | | | 3 |
| Stonework | | | X | | | X | | | 0 |
| Woodwool slabs | X | | X | | | X | | | 1–0 |

**Fig. 3.4** Porous material

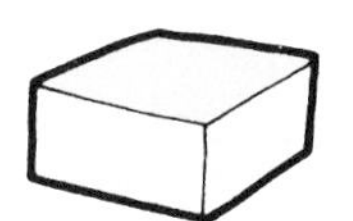

**Fig. 3.5** Non-porous material

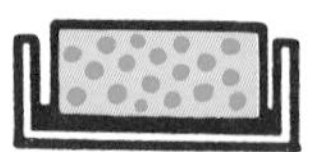

**Fig. 3.6** Very absorbent material

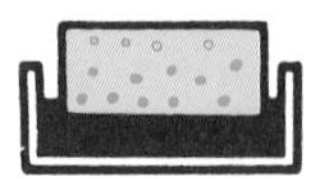

**Fig. 3.7** Absorbent material

**Fig. 3.8** Non-absorbent material

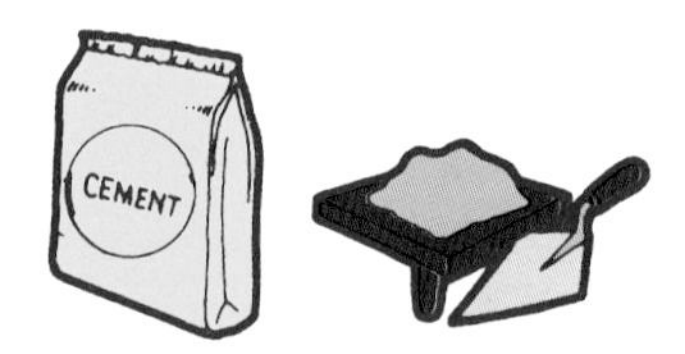


**Fig. 3.9** Chemically active material

**Fig. 3.10** Corrodible material

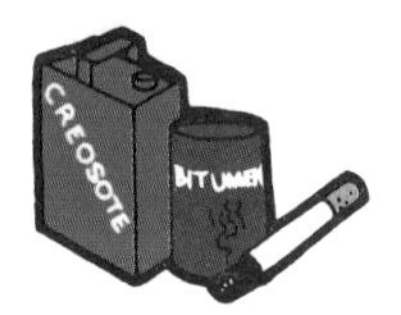


**Fig. 3.11** Non-drying or bleeding material

***Preparation***

(i) Ensure surface is dry and free from dust and grease.

(ii) Punch nails below the surface and spot-prime.

(iii) Caulk joints and nail holes (see 3.2).

(iv) *Do not* abrade, as this will cause surface damage.

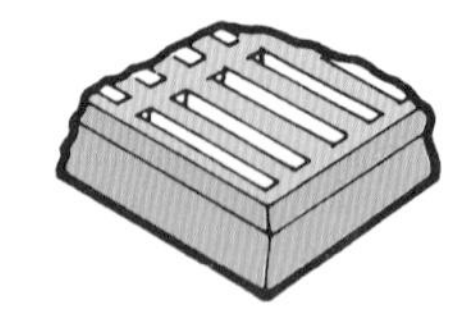

**Fig. 3.12** Acoustic board

***Primer***

(a) Low-pigmented oil sealer to penetrate and reduce absorption.

(b) Thinned emulsion paint.

(c) Acrylic primer/undercoat.

(d) Hardboard primer.

(e) Glue size solution.

*Note:* Water-borne primers tend to raise the surface fibres of the board and corrode nail heads.

## Acoustic board (Fig. 3.12)

Sheets or tiles of insulating board with a variety of holes or slots covering the surface. The surface texture increases the surface area, so improving the sound-insulating properties of the board.

*Types, Preparation, Primer*, all as insulating board.

*Note:* Short-pile paint roller will prevent the textured surface being filled with paint, which would reduce the acoustic properties of the board.

## Paper and fabrics

With the possible exception of wallpaper, hessian and canvas, paper and fabrics are not usually painted.

***Preparation*** Ensure that the surface is dry and free from dust.

***Primer***

(a) Glue size solution.

(b) Thinned emulsion paint.

(c) Acrylic primer/undercoat.

Oil-based primers must not be used because they will either rot or discolour the vegetable fibres of the material.

## Plasterboard (Fig. 3.13)

Building board of hard gypsum plaster sandwiched between two sheets of stout paper or card. Two usable

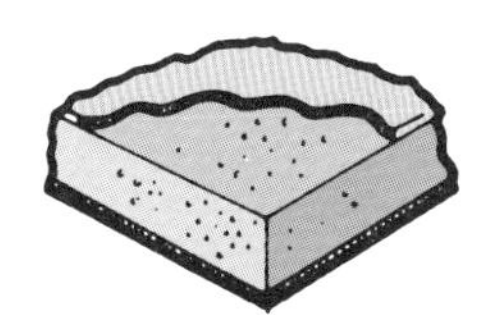

Fig. 3.13 Plasterboard

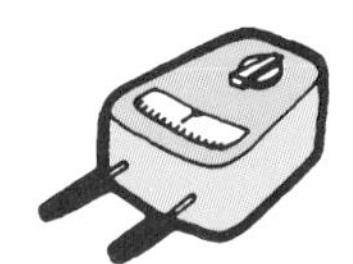

Fig. 3.14 Moisture meter

sides; grey for plastering and cream for decorating. Used as dry lining.

*Preparation, Primer*, both as insulating board.

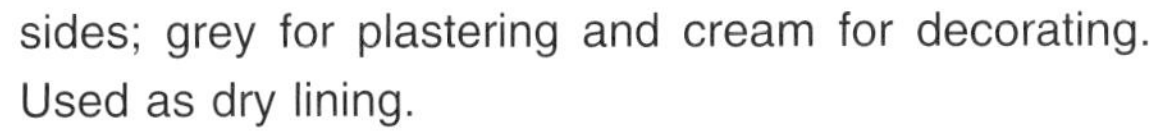

## 3.4.3 Absorbent surfaces

### Timber

#### Moisture content

Timber contains moisture in varying amounts, and it is the drying out and absorption of water which cause wood to expand and contract, so damaging paint films. Before painting wood, ensure that moisture content does not exceed 18 per cent. Use a moisture meter to check this (Fig. 3.14). Painting timber having over 20 per cent moisture content may result in blistering and flaking (see 3.6).

#### Preservatives

Timbers with high moisture content can rot even under paint systems. To avoid this, special preservatives can be applied. Such treatment is recommended for exterior softwood joinery and cladding. Applied to all surfaces before fixing. It is common practice for self-coloured preservatives to be left unpainted.

#### Back-priming

Where wood is to be fixed to structures such as brickwork or plaster, the entire surface, particularly end grain, should be primed before fixing to prevent any future penetration of water (Fig. 3.15). Timber with over 20 per cent moisture content is open to attack by dry rot.

#### Primer application

Good adhesion is obtained by brush application which ensures intimate contact between paint and substrate.

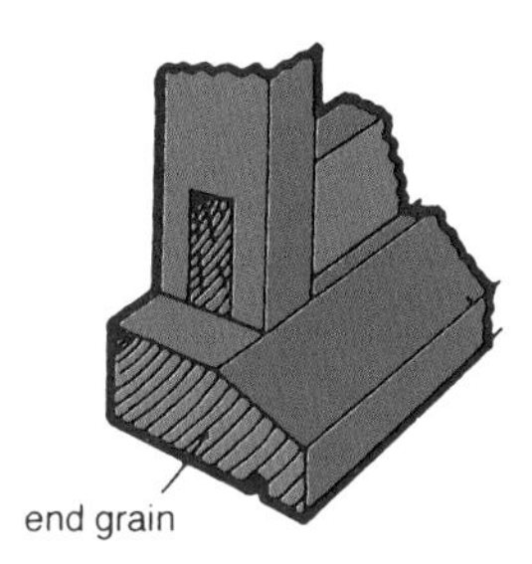


Fig. 3.15 End grain requiring back-priming

#### Shop-primed timber

Some timbers and building boards can be obtained already primed, but the priming paint and application are often inferior to 'on-site' priming. If in doubt, dust off the surface and reprime with suitable primer.

#### Softwood

Softwood timbers are obtained from fast-growing trees. The main timber used in construction is from coniferous trees (evergreen, with needle leaves). Softwoods require protection by surface coatings. Common types are pine, cedar, spruce (whitewood), redwood (deal).

***Preparation***

(i) Ensure that moisture content is low.
(ii) Punch nails below surface.
(iii) Dry-abrade to a clean, smooth surface.
(iv) Prevent resin staining by coating knots and resinous parts with shellac knotting (see 3.6).
(v) Apply preservatives when specified.

***Primer***

(a) Prink wood primer.
(b) White wood primer.
(c) Lead-free primer.
(d) Aluminium wood primer.
(e) Acrylic primer/undercoat. Emulsion paint is unsatisfactory.

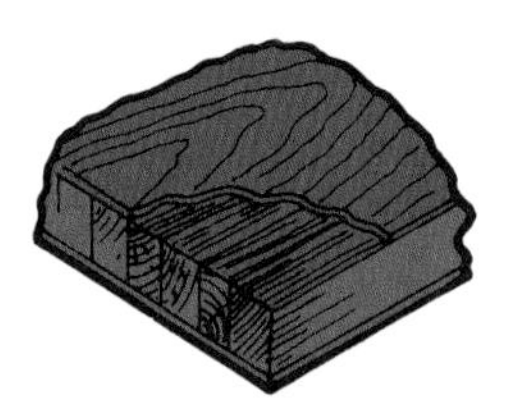

**Fig. 3.16** Blockboard

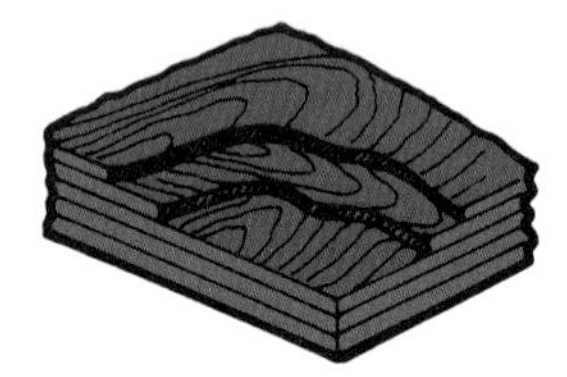

**Fig. 3.17** Plywood

(f) Calcium plumbate primer (composite wood/metal components).

### Hardwood

Hardwood timbers are obtained from slow-growing trees, deciduous (leaf-shedding), with broad leaves. They are not usually painted, but often given a clear finish to enhance the grain. Common types are ash, oak, beech, mahogany, sapele.

***Preparation*** As softwood.

***Primer*** As softwood. Better penetration of the usually tighter grain may be obtained if the primer is thinned.

### Blockboard (Fig. 3.16)

Strips of softwood sandwiched and stuck between double ply boards. Rigid, smooth-faced building board 16–25 mm thick. No knots or resinous parts. Available: (a) standard, (b) veneered, usually clear-finished.

*Preparation, Primer*, both as softwood.

### Plywood (Fig. 3.17)

Thin veneers of wood stuck together with the grain running at right angles one to another. Rigid and of varying texture. Thickness 3–25 mm depending on number of veneers or plies. Grain often difficult to smooth by filling. No knots or resinous parts. Available:

(a) interior grade
(b) exterior grade
(c) marine (resin-bonded)
(d) veneered, usually clear-finished.

*Preparation, Primer*, both as softwood.

### Hardboard/fibreboard

Wood fibres and resin firmly compressed at high temperature. Rigid, smooth-faced, even-textured building board. Available:

(a) standard, 3–12 mm thick. Smooth surface on one side, mesh texture on reverse.
(b) medium density fibreboard (MDF) from 2–25 mm thick. Less dense than standard hardboard.
(c) oil-tempered, with increased resistance to moisture. Very smooth surface.
(d) perforated: variety of holes and slots.
(e) prepared or ready-primed.
(f) fireproofed (see 3.4.5).

***Preparation***

(i) Ensure that the surface is dry and free from dust and grease. Degrease with white spirit if necessary.
(ii) Punch nails below the surface. Caulk joints and nail holes (see 3.2).
(iii) *Do not* abrade as this will cause surface damage.
(iv) Prime *before* making good, paying special attention to edges.

***Primer***

Internal: (a) thinned emulsion paint; (b) acrylic primer/undercoat; (c) hardboard primer.

*Note:* Pink primer is unsatisfactory.

External: (a) hardboard primer; (b) aluminium wood primer.

*Note:* Both thinned emulsion paint and pink primer are unsatisfactory. Extra durability can be obtained by back-priming.

### Chipboard

Wood chips and resin firmly compressed at high temperature. Rigid, dense, even-textured building board 9–38 mm thick. Available: (a) standard, (b) medium, for furniture, (c) flooring quality. *Preparation, Primer*, both as softwood.

## 3.4.4 Non-absorbent surfaces

Usually very dense, smooth and shiny, so offering very little adhesion for paints.

### Glass, glazed tiles, glazed bricks

***Preparation***

(i) *Do not* abrade as the surface may be damaged.

(ii) *Glass* Degrease by washing with methylated spirit and whiting, allowing to dry to a powder and polishing with clean rag.

(iii) *Glazed tiles* Wash with strong sugar soap or detergent; rub joints between tiles with cellulose thinners.

(iv) Prime as soon as surface is dry. Do not allow further contamination, even with fingerprints.

***Primer*** Must possess good adhesive properties: thinned alkyd gloss paint or special tile-bonding coat based on PVA or acrylic resin.

### Plastics

A number of synthetic materials either in sheet form or capable of being moulded by a variety of methods. Usually self-coloured, therefore not often painted. Those most commonly encountered in painting are:

(a) PVC (polyvinyl chloride). Usually coloured white, grey or black. Used extensively for general soil, waste and water services.

(b) Glass-reinforced plastic (fibreglass). Plastic resin reinforced with layers of glass-fibre matting. Used for a large variety of moulded items. Often self-coloured.

(c) Melamine laminates. Decorative sheeting used for panelling, door covering, and table and counter tops.

(d) Acrylic (Perspex, Oroglas). Clear or coloured sheeting or formed units such as baths, storage tanks and signs.

***Preparation***

(i) Degrease with detergent and rinse.

(ii) Paint as soon as surface is dry with specified finishing paint. Do not allow further contamination.

### Expanded polystyrene

Polymerised styrene resin expanded into an extremely lightweight, porous, insulating material. Used extensively for ceiling tiles and wall insulation sheets. Available:

(a) standard: highly flammable, the most common type.

(b) self-extinguishing: more dense than standard, slow to ignite so reducing surface spread of flame, provided it is fixed with a full film of adhesive and not 'spotted'.

***Preparation*** Make good with plaster-based filler only, as oil filler may dissolve polystyrene.

***Primer*** Emulsion paint or acrylic primer/undercoat.

## 3.4.5 Chemically active surfaces

Most chemically active surfaces are mixed with a considerable amount of water when being manufactured or used, and this water may take many months to dry out. If painting is essential before complete drying out (e.g. in new buildings), only permeable water-borne paints should be used.

Chemically active surfaces are those which contain *lime* or *cement*. They become active alkalis only when wet. Suspect surfaces can be checked for alkalinity by dampening with distilled water and applying red litmus paper. Alkali will change the colour of the paper to blue.

*Saponification:* the effect of alkali on most drying oils used in paints. The alkali changes the drying oil into a non-drying, water-soluble soap (see 2.6).

*Efflorescence:* most chemically active surfaces contain water-soluble salts which rise to the surface in solution and crystallise as the water dries out (Plate 8). Remove by dry brushing only, as washing will aggravate efflorescence.

***Primer***

***Wet surfaces*** Emulsion paint or cement paint – permeable coatings which allow the water to dry out through the paint film without breakdown of the film.

**Plate 8** Efflorescence on brickwork

***Dry surfaces*** (a) alkali-resisting primer/sealer (see 2.2.2); (b) thinned emulsion paint.

## Gypsum plasters

Plasters produced by the burning of natural gypsum (calcium sulphate). Although they are almost lime-free, they are often applied to strongly alkaline surfaces which make them chemically active.

***Classes according to BS 1191: Parts 1 and 2: 1973***

***Class A*** Hemi-hydrate (plaster of Paris), used for fibrous plastering.

***Class B*** Retarded hemi-hydrate ('Thistle').

***Class C*** Anhydrous ('Sirapite').

***Class D*** Keene's cement.

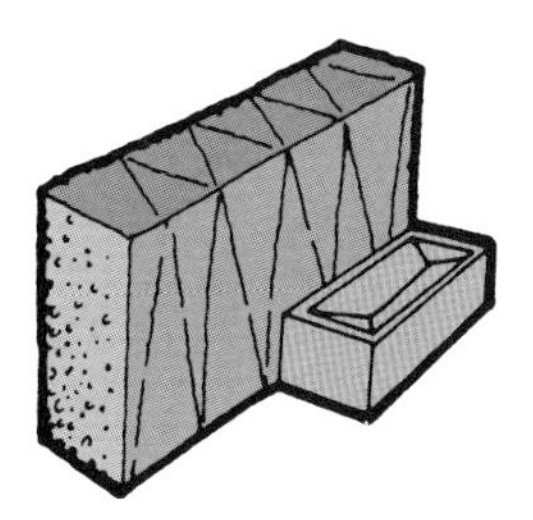

**Fig. 3.18** Building block

Premixed: Class B plaster with the addition of exfoliated vermiculite and expanded perlite to produce a lightweight, insulating, fire-retardant plaster ('Carlite').

***Preparation***

(i) Ensure that the moisture has dried out.
(ii) Remove any efflorescence.
(iii) Denib with stripping knife. *Do not* abrade as the surface will be damaged.
(iv) Make good large holes and cracks with a similar plaster or plaster filler.

***Primer***

(a) Thinned alkali-resisting primer. Thinning improves the primer's penetrative power.
(b) Thinned emulsion paint for emulsion paint finish.

### Lime plaster

Plaster produced by the burning of natural limestone to make 'quicklime' which is then steeped in water to form 'lime putty'. Not often used today, but older buildings would certainly contain it. Hardens very slowly but often speeded-up by adding gypsum plaster (gauged lime plaster).

***Preparation*** As gypsum plaster.

### Building blocks (Fig. 3.18)

Pulverised fuel ash or clinker (burnt coke) and cement slurry. Lightweight, even-textured, load-bearing and non-load-bearing blocks. Used extensively for partition and cavity walls. Sometimes incorrectly called 'breeze blocks'.

***Preparation***

(i) Ensure that the moisture has dried out.
(ii) Clean and remove any efflorescence with a wire brush.
(iii) Clean and sterilise any areas of mould growth or lichen (see 3.6).

### Brickwork, stonework, concrete, cement rendering

***Preparation*** Treat as building blocks.

### Glass-reinforced cement

Glass fibres and cement slurry pressed to form sheets or cast in moulds. Dense, even-textured, fireproof. Used extensively for rainwater goods, fireproofing doors and walls, and corrugated sheeting for roofs.

***Preparation*** Dry-brush to remove dust. Abrading causes surface damage and is a health risk (see 6.6). Back and edges should be coated with bitumen or other water-resistant coating to prevent moisture penetrating and damaging the finish.

### Woodwool slabs

Long-tangled wood fibres and cement slurry pressed into a lightweight building slab. Good insulator, very coarse surface texture resembling straw. Commonly used in partition walls.

***Preparation*** Dry-brush to remove dust.

*Note:* Because of their open texture, these surfaces are best coated by spray.

### Fireproofed boards

Building boards, especially insulating board and hardboard, which have been treated with water-soluble fire-retardant solutions to lower their class of flame spread (see 3.4.1).

***Preparation*** According to type of board (see 3.4.2).

***Primer*** Alkali-resisting primer/sealer only.

***Do not*** use water-borne paints as they may promote efflorescence of the fire-retardant salts.

## 3.4.6 Corrodible surfaces

### Ferrous metals

All metals which have iron in their composition are classed as ferrous metals. Cast iron, wrought iron, mild steel, high tensile steel and stainless steel are the most common examples. The primary difference

**Plate 9** Rust running down a wall

between them is the amount of iron in the metal, which ranges from a very large amount (cast iron) to a very small amount (stainless steel).

When iron and steel are exposed to oxygen and water (usually in the atmosphere), they quickly corrode: reddish-brown deposit known as rust forms on the surface (Plates 9 and 10). It is the iron content which produces the rust deposit; metals with a large iron content corrode more readily than those with a small iron content. Rusting weakens the metal.

***Chemistry of corrosion* (Fig. 3.19)** Corrosion is a chemical change brought about by an electrochemical process called electrolysis. When metal is immersed in an electrolyte (water contaminated with compounds such as acids or salts), an electrolytic cell is set up. An electrical current will flow between certain areas of the metal surface known as anodes and cathodes. The current flows from the cathode through the metal to the anode and returns through the electrolyte, which is capable of conducting electricity. The current results in destructive corrosion of the anode area with the formation of rust. While the anode corrodes, the cathode is protected against corrosion.

**Plate 10** Rust damage

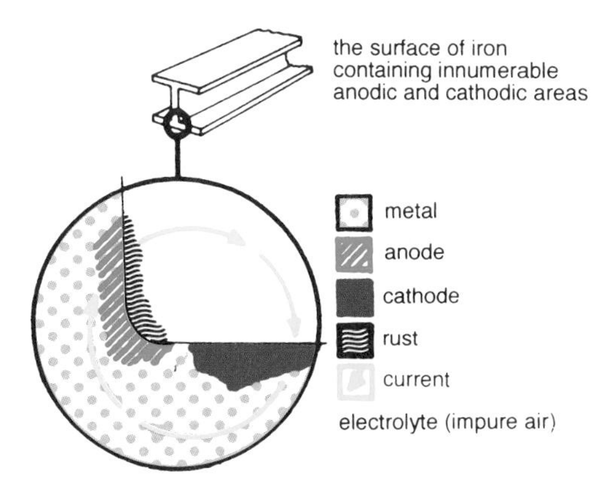


**Fig. 3.19** The corrosion process

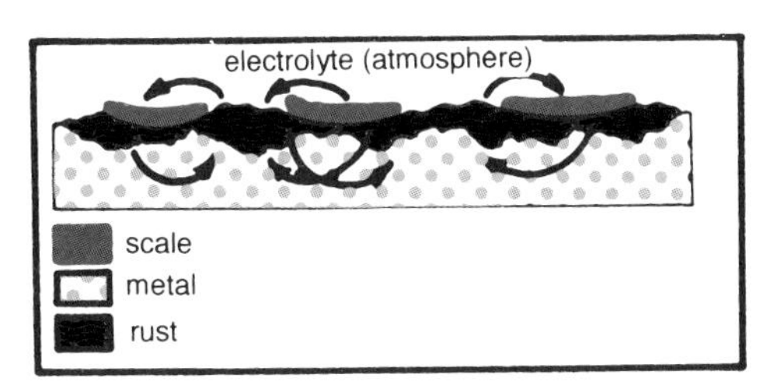


**Fig. 3.21** Millscale

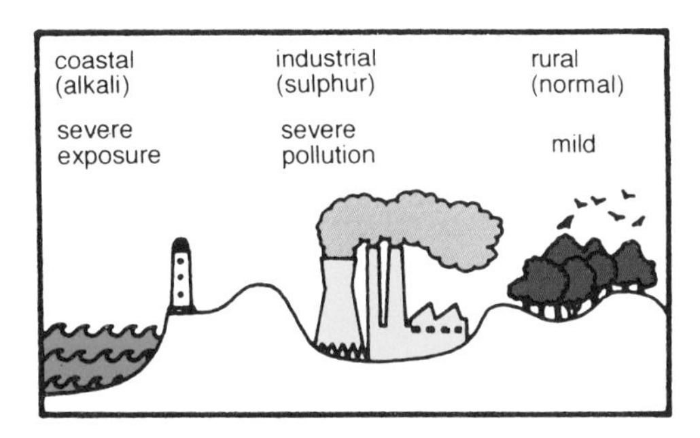


**Fig. 3.20** Atmosphere as the electrolyte

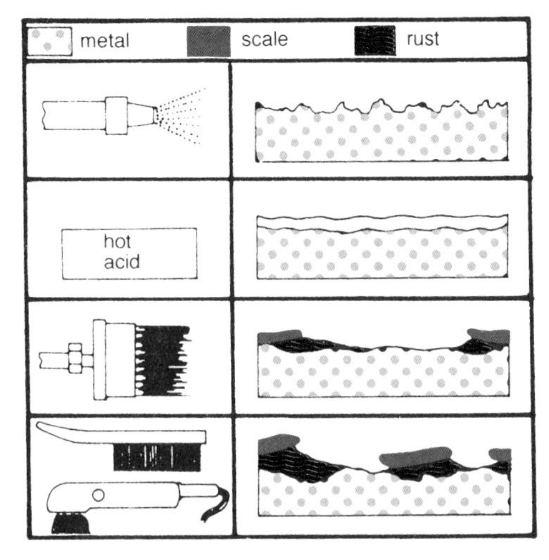


**Fig. 3.22** Removal of rust and millscale

When metals are exposed to the atmosphere, they are actually immersed in an electrolyte, since the air contains water, dilute acids and salts, especially in industrial and coastal areas. Corrosion is more severe in a strong electrolyte, which explains why metals rust more rapidly in industrial and coastal atmospheres (Fig. 3.20).

***Millscale* (Fig. 3.21)** When ferrous metal is forged into flat plates, beams and girders, etc., it is worked at temperatures often exceeding 1000°C. While the metal is in this white-hot state, it is in contact with the cool air and rapidly oxidises. This results in the formation of a thin, flaky layer of black, oxidised iron called millscale. The millscale is cathodic to the bare steel, and when it is broken, anodic areas are exposed.

***Preventing corrosion*** To prevent rusting, ferrous metals can be completely enveloped in a thick, elastic paint system which must adhere firmly to the surface. This can only be accomplished by:

(i) The complete removal of all rust and millscale.
(ii) Use of a suitable rust-inhibitive primer.
(iii) Use of a weather-resistant paint system of adequate thickness.

(i) *Surface preparation (Fig. 3.22)* Millscale and rust promote further corrosion and prevent the firm adhesion of paint films. Methods of removal, in order of effectiveness, are:

(a) Abrasive cleaning: a stream of abrasive particles is directed against the metal at a high velocity to dislodge all rust and millscale, leaving a clean, white metallic surface (see 1.2.3).
(b) Acid pickling: a factory method of treating metals by use of acids. One of the most common is the 'Footner process', which

involves first dipping the metal in dilute sulphuric acid, then rinsing in hot water and finally dipping in dilute phosphoric acid. This leaves a black, phosphate film on the surface, ideal for painting.

A less effective site process is to use an acid solution, mainly phosphoric, applied by brush during preparation of the surface.

(c) Flame-cleaning: a high-temperature oxy-acetylene flame which causes rapid expansion of the metal, loosening the scale and dehydrating the rust, which is brushed off (see 1.2.2).

(d) Mechanical and hand methods: preferably carried out after weathering. Loosened rust and millscale are partially removed using chipping hammers, derusting pistols and wire brushes (see 1.1.3).

*Removal of grease and ferrous sulphate* The surfaces of metals are naturally greasy, especially around welded joints where oily fluxes are used during welding operations. To prevent paint films from flaking, the surface of the metal must be degreased before painting by washing with either naphtha, white spirit or an emulsifying agent. Also, washing with hot, clean water removes ferrous sulphate salts which promote corrosion by attracting and absorbing water.

(ii) *Priming* Immediately after preparation, the metal should be primed with a special rust-inhibitive primer. Because of the effects of surface tension, arrises and bolt heads should receive two coats. Where possible, apply by brush or roller. If spray application is necessary, airless spraying is superior to conventional methods. Factory-primed metal should be washed with detergent, rinsed, dried and reprimed after fabrication.

*Rust-inhibitive primers* The pigments used in these paints actually prevent the formation of rust even if the film is slightly damaged. They include red lead, metallic lead, calcium plumbate, zinc chromate, zinc phosphate and zinc dust.

(iii) *Paint system (Fig. 3.23)* The whole surface must be protected with a weather-resistant paint system which must not be less than 125 microns in thickness. This ensures that all peaks or high spots on the surface are adequately covered.

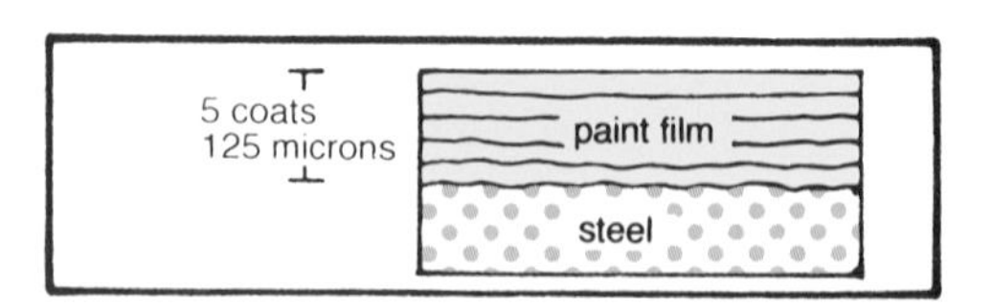


**Fig. 3.23** The paint system

*Note:* 125 microns represent approximately four to five coats of conventional paint.

## Non-ferrous metals

All metals which have no iron in their composition are classed as non-ferrous metals. Aluminium, zinc, copper, brass, bronze, lead, chromium, tin and cadmium are commonly used in buildings. Coated iron and steel may also be classed with non-ferrous metals, e.g. galvanised iron.

***Corrosion of non-ferrous metals*** Non-ferrous metals, like ferrous metals, corrode by the chemical process of electrolysis, with the same influencing factors. The main difference is that, on the surfaces of non-ferrous metals, firmly-adhering corrosion deposits are formed which inhibit and slow down further corrosion. Corrosion can be accelerated by contact with dissimilar metals, acid or alkaline materials.

***Surface preparation*** Generally speaking, the surface must be dry and free from grease, dirt, corrosion deposits and loose paint. Methods of pretreatment are shown in Table 3.9.

## Sacrificial coatings

One method of reducing corrosion is to apply a coating containing a metal which corrodes more rapidly than the substrate metal it is in contact with.

Metals are classified by their activity in the galvanic series (Table 3.10). Metals at the top of the series are more active than those lower down. If a coating containing a metal is applied to a substrate which is lower on the galvanic series, the metal in the coating will corrode rapidly, so 'sacrificing' itself for the substrate metal.

Galvanising is a zinc coating, and when applied to iron, it is sacrificed and protects the iron from

**Table 3.9** Non-ferrous metals – surface preparation and primers

| *Metal* | Preparation | | *Primer* |
|---|---|---|---|
| | *Factory* | *On site* | |
| Aluminium sheeting extrusions | Degrease in acid bath | Etch and degrease with emery cloth and white spirit | Etch primer<br>Zinc chromate (No lead paints) |
| Galvanised iron (steel dipped in molten zinc) | Degrease in acid or alkali bath | (i) Allow to weather for several months<br>*or*<br>(ii) Degrease | (i) Zinc chromate<br>(ii) Calcium plumbate or etch primer |
| Zinc sprayed sherardised | Denib with emery cloth and degrease | Denib with emery cloth and degrease | Etch primer followed by zinc chromate |
| Copper<br>Brass<br>Bronze | Degrease in solvent bath | Emery cloth and white spirit (do not weather or dry-abrade) | Etch primer |
| Lead | Not usual | Allow to weather for several months<br>*or*<br>Emery cloth and white spirit (do not dry-abrade for health reasons) | Etch primer |

**Table 3.10** Galvanic series

| |
|---|
| Aluminium (unstable or active metals) |
| Chromium |
| Zinc |
| Iron |
| Lead |
| Copper |
| Silver |
| Gold (stable or noble metals) |

corrosion. Alternatively, if an aluminium surface is coated with a lead-based coating, the aluminium will corrode, which is the reason why lead-based primers must not be used on aluminium or zinc.

## 3.4.7 Bleeding and non-drying surfaces

(a) Surfaces coated or contaminated with bitumen, coal-tar oils, nicotine or mineral oil and grease.

(b) Surfaces or paints containing certain lake pigments or dyestuffs.

When a surface contains material which is soluble in the medium of the paint applied to it, the result may be discoloration by bleeding (see 3.6) or possible slow drying or non-drying of the finish material.

***Preparation***

(i) *Do not* abrade as the resultant scratches may increase the degree of bleeding.

(ii) Ensure that the surface is free of dust and grease.

***Primer*** Aluminium sealers or proprietary anti-bleed sealers. Shellac or 'Stop Tar' knotting is satisfactory only for small areas.

### Resinous and oily woods

Softwoods and hardwoods containing an excessive amount of natural resin or oil which may slow the drying of paints applied to them and cause discoloration by bleeding (see 3.6). Common types are: teak, pitch pine, Oregon pine, Douglas fir.

***Preparation***

(i) Ensure that the surface is dry.

(ii) Punch nails, etc. below surface.

(iii) Wash the surface free of excessive resin or oil with cellulose thinner or white spirit.
(iv) Dry-abrade to a clean, smooth surface.

***Primer*** Aluminium wood primer will prevent:

(i) discoloration by the resin (see 3.6).
(ii) natural oils affecting the hardening of the coating.

*Note:* Aluminium paint will prevent bleeding. It will not prevent exudation of resin.

# 3.5 Natural wood finishes

Because of their attractive grain and natural colour, many hardwoods and some softwoods are not required to be painted. Their natural beauty can be preserved by protecting them with transparent finishes.

## 3.5.1 Preparation of natural wood surfaces

### Previously treated sound surfaces

(a) Previously varnished, lacquered or coated with a catalysed finish, in good condition and to be recoated with similar material.
(b) Previously lacquered or coated with a catalysed finish, in good condition and to be coated with an oil or alkyd varnish.

***Treatment*** Wash and wet-abrade (see 3.3.2).

### Previously treated unsound surfaces

(a) Having old coatings which are cracked, discoloured, flaking or otherwise defective.
(b) Previously coated with a spirit varnish or wax, to be recoated with similar or other material.
(c) Previously varnished, to be recoated with cellulose lacquer or a catalysed material.

***Treatment*** Complete removal with spirit paint remover (see 3.3.1). Wax polish can be removed by scouring with steel wool and white spirit.

### New or stripped surfaces

Dry-abrading by hand or orbital sander is necessary to remove possible plane marks, or to smooth raised or rough grain. It is essential when hand-abrading to rub with the grain to avoid scratching.

Stained surfaces may require either scraping, using a universal scraper, followed by abrading, or, if stains are very deep, bleaching, for which two types of material are available:

(a) Proprietary bleaching solutions which require to be left for from 6 to 24 hours, depending on type, timber or stain.
(b) Bleach site-mixed by dissolving 50 g oxalic acid crystals in $\frac{1}{2}$ litre water (fast-acting but poisonous and must be used with care).

Surfaces must be thoroughly dry-abraded after bleaching and rinsing.

### Making good

*Stopping* under a spirit-carried finish or spirit stain-tinted hard stopper, plastic wood or tinted stopper (see 3.2). If water-carried stain or finish is to be used, only tinted stopper is suitable.

*Filling,* when necessary, should be carried out with an oil-based rubbing filler (see 3.2) or vinyl-based grain filler.

### Staining

Staining may be necessary if

(i) the grain of the timber is required to be exaggerated or darkened;
(ii) the natural colour of the timber is not desirable;
(iii) the timber requires to be colour-matched to existing joinery.

Six principal stains are available:

(a) Chemical stains – the application of warm solutions of Epsom salts, washing soda, lime or ammonia. The effect varies with each type of wood, and these stains are difficult to control to produce a specific effect.

(b) Water stains.
(c) Spirit stains (see 2.5.3).
(d) Oil stains.
(e) Tinted catalysed finishes – which combine stain and finish in one operation. Tend to obscure grain.
(f) MVP coatings (see 2.5.2).

## 3.5.2 Finishes

### Preservative

Material having toxic properties that resist wood rot, wood-boring insects and general decay.

(a) Creosote, dark brown and cannot be subsequently coated with paint or varnish (see 2.5.1).
(b) Non-bleeding types, considerably more expensive than creosote but available in a large range of colours (see 2.5.1).
(c) MVP stain/varnish (see 2.5.2).

### Shellac varnish

The most common types are button polish and French polish. They provide a high-sheen surface which is brittle and offers limited resistance to moisture (see 2.5.8).

### Oil, alkyd or one-pack polyurethane varnish

Slow-drying coating having excellent protective qualities for both interior and exterior woodwork. Tends to yellow on ageing (see 2.5.5 and 2.5.6). A minimum of four coats required on external surfaces.

### Oil sealer

Prepared principally for floors.

### Nitrocellulose lacquer

Sprayed material which dries rapidly and produces very clear, water-resistant films for interior and exterior woodwork (see 2.3.8).

### Clear finish

Usually two-pack material based on polyurethane, epoxy, melamine or urea. Produces very hard finish which is resistant to abrasion, water and many chemicals. Most types are produced in qualities suitable for floors (see 2.4.7 and 2.5.6).

### Emulsion glaze

Very clear, quick-drying finish for interior woodwork. Tends to raise the grain (see 2.5.4).

### Wax polish

Beeswax polish, sometimes containing silicones to improve water-resisting properties. Produces a matt finish which does not resist abrasion or solvents, and is readily marked by water.

# 3.6 Surface and paint film defects

**Table 3.11** Surface defects

| *Defect* | *Cause* | *Treatment* |
|---|---|---|
| Efflorescence (Plate 8) | See Chemically active surfaces (3.4.5) | |
| Lichen, moss and algae<br>Growths of small, thickly growing, non-flowering plant life | Plant life thriving on damp surfaces | (i) Sterilise with proprietary toxic solution<br>(ii) Remove by scraping<br>(iii) Sterilise surface with proprietary toxic solution (N.B. avoid splashes on skin and accidental consumption by children and animals) |
| Moulds and fungus (Plate 11)<br>Multi-coloured spots or patches on infected areas | A variety of airborne spores which multiply and feed on the organic matter in paints, papers, pastes and surface deposits<br>Promoted by dampness and poor ventilation. Common in bakeries, breweries and damp buildings and under vinyl wall coverings hung with unprotected paste. Often found behind curtains and in cupboards where ventilation is poor | (i) Trace and remedy cause of dampness<br>(ii) Sterilise area with proprietary fungicidal solution or 3:1 mixture of water and household bleach<br>(iii) Remove mould and fungus by scraping and washing<br>(iv) Sterilise area with proprietary fungicidal solution or 3:1 mixture of water and household bleach<br>(v) Observe for 7 days for any reappearance of the mould<br>(vi) Coat with paints containing a fungicide (N.B. avoid splashes on skin and accidental consumption by children and animals) |

**Table 3.11** Surface defects (cont'd)

| *Defect* | *Cause* | *Treatment* |
|---|---|---|
| Dry rot (Plate 12)<br>A fungus which destroys timber<br>Early stages: wood excessively wet, spongy and covered with silvery white branching strands (hyphae), which in extreme dampness may form a fluffy white mass<br>Characteristic musty, damp smell<br>Advanced stages: wood totally dry and brittle, with no strength, cracked in a cube pattern across the grain, and darkened<br>At a very advanced stage: the fungus develops a fruiting body with a brick-red centre | Airborne spores of dry rot germinate on wood under the following conditions:<br>(a) Wood with more than 20% moisture content (see 3.4.3)<br>(b) Humid atmosphere<br>(c) Poor ventilation<br>*Life cycle:*<br>(i) The spores send out hyphae in search of nourishing moisture and cellulose which is sucked from the wood until it is dry and brittle<br>(ii) The hyphae carry moisture to dry, sound wood to enable fresh spores to germinate and so spread the growth<br>(iii) A plate fungus forms with a red centre which contains millions of fresh spores which further contaminate damp wood | (i) Trace and remedy the cause of dampness Increase ventilation<br>(ii) Cut out all infected timber 1 metre past the termination of the hyphae and *burn immediately*<br>(iii) Treat adjacent timber with wood preservative Treat all replacement timber with preservative<br>(iv) Wall plaster through which 1 hypha may have travelled must be removed and the brickwork treated with dry-rot killer and thoroughly scorched with a blowlamp before replastering. This is a job for the specialist – rarely carried out by the painter.<br>(i) Trace and remedy cause of dampness<br>(ii) If attack is not too severe, the removal of the source of dampness is often sufficient to kill the rot<br>(iii) If attack is severe, cut out infected timber and treat as for dry rot<br>(i) Sterilise with proprietary toxic solution<br>(ii) Remove by scraping<br>(iii) Sterilise surface with proprietary toxic solution (N.B. avoid splashes on skin and accidental consumption by children and animals) |
| Wet rot (Plate 13)<br>A fungus which destroys timber<br>Wood is darkened and cracked with the grain. It becomes spongy and excessively wet beneath a seemingly sound veneer | Airborne spores of wet rot germinate on very wet wood (see 3.4.3)<br>*Moisture penetration:*<br>Commonly found in window sills and the bottom of doors, windows and gates | (i) Trace and remedy cause of dampness<br>(ii) Sterilise area with proprietary fungicidal solution or 3:1 mixture of water and household bleach<br>(iii) Remove mould and fungus by scraping and washing<br>(iv) Sterilise area with proprietary fungicidal solution or 3:1 mixture of water and household bleach<br>(v) Observe for 7 days for any reappearance of the mould<br>(vi) Coat with paints containing a fungicide (N.B. avoid splashes on skin and accidental consumption by children and animals) |
| Pattern-staining (Fig. 3.24)<br>Seen as a light and dark pattern in rectangular form, corresponding with the beams, rafters and joists of ceilings or mortar joints on building-block walls. Also seen on walls above radiators. | Air in a room always contains floating dust. The warm air rises by convection current carrying the dust and depositing it on the ceiling. Most dust is deposited on the cooler areas. | There is little a decorator can do to cure this problem but any form of heavy covering will help to insulate the surface and reduce the defect, e.g. lining and anaglypta or polystyrene tiles. Much better results can be achieved by fitting up to 50 mm of fibreglass or other insulating material between rafters |
| | *Steel:* dark areas will be immediately under steel beams as these are colder than the space between | Insulate the ceiling surface with a heavy covering such as 'Superglypta' or expanded polystyrene |
| | *Wood:* dark areas will be between timber rafters because the timber presents a warmer surface than the space between | As above, or fill the space between the rafters with an insulating material such as 50 mm of fibre glass, expanded polystyrene or vermiculite, to even up the surface temperature |

**Plate 11** Mould

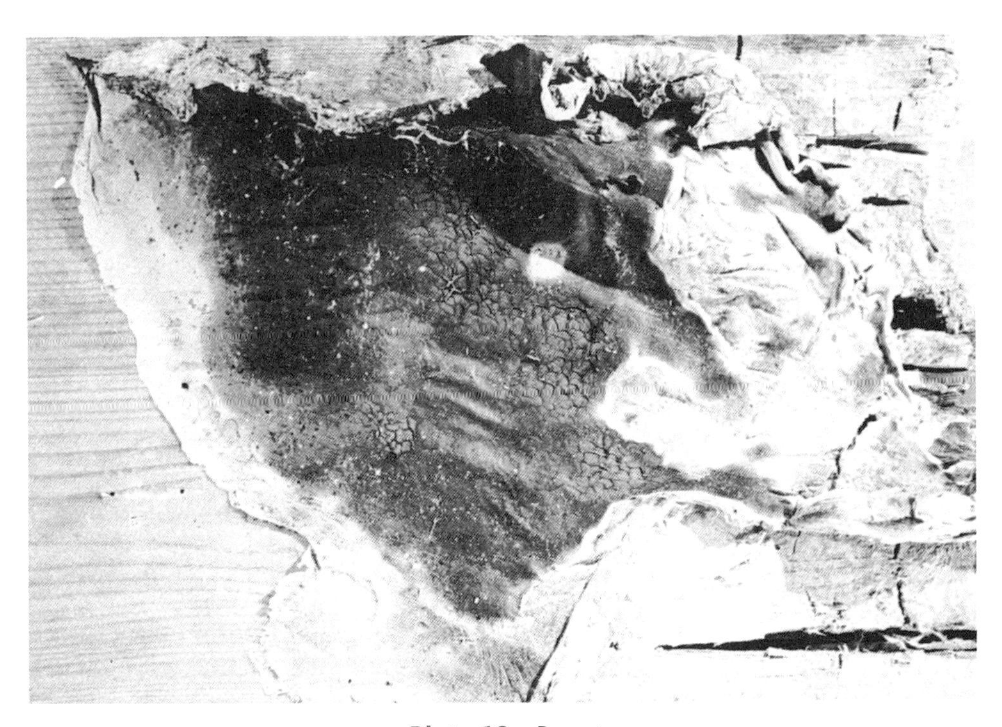

**Plate 12** Dry rot

**Plate 13** Wet rot

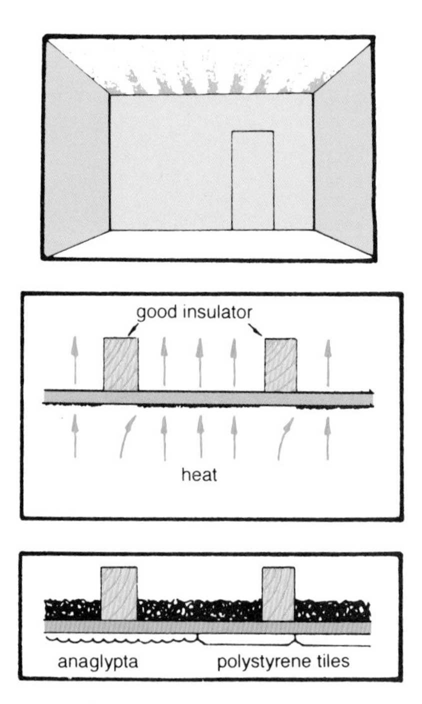


**Fig. 3.24** Pattern-staining

**Table 3.12** Paint film defects

| *Defect* | *Cause* | *Prevention* |
|---|---|---|
| Bleaching<br>A 'whitening' or a complete change of colour of the paint film due to chemical action | Action of acid fumes upon pigments such as ultramarine blue | Correct choice of paint and colours to suit the environment |
| | Action of alkali upon pigments such as Prussian blue and Brunswick green | Use of alkali-resisting primer on alkaline surfaces |
| Bleeding<br>Discoloration of the finishing paint by some constituent of the surface being dissolved by the paint medium | Painting surfaces coated or contaminated with bitumen, oils or nicotine | Use aluminium paint as a sealer or sizing and lining |
| | Painting surfaces or paints containing lake pigments or dyestuffs | Use aluminium sealer |
| | Painting knotty or resinous woods | Use shellac or aluminium paint as a sealer |
| Blistering (Plate 14)<br>Eruption of a paint or varnish film forming bubbles on the surface | Trapped moisture in wood<br>Made worse by heat, particularly dark colours | Use moisture meter to make sure timber is dry |
| | Resin exuding from knots due to heat or sunlight | Replace resinous knots with sound timber |
| | Trapped moisture in plaster and rendering | Use permeable coatings or wait until surface is dry |
| | Using non-heat-resistant paints adjacent to sources of extreme heat | Use heat-resistant paint. Light colours have more resistance to blistering by heat than dark colours |
| Chalking (Plate 15)<br>Formation of a powdery deposit on the surface of dry paint film, the powder being unbound pigment | Paint deficient in binder because of overthinning | Thin paints only when specified |
| | Painting insufficiently sealed absorbent surfaces | Seal absorbent surfaces and touch-up filling |
| | Highly pigmented spirit-borne paints on exterior exposure | Correct choice of paint |
| | Use of interior paints on exposed surfaces | Correct choice of paint |
| Cissing (Plate 16)<br>Failure of paint or varnish coating to form a continuous film on the surface<br>The film rolls back in globules, leaving small round bare patches | Painting greasy surfaces | Thorough washing and rinsing of the surface |
| | Painting very smooth, shiny surfaces | Abrade to remove shine |
| | Finishing over an oil undercoat | Do not adulterate undercoats<br>Lightly wet-abrade |

**Table 3.12** Paint film defects (cont'd)

| *Defect* | *Cause* | *Prevention* |
|---|---|---|
| Cracking or crazing (Plate 17)<br>Splitting of the surface due to the top coat of paint or varnish being unable to expand to the same degree as the previous coatings | Applying hard, drying coatings over soft, elastic coatings | Choice of correct paint system |
| | Recoating before undercoat has dried | Allow specified drying time before recoating |
| | Adding excessive driers | Add only minimum amount of driers |
| | Paste and size over new paintwork | Remove all paste and size from paintwork immediately after paperhanging |
| Flaking or peeling (Plate 18)<br>Breakdown of adhesion resulting in paint or varnish lifting away from the surface in flakes | Painting damp surfaces, especially wood | Use moisture meter to make sure timber is dry |
| | Painting powdery, friable surfaces, e.g. distemper | Correct preparation and sealing |
| | Painting in humid conditions which promote condensation | Do not paint in foggy, wet or frosty conditions |
| | Expansion and shrinkage of painted surfaces | Select paints to suit nature of surface |
| | Formation of rust under paints | Ensure thorough preparation and correct primer |
| | Lack of adhesion on smooth, shiny surfaces | Wash and wet-abrade surface |
| | Efflorescence forming under the paint film (see 3.4.5) | Remove all efflorescence by dry-brushing before painting |
| Loss of gloss and sinking<br>Failure of a paint or varnish film to maintain its potential or original sheen or shine | Painting insufficiently sealed absorbent surfaces | Seal absorbent surfaces and touch-up filling |
| | Use of wrong paint system | Use correct primers and undercoats |
| | Wet paint film affected by damp weather or condensation | Plan the work so that paint will dry in correct conditions |

**Plate 14** Blisters on a door panel

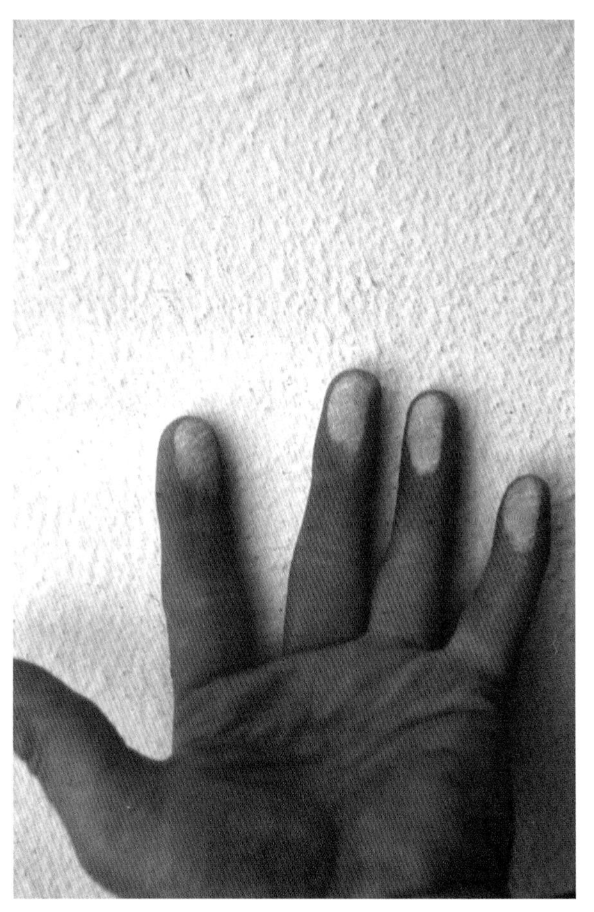

**Plate 15** Chalking

**Plate 16** Cissing

**Plate 17** Cracking or crazing

**Plate 18** Flaking or peeling

# 3.7 Preparation of surfaces for graining, marbling and surface texturing

All surfaces to be surface textured must be:

(i) Hard. Allow at least 2 or 3 days for the ground coat to dry.

(ii) Free from surface indentation. The surface must be thoroughly filled, otherwise scumble will collect in the holes and show up as darker areas.

(iii) Free from grit. All coatings must be strained and the dried ground coat-abraded with 400 grade abrasive paper.

(iv) Free from brushmarks.

(v) Ground coat of slight sheen. An eggshell finish is ideal.

(vi) Full ground coat. No misses or scuffs in the ground coat which will be accentuated by the scumble.

(vii) Ground coat of correct colour. This should usually be a tone lighter than the lightest part of the desired finish.

(viii) Free from grease. Wash with detergent and rinse. This is essential when using water colour. If cissing still occurs, a slurry of fuller's earth, whiting, or fine pumice and water, rubbed in over the surface, will ensure good wetting.

# 3.8 Preparation of surfaces for gilding and bronzing

Metallic leaves and powders can be applied to any dry and hard surfaces:

***Ferrous metal***

(i) Degrease, remove all rust and coat with a rust-inhibitive primer (see page 150).

(ii) Allow the primer to dry for at least 48 hours before sizing.

***Non-ferrous metal***

(i) Degrease and wipe clean (see page 151).

(ii) Priming is not essential, but if the surface is exposed, one coat of etch primer is recommended.

***Timber***

(i) Thoroughly dry-abrade to remove all fibres and to smooth rough grain.

(ii) Because gold leaf will not bridge over open pores, it is necessary to fill the grain.

(iii) A minimum of one coat of oil-based primer and one coat of oil/resin undercoat are necessary to ensure a non-absorbent surface suitable for sizing.

***Plaster***

(i) Must be thoroughly dry and free from dust and residue.

(ii) A minimum of one coat of plaster sealer and one coat of oil/resin undercoat to ensure a non-absorbent surface suitable for sizing.

***Painted surfaces***

(i) Must be clean by washing with detergent and rinsing.

(ii) Free from texture by lightly abrading with 400 grade waterproof abrasive paper.

(iii) Thoroughly hard, without surface defects, and free from moisture.

(iv) Surfaces which have recently been gloss painted or varnished and are to be sign-written or painted with decorative motifs should be treated also by one of the following processes:

***Pouncing*** Dabbing over the entire surface to be gilded with a small muslin bag (pounce bag) containing French chalk. This absorbs any slight tackiness on the surface that may adhere the gold outside the sized areas. Surplus chalk should be brushed off before sizing.

***Glair coating*** Glair is the white of an egg mixed with 0.5 litres of warm water. It is applied to the whole area to be gilded or bronzed, without misses, and allowed to thoroughly dry. The gold size is applied to the glaired surface. Immediately the gilding or bronzing process is completed, the surface must be washed with warm water to remove all traces of glair and any surplus metallic leaf or powder.

# PART 4

# Surface Coverings

# 4.1 Preparation of surfaces for surface coverings

### Stripping old paper

All previous covering and all traces of adhesive and size must be removed to reveal a clean, sound surface by either dry-scraping, soaking and scraping, or steam stripping.

### Sizing

Increased adhesion of coverings to a surface can be obtained by the application of either a glue size solution, when using a starch paste, or a thin coat of adhesive, when using cellulose or polyvinyl acetate (PVA) adhesive. Sizing also reduces the absorbency of porous surfaces, so preventing 'snatch'.

### Nail or screw fixings

Timber building boards, plasterboards and hardboards are often fixed with nails or screws. To prevent these fixings corroding they should be touched-up with metal primer before the surface is primed and sized.

### Oil priming paints

Some surfaces require an oil primer before sizing and hanging coverings. The reasons are:

(i) to reduce surface porosity;
(ii) to bind down any existing emulsion or oil-bound distemper coatings;
(iii) to make subsequent stripping easier.

### Degreasing

Previously painted, non-porous and new concrete surfaces require degreasing with detergent before wallpaper is hung. New concrete surfaces are greasy because of the use of release oil when casting from moulds or shuttering.

**Table 4.1** Preparation of bare surfaces

| | Preparation | | | | | | | | |
|---|---|---|---|---|---|---|---|---|---|
| *Surface* | *Ensure material is thoroughly dry* | *Wash with detergent and rinse* | *Dry-scrape* | *Dry-abrade* | *Touch-up nail heads* | *Fill joints, nail holes and cracks* | *Apply linen tape over joints and fill edge* | *Apply appropriate primer/sealer* | *Size* |
| Concrete | X | X | X | X | | X | | | X |
| Expanded polystyrene | X | | | | | X | | | |
| Gypsum plaster | X | | X | X | | X | | | X |
| Hardboard | X | | | | X | X | X | X | X |
| Insulating board | X | | | | X | X | X | X | X |
| Lime plaster | X | | X | X | | X | | | X |
| Plasterboard (dry-lining) | X | | | X | X | X | X | X | |
| Wood building board | X | | | X | X | X | X | X | X |

**Table 4.2** Preparation of previously treated surfaces

| | Preparation | | | | | | | | | |
|---|---|---|---|---|---|---|---|---|---|---|
| ***Surface*** | ***Wash with detergent and rinse*** | ***Soak and wash off surface coating*** | ***Strip wallpaper and wash wall*** | ***Dry-scrape and dry-abrade*** | ***Wet-abrade*** | ***Fill joints, cracks and imperfections*** | ***Apply penetrating sealer*** | ***Size*** | ***Apply lining paper*** | ***Apply reinforced lining paper*** |
| Emulsion painted | | | | X | | X | X | | | |
| Oil painted | X | | | | X | X | | X | X | |
| Painted tongued-and-grooved boarding | X | | | | X | X | | X | | X |
| Papered plaster | | | X | X | | X | | X | | |
| Size bound distempered | | X | | X | | X | | X | | |

## Etching or abrading

Previously painted and other non-porous surfaces should be roughened by abrading to provide extra mechanical adhesion for the wallpaper adhesive. Common practice is to abrade with soda block or Grade 150 waterproof silicon-carbide paper ('Wetordry') during the washing process.

## Jointing tape

A 50-mm or 75-mm linen-reinforced or paper tape is stuck over the joints of timber building boards, plasterboards and hardboards before hanging wallpaper. This prevents joints from opening which might cause the wallpaper to split. Some linen types are self-adhesive.

## Lining paper

Lining paper adheres firmly to the surface and provides a ground of even porosity and absorption for the wallpaper. Surfaces should be cross-lined before hanging heavy wallpapers or hand-printed papers, or hanging to previously oil-painted surfaces. Heavy lining should be used on badly cracked surfaces, and reinforced lining on surfaces likely to move, e.g. tongued-and-grooved boarding.

## Special surface characteristics

*Dry-lining* is a timber framing clad with plasterboard with its decorative side exposed. Joints should be caulked and taped (see 3.2). The surface should be sealed with an oil primer to prevent damage to the paper surface when stripping subsequent coverings.

***Damp walls*** Hanging wallpaper to damp walls may cause bleaching of the wallpaper colours, disintegration of the paper or mould growth. The treatment is:

(i) trace and cure source of dampness;
(ii) if cure is not possible and dampness is not severe, apply either pitch paper, metal foils or proprietary damp-proofing solution.

*Efflorescence* is a white, fluffy, crystalline salt deposit on the face of alkaline materials (see 3.4.5) which must be rubbed off.

*Mould growth* is seen as small, multi-coloured spots or patches on infected wall or surface covering (see 3.6).

The treatment is:

(i) remove infected wallpaper and burn;
(ii) apply a fungicidal solution (see 3.6);
(iii) use fungicidal adhesive to hang surface coverings, particularly vinyls.

# 4.2 Adhesives

The range of adhesives for surface coverings can be broadly divided into four groups.

## 4.2.1 Starch paste

Swellable starches from maize, corn, potato, wheat, tapioca or manioc. Available in powder or ready-to-use form.

***Properties***

High-solid–low-water content.

Opaque.

***Advantages***

Long open time.

Good bonding properties to substrate and material.

***Disadvantages***

High-marking property (easily stains face of paper).

Encourages and supports mould growth. Many contain preservative to reduce risk of early putrification.

### Cold-water paste powder

Easily prepared and suitable for all types of wall and ceiling papers except vinyls. Should be mixed fresh daily, as stale paste is more inclined to stain.

### Hot-water paste powder

Inconvenient where boiling or very hot water is not readily available. Must be allowed to cool before use as hot paste will oversoak the paper. Very glutinous. Suitable for all types of wall and ceiling papers, hessian, corks, grasscloths, but not vinyls. Must be used fresh.

### Prepared or tub paste

Requires no preparation other than dilution to suit weight of paper. A very smooth paste. May be acidic, affecting metallic prints; even if washing soda is added, some discoloration may occur. Similar in use to hot-water paste.

### Dextrine

A very stiff, yellow maize glue. Has great strength. Available in powder or ready-prepared form. Suitable for anaglypta panels, lincrusta mouldings, or for strengthening other pastes.

## 4.2.2 Cellulose paste

Combination of various cellulose ethers obtained from cotton or wood pulp. Thin, quick-setting paste, suitable for lightweight papers only.

***Properties***

Low-solid–high-water content.

Transparent.

***Advantages***

Good affinity with paper, as it is of similar origin.

Low-marking quality (unlikely to stain face of paper).

Does not encourage or support mould growth.

***Disadvantages***

Higher wetting properties than starch pastes.

Lower adhesive properties than starch pastes.

Insufficient *immediate* adhesion to anchor vinyls to surface, therefore not recommended for these papers.

## 4.2.3 Starch ethers or starch-based

Chemically produced starch ethers combined with alkali-treated cellulose.

***Properties***

Medium-solid content, more than cellulose and less than starch.

Whitish paste which is partially transparent.

Usually available in fine flake form.

***Advantages***

Mixes easily to a smooth paste.

Easy to apply.

Similar setting time to starch pastes.

Better adhesive qualities than cellulose.

***Disadvantages***

Greater marking properties than cellulose, but less than starch.

Can support mould growth although most contain fungicides to reduce risk.

### All-purpose or multi-purpose pastes

The most common name for starch-based pastes.

When mixed at different strengths they are suitable for all types of wallpapers, vinyls and expanded polystyrene.

*Note:* Fungicide can be dangerous if swallowed. Particular care must be taken when using adhesives containing fungicide where there are children or animals. It is essential to wash hands after use.

## 4.2.4 PVA and acrylic adhesives

Ready-made adhesives containing PVA and/or acrylic resin.

### Ready prepared

Available in either brushing consistency or very thick state for fixing of expanded polystyrene. Has very strong adhesive strength for applying vinyl fabric or thick expanded polystyrene sheet, and most laminates, e.g. hessian, grasscloths and fabrics.

### Overlap adhesive

Particularly strong copolymer resin glue for bonding vinyl coverings when it is necessary to lap them, and for sticking seams in high-condensation areas and on hot areas such as behind radiators.

# 4.3 Preparatory papers

## 4.3.1 White lining paper

Texture-free white wood pulp paper, usually rolled with smoother face on outside. Available in various weights, e.g. 600 (110 g/m$^2$), 800 (130 g/m$^2$), 1000 (150 g/m$^2$), 1200 (170 g/m$^2$) and 1400 (190 g/m$^2$). Also sold in two qualities: white (w) to go under wallpaper and extra white (xw) which is wood-free and more suitable for emulsion painting.

***Size*** 555 mm wide × 11 m long. Also available in double, triple and quadruple length rolls.

***Use***
(i) Provides uniformly absorbent surface for most surface coverings, especially heavy papers and materials.
(ii) Essential on non-porous surfaces because its absorbency enables subsequent papers to dry quickly.
(iii) Evens porosity on surfaces of varying absorbency, e.g. over large areas of making good.
(iv) Helps to mask surface irregularities, e.g. on surfaces that have received considerable making good, whether painting or paper-hanging is specified as the finish.
(v) Class O grade available.

***Hanging*** If hanging under wallpaper, you should cross-line in the opposite direction to the final surface covering. If used as a foundation for painting, then hang vertically in the traditional manner.

### Non-woven lining paper

Incredibly strong, tougher than glass fibre but feels like paper. The weight is comparable to 1200–1400 grade traditional white lining.

***Size*** 750 mm wide × 25 m long

***Use*** As per lining paper but has the following advantages.
(i) It is much stronger
(ii) Offers specifiers a super smooth, even finish (like plaster even after painting)
(iii) No shrinkage or stretching
(iv) Use manufacturer's recommended adhesive
(v) Paste the wall hanging technique
(vi) No soaking required
(vii) High stability and crack resistance
(viii) Dry strippable

*Note:* (i) Replaces linen/scrim-backed lining paper.
(ii) There are at least two qualities available.

### Thermal wall liner

Energy-saving interior wall covering, allowing up to 65 per cent quicker heat-up of a room. It is permeable, sound-absorbing and mould-inhibiting.

***Size*** 750 mm × 10 m.

***Use*** As a substrate for wallpaper.

***Hanging***
(i) Use manufacturer's adhesive.
(ii) Paste the wall.
(iii) Do not overlap at corners.
(iv) Clean off excess at ceiling/floor using a sharp knife.

## 4.3.2 Glass fibre

Made from very fine fibres of glass either by weaving into coarse, medium or fine open woven material, or by pressing into a fine random mesh called chopped strand, or fine surfacing tissue.

***Size*** 50 m long × 1.0 m wide and purchased by the metre run.

***Use*** A fire-resistant material used (i) to reinforce cracked and imperfect surfaces, and applied to shuttered concrete to provide a flat, even texture; (ii) as a finishing material over coloured grounds to produce a two-coloured relief texture.

***Hanging***

(i) PVA adhesive applied to the wall a little short of the width of the material.

(ii) The joints are lapped and cut dry with knife and straight-edge. The material is then peeled back and the wall pasted beneath the joint.

(iii) Smooth with felt roller, plastic or metal spatula.

(iv) The edges can be trimmed dry before hanging, and the walls pasted and material hung in the usual way.

# 4.4 Flat papers

## 4.4.1 Grounds

Paper completely coated with a casein-bound paint before being printed with a repeating pattern or texture. Non-washable surface.

### Machine prints (Fig. 4.1)

Repeating pattern, overall surface textures or stripes printed by rollers and containing up to 22 colours. The method by which most papers are produced.

Some grounds have a particular character and may be known by other titles. The most common are:

***Jaspe*** A fine, irregular, vertical texture similar to that of scumble texture dragging (Fig. 4.2).

***Chintz*** Pattern based on natural forms treated in a realistic manner to imitate the cotton furnishing fabric of the same name (Fig. 4.2).

***Satin*** Ground colours polished or glazed to produce a sheen before being printed.

***Moire*** Satin or satinette ground, finely textured to produce a watered silk effect, most commonly produced as vinyl.

***Satinette*** Fine mica flakes incorporated in ground colour to produce a sheen.

***Metallic*** Bronze or aluminium powder included in ground colour or print to produce metallic sheen.

### Hand prints

Good-quality, ground-coated papers printed by a hand process. Considerably more expensive than machine prints. Some have a selvedge (see Glossary of terms, 4.10). There are two principal types:

***Block prints* (Fig. 4.3)** Printed by a method using hand-operated carved wooden blocks.

***Screen prints* (Fig. 4.4)** Each colour of a design is squeezed out through a stencil design which is mounted on nylon material or metal mesh stretched

**Fig. 4.1** Machine printing

**Fig. 4.2** Machine prints

**Fig. 4.3** Block printing

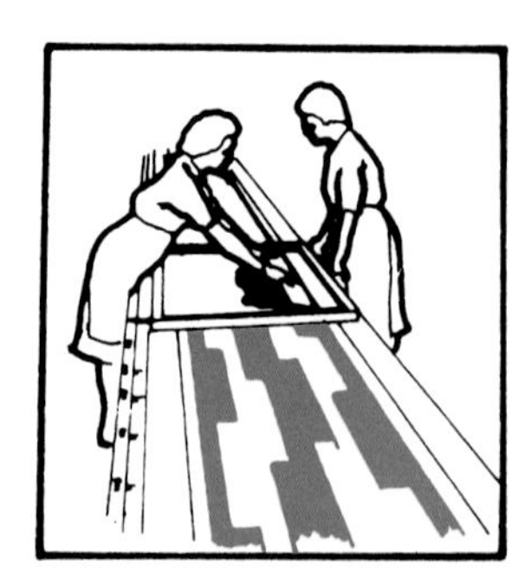

**Fig. 4.4** Screen printing

tight over a printing frame. This method is also carried out mechanically and may be classified as a machine print.

***Size*** BS EN 233: 1999 states that wallpaper shall be 10.05 m long × 530 mm wide. A plus or minus tolerance of 5 per cent is permitted on the width, which has resulted in many standard papers being made available in 520 mm widths.

***Use*** Because the size of the pattern repeat is restricted by machine (maximum about 500 mm), machine prints constitute the bulk of wall and ceiling papers used for domestic purposes.

Hand prints are available in longer repeats, therefore more adaptable for use on very large areas and where cost does not restrict their use.

***Hanging***

(i) Applied with any stout starch or starch ether paste. Cellulose pastes can be used with lightweight papers.

(ii) Joints should be butted.

(iii) May be smoothed with brush or felt roller.

(iv) If paste penetrates through paper or accidentally gets on to the face of some dark prints, it can result in a patchy sheen or colour change.

# 4.5 Embossed papers

Papers which have been pressed into a relief texture so that the back of the paper is hollowed.

## 4.5.1 Dry embossed paper

Printed paper passed between embossing rollers: either a steel roller with the pattern or texture in relief presses the paper into a soft roller, or male and female rollers are used (Fig. 4.5).

(a) *Embossed* The most common and cheapest range, either with all-over random relief texture or with the pattern in relief. Depth of relief may decrease considerably after hanging.

(b) *Duplex* Two papers bonded together before being embossed, to produce a more pronounced relief. Usually the pattern only is embossed.

(c) *Low relief anaglypta-type* A duplex paper, embossed while the adhesive bonding the two papers is still wet, which helps to retain maximum relief after hanging. The top paper is a craft type for strength and whiteness. These papers differ from those described above as they are not coloured or printed with a pattern. Generally available either in textures such as pebbledash, plaster swirl, broken glass or in geometrical patterns. Available in white only and can be painted after hanging.

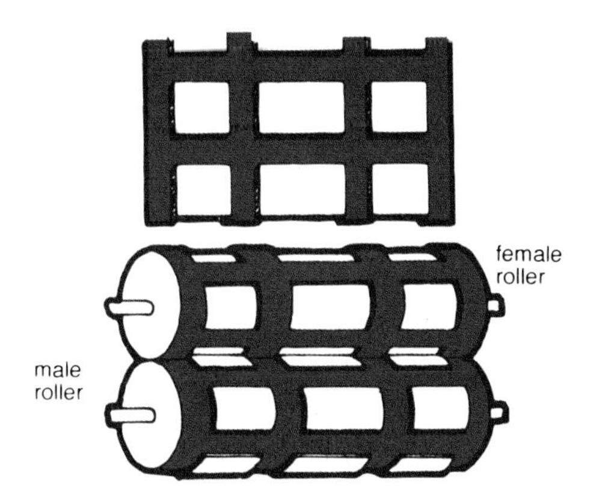


**Fig. 4.5** Embossing machinery

***Size*** 10.05 m long × 520 mm or 530 mm wide.

***Use*** Similar use to grounds. Texture tends to mask slight irregularities on the surface which is a common use for anaglypta.

***Hanging***

(i) Applied with stout starch or starch ether paste.

(ii) The relief should not be filled with paste.

(iii) Duplex papers require soaking until supple. Oversoaking may cause delamination.

(iv) Use of seam roller will flatten the texture and spoil the result.

## 4.5.2 Wet embossed paper

A much heavier-quality paper than anaglypta, prepared from cotton linters, rosin size, china clay and alum. While in a wet state, it is moulded between a steel roller and its counter made of gutta-percha. Thc wet moulding ensures that its relief is retained after hanging. Usually available in white only, and suitable for subsequent painting in oil- or water-based paint.

(a) *Supaglypta-type* Available in rolls and a range of textures similar to anaglypta.

(b) *High relief panels* Very heavy, moulded paper available only in panels, usually in imitation of brick or stonework. Relief may be up to 25 mm in depth. The hanging technique is entirely different from that of other embossed papers.

***Size***

***Supaglypta*** 10.05 m long × 520 mm wide.

***Panels*** Vary according to design: approximately 0.5–0.75 $m^2$.

***Use***

***Supaglypta*** Often used in domestic or industrial situations to mask irregularities on badly marked surfaces. Subsequent painting provides durable and washable finish.

***Panels*** Used for decorative effect domestically, in restaurants, hotels, and film and television sets.

***Hanging***

***Supaglypta***

(i) Applied with stout starch or starch ether paste, or prepared paste.
(ii) Requires to be soaked until supple before hanging.
(iii) Joints should be butted.
(iv) Smooth with paperhanging brush.
(v) Seam roller will flatten relief and spoil the result.

***Panels***

(i) Soaked with water before adhesive is applied.
(ii) Fixed with dextrine, which is knifed on to areas which come in contact with the surface.
(iii) Held to surface with steel pins until adhesive has set.
(iv) Usually designed to overlap or interlock.

# 4.6 Relief papers

Papers having a relief texture on the face, but with a smooth backing. There are two principal types: ingrains, and relief or blown vinyls.

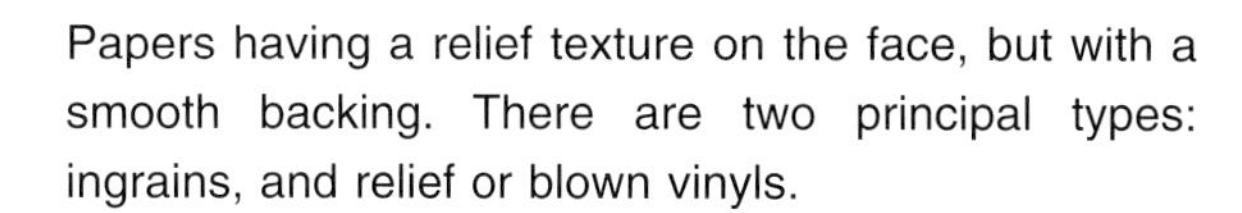

## 4.6.1 Ingrains: woodchip, oatmeal

Paper which has small particles of wood or cork sandwiched between a heavy backing paper and a thin surface paper.

***Size*** 10.05 m long × 520 mm or 530 mm wide.

***Use*** Grades having very fine particles may be machine- or hand-printed and have uses similar to grounds.

The coarser grades (woodchip) are supplied in natural colour and subsequently painted with oil- or water-based paints. Their texture tends to mask irregular surfaces.

***Hanging***

(i) Applied with any stout starch or starch ether paste.
(ii) Oversoaking may cause delamination, or make paper difficult to handle without damage.
(iii) Joints should be butted.
(iv) May be smoothed with felt roller or brush.

## 4.6.2 Relief or blown vinyls

A paper-backed material having an expanded coating of polyvinyl chloride (PVC) which is textured in a variety of random or repeating patterns. Usually available in a neutral colour suitable for painting with alkyd or emulsion paint after hanging.

***Size*** 10.05 m long × 530 mm wide.

***Use*** Domestic or industrial surfaces to provide interesting textures with good durability.

***Hanging*** As vinyl papers.

# 4.7 Washable papers

Papers which have a surface that can be cleaned without damaging the pattern.

## 4.7.1 Ordinary washable paper

A ground or embossed paper which has been machine-coated with a clear glaze based on PVA. It will withstand regular wiping with a damp sponge, but not usually saturation washing or scrubbing with brush or abrasive.

***Size*** 10.05 m long × 520 mm or 530 mm wide.

***Use*** For surfaces which may be subject to handling or rubbing, or where condensation is not excessive, e.g. ventilated kitchens or bathrooms. Principally used for domestic purposes.

***Hanging***

(i) Applied with any stout starch or starch ether paste which contains a fungicide.

(ii) Joints should be butted.

(iii) May be smoothed by brush or felt roller.

*Note:* Some types of washable paper are manufactured as loosely bonded duplex papers to facilitate removal prior to redecoration. The top coated paper can be pulled away, leaving the plain backing which can be stripped off by soaking and scraping or left as a lining paper for subsequent coverings.

## 4.7.2 Vinyl paper

Smooth or textured thin film of PVC bonded to a paper backing and printed with PVC inks heat-fused on to the surface. Provides a completely washable surface which is also resistant to abrasives and hard wear. Has anti-static properties, therefore tends to repel dust.

***Size*** 10.05 m long × 520 mm or 530 mm wide.

***Use*** For surfaces which are subject to condensation, considerable handling and rubbing, or splashing and staining. The wide range of patterns available makes them suitable for both domestic and industrial use.

***Hanging***

(i) Because of its impervious surface, pastes dry very slowly; unless they contain a powerful fungicide, mould growth will result. It also has a tendency to pull away from the surface when drying, and the paste must be particularly strong to resist this pull. Special adhesives are available which incorporate both these properties.

(ii) Joints should be butted.

(iii) May be smoothed with brush, rubber roller or plastic squeegee.

(iv) Special adhesive is necessary where paper has to be lapped, as paste will not adhere to vinyl.

*Note:* Removal of vinyl paper prior to redecoration is carried out by pulling away the vinyl film from the paper backing. The paper can be left as a lining for subsequent coverings, or removed by soaking and scraping.

## 4.7.3 Paper coated after hanging

Most pulps, grounds and embosses which are not produced with a washable finish can be coated after hanging has been completed and the paste has dried.

The coating that is used is based on PVA (see 2.5.4) and is brushed, rolled or sprayed on to the paper. Some papers are printed with poorly bound colour which may lift or smudge if coated. Before sealing, the paper should be tested for its suitability for this treatment.

Usually the sheen produced is greater than that of a washable paper, although the degree of gloss can be controlled by thinning the glaze with water.

## 4.7.4 Pre-pasted (ready-pasted) paper (Fig. 4.6)

Vinyl or washable paper which has been coated on the back with a water-activated fungicidal paste. More expensive than similar non-pasted papers, but this may be offset by the saving of pasting time.

***Size*** 10.05 m long × 520 mm wide.

***Use*** Similar to vinyl and washable papers.

***Hanging***

(i) After being cut into lengths, the paper is soaked in a trough of water.

(ii) Joints should be butted.

(iii) Smooth with a sponge.

**Fig. 4.6** Pre-pasted paper

## 4.7.5 Photo murals

Very large photographs often consisting of several panels to obtain the full picture. Each panel has an overlap to allow the joints to be trimmed and butted.

Some are normal bromide photographic paper, others are on special resin-coated paper.

***Size*** Single or multi-panelled units up to 1.270 m wide.

***Use*** Feature panels in a range of commercial and domestic offices and buildings.

***Hanging***

(i) Applied with tub paste or PVA adhesive.

(ii) Plan layout and joints before hanging multi-panelled murals.

(iii) Trim surplus leaving a gap of about 10–15 mm.

(iv) Important that light source shines into the joint.

(v) Trim with knife and straight-edge.

(vi) Smooth with screen printer's squeegee working from the centre outwards.

(vii) Bromide papers need soaking in water for 5–10 minutes until pliable.

(viii) Always keep surface wet, otherwise the squeegee will damage the surface.

# 4.8 Decorative laminates

Paper, fabric and fibre-glass backings on to which is laminated a range of natural and synthetic decorative materials.

*Note:* Cross-lining is recommended as the preparation for all paper-backed decorative laminates.

## 4.8.1 Flock papers

Originally made to look like brocade or cut velvet fabrics. Many are still made with traditional patterns.

***Hand made*** The pattern is block printed or screen printed onto a paper ground using a slow-setting adhesive. Short fibres of wool, silk or synthetic fibre about 2 mm long (flock) are blown on, sticking to the adhesive. This gives a raised pattern known as the pile.

***Machine made*** The majority of flocks are now machine printed onto a variety of backings including vinyl and metallics. By using synthetic adhesives and fibres, washable flock papers can be produced.

***Double flocks*** Higher relief, sometimes in two tones of colour. Produced by applying one tone of flock onto parts of another. Patterns can also be made by using different coloured fibres or different coloured adhesives.

***Storage*** The pile can be crushed if stored laying down, giving a shady finish when hung. Always stand on end to store.

***Size*** 10.05 m long × 520 mm wide.

***Use*** Decorative appearance used in domestic buildings, restaurants and older-type manors and large houses.

***Hanging***

(i) Applied with tub paste or PVA adhesive.
(ii) Trim by undercutting with knife and straight-edge, ensuring that the straight-edge is placed on the selvedge of the paper so as not to damage the flock.
(iii) Do not crease after pasting.
(iv) Joints should be butted.
(v) Smooth with felt roller, rolling downwards.
(vi) Paste on the face can cause matted fibres which are impossible to rectify.
(vii) After hanging, brush the pile into one direction with a soft brush to give an even-coloured appearance.

## 4.8.2 Wool strand and weftless materials

Separate strands or threads of wool or other material laminated to a paper backing. There is no weft, and they can be of an overall plain colour or produced in a variety of coloured stripes.

***Size*** (i) 10.05 m long × 525 mm wide.
(ii) 50 m long × 1.0 m wide.

***Use*** General decorative finish mainly in domestic buildings and offices.

***Hanging***

(i) Applied with stiff tub paste or PVA-reinforced paste.
(ii) Very little soaking necessary.
(iii) Trim with knife and straight-edge.
(iv) Joints should be butted.
(v) Smooth with felt roller.
(vi) Cleanliness is important. Paste removed from the face can damage the fibres.

## 4.8.3 Paper-backed hessian

Made from jute which is either dyed and woven into a cloth, or woven into a cloth and then dyed. The hessian cloth is laminated on to the paper. Some types have a very even texture. Others have random joints in the warp and weft known as slubs to give more texture and a pronounced feature to the finish.

***Size*** (i) 10.05 m long × 520 mm wide.
(ii) 50 m long × 910 mm wide.

***Use*** General decorative finish in offices, public and domestic buildings.

***Hanging***

(i) Applied with stiff tub paste or PVA-reinforced paste.
(ii) Stiff paste and short soaking time prevent undue stretching and reduce the risk of delamination.
(iii) Trim with sharp knife and straight-edge to achieve a clean first-time cut. A blunt knife or poor trimming will cause frayed edges.
(iv) Reverse alternate lengths to reduce the effects of edge-to-edge shading.
(v) Smooth with felt roller.
(vi) Joints should be butted.
(vii) Cleanliness is important. Paste can mark the face of the cloth.

## 4.8.4 Grasscloth

Originally made in Japan but is now produced in other Far Eastern countries.

Made from dried grasses, reeds, split bark and split canes which are sometimes dyed. The grasses are woven with silk or cotton thread to form a random design which is adhered to a paper backing. Some types have the background showing through as part of the design, others have the whole surface dyed.

***Size*** 7.135 m long × 910 mm wide. They can vary in size.

***Use*** Expensive and exclusive wall coverings, very fragile and difficult to clean. Mainly used in domestic buildings and offices.

***Hanging***

These papers vary a great deal, and manufacturer's instructions should be carefully adhered to.

(i) Applied with a stiff tub paste.
(ii) Do not soak. If the cloth becomes wet, it will delaminate from the paper.
(iii) Trim with knife and straight-edge or a Ridgley track and wheel trimmer.
(iv) Matching is impossible and joints will show. Plan the area to achieve a balanced effect.
(v) Joints should be butted.
(vi) Smooth with a felt roller.
(vii) Soft grasscloths will turn into and around angles like other wall hangings. Stiff grasscloths must be cut to fit exactly into internal angles, and partially cut through from the face on external angles.

## 4.8.5 Cloth-backed vinyl

Plain or textured wall coverings made by coating or laminating PVC on to paper or scrim cloth. They can be self-coloured, textured to imitate fabrics, and have designs printed with special inks. They generally provide a tough, waterproof and easily cleaned surface.

***Size*** Usually 30 m long × 1.3 m wide, but can be purchased by the metre run.

***Use*** For decorative effects and surfaces which have to take hard wear, such as corridors, staircases, offices and public buildings.

Light colours are discoloured by fumes from gas fires and tobacco smoke.

***Hanging***

These materials are waterproof and the adhesive must dry out through the wall.

(i) Applied with PVA adhesive. All pastes must contain a fungicide, or be of a type which does not support mould growth.
(ii) The paste can be applied either to the back of the material, or to the wall, depending on manufacturer's instructions.
(iii) Joints should be overlapped and cut through, or the material can be trimmed with knife and straight-edge and butt jointed.

(iv) Smooth with a plastic or flexible metal spatula.
(v) Where overlaps cannot be avoided, use special overlap adhesive to adhere edges.
(vi) Any paste on the surface is best left to dry and then removed by stiff, dry brushing.

## 4.8.6 Silk

Silk yarn dyed and woven with an extremely fine warp. The weft is usually thicker and joints in the yarn are rolled together producing a pronounced marking in the weave called a slub. The silk is then laminated on to a paper backing. The natural shimmer of the silk creates a unique sparkling effect.

***Size*** (i) 7.30 m long × 910 mm wide.
(ii) 6.85 m long × 760 mm wide.

***Use*** Expensive and exclusive wall covering, very fragile and difficult to clean. Used mainly in domestic buildings and offices where it will receive little wear and tear.

***Hanging***

(i) Surface preparation is critical as any surface defects will be highlighted by the sheen of the silk.
(ii) Applied with tub paste used sparingly.
(iii) Trim with knife and straight-edge.
(iv) Matching is not possible and the joints will show. Plan the area to achieve a balanced effect.
(v) Do not soak. If the material becomes wet, it will delaminate from the paper backing.
(vi) Smooth with a felt roller.
(vii) Joints should be butted.
(viii) Trim around obstacles with scissors not knives.
(ix) Cleanliness is vital as paste on the surface cannot be sponged off. Thin paste, oversoaking and water on the face will cause delamination.

## 4.8.7 Lincrusta

Lincrusta has a thick backing paper coated with a putty-like compound which is rolled with textured rollers to produce a range of wallcoverings, friezes, dado panels and borders.

***Surface preparation*** Thorough preparation of the surface is essential when installing Lincrusta. Remove old paper, wash down, fill all imperfections and allow to dry. Rub down before applying a coat of size.

***Lining*** We recommend cross-lining all non-porous surfaces (e.g. solvent-painted walls) to give perfect adhesion. Apply lining paper using a ready-mixed adhesive containing a fungicide. Allow to dry thoroughly before applying Lincrusta.

***Recommended adhesive*** Lincrusta Adhesive has been specially formulated. If the Lincrusta Adhesive is a little thick, stir well – *do not dilute*.

***Hanging***

(i) Plan your start and finish points in the room to minimise wastage, and plumb your first length.
(ii) Match each length where necessary. Cut individual lengths to size adding 50 mm (2″) at the top and bottom for trimming.
(iii) Trim off the selvedge by undercutting the edges using a knife and straight edge, taking care to avoid edge damage.
(iv) Sponge the back with warm water and leave to soak for 20–30 minutes. For best results place lengths back to back while soaking. Soaking allows for expansion and aids application of adhesive.
(v) After soaking, wipe the back with a dry sponge to remove excess water.
(vi) Apply Lincrusta Adhesive to the back using a 50–75 mm (2–3″) synthetic bristle paintbrush. A roller can be used for larger areas.
(vii) Apply length to the surface and smooth down using a 178 mm (7″) felt roller, working from the middle to the outer edges expelling all air bubbles.
(viii) To form lower edge, mark Lincrusta at each side, level with top of skirting or dado. Place cutting board behind Lincrusta leaning against wall, and cut between pencil marks using knife and straight edge. Repeat at top. See Fig. 4.7.
(ix) Sponge off any surplus adhesive from the surface.
(x) Leave to dry for at least 24 hours prior to degreasing and decorating.

*Internal corners* – cut to fit into the corner. Hang the remaining off-cut to a plumb line on the adjacent wall,

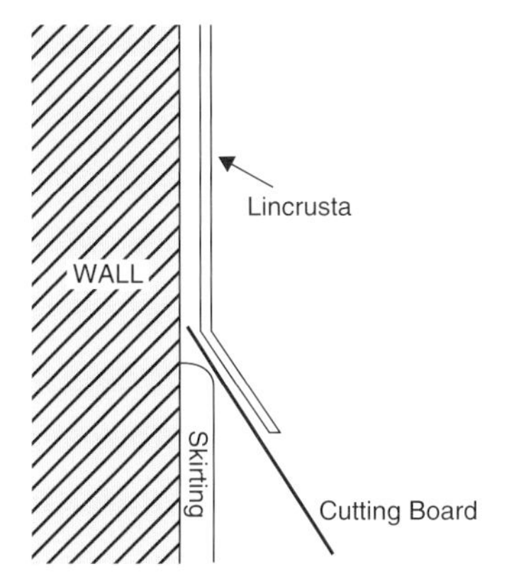


**Fig. 4.7** Lincrusta – forming the lower edge

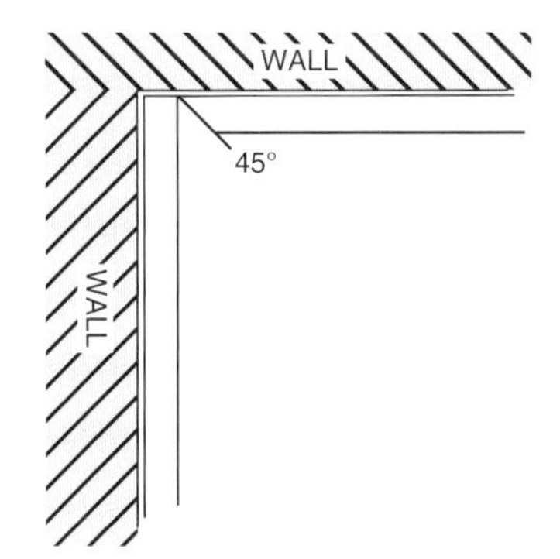


**Fig. 4.8** Lincrusta – internal corner

trimming to fit in corner. Trim any areas of high relief in the corner at 45° angle to fit. See Fig. 4.8.

*External corners* – for gently rounded corners, Lincrusta can be smoothed round in one piece. Where the angle is sharp, cut the Lincrusta to finish flush at the corner. Hang the remaining off-cut to a plumb line on adjacent wall, equal to the minimum width of off-cut. Fill any gaps at the external corner using linseed putty. See Fig. 4.9.

***Application of friezes***

1. Cut the Lincrusta into required lengths, ensuring the design is centred at the mid-point of the main wall. Continue matching the pattern in the corner with subsequent lengths, finishing at the least obvious corner.
2. If your wall is more than 2 m (6′) long, cut the frieze (across the depth) into manageable lengths, no more than 2 m (6′) long. For lengths in excess of this it is recommended that two people hold and support the frieze during application.
3. Trim off the selvedge by undercutting the edges using a knife and straight edge.

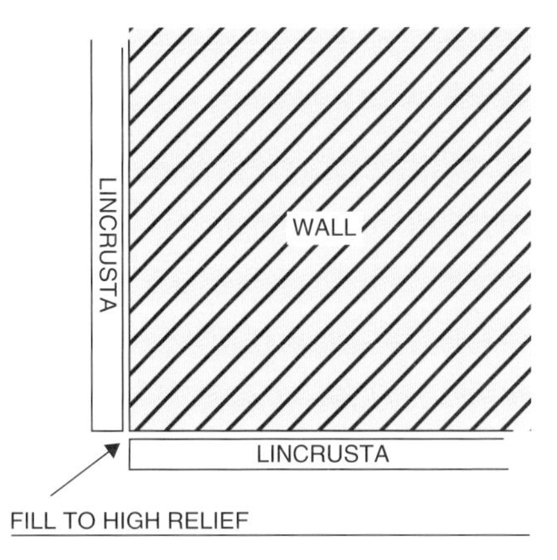


**Fig. 4.9** Lincrusta – external corner

***De-greasing Lincrusta*** Before painting the surface, the Lincrusta should be brushed with white spirit or turps substitute to remove any grease, dried with a lint-free cloth, then allowed to dry overnight.

Lincrusta can be painted using both oil-based and water-based paint systems.

*Oil-based* – apply an oil-based primer prior to application of 2 coats of oil-based eggshell or gloss to create base coat. Apply topcoat to achieve required paint effect if necessary.

*Water-based* – apply a water-based acrylic primer prior to application of 2 coats of water-based eggshell or matt emulsion to create base coat. Apply topcoat to achieve required paint effect. Add a decorator's varnish for extra protection if necessary. As an alternative to painting, many decorative effects can be used in accordance with manufacturer's instructions.

## 4.8.8 Metallic foils

Made by either laminating a paper backing with tarnished leaves of various metals, or by spraying metallic or coloured polyester resin on to a paper backing. They can be overprinted with a range of designs.

***Size*** 7.30 m long × 910 mm wide. Other sizes also available.

***Use*** Rich, decorative effects in exhibition display and feature areas of domestic and office decoration.

***Hanging***

(i) Applied with PVA adhesive.
(ii) Advisable with some polyester-coated papers to paste the wall twice to obtain a better surface.
(iii) Trim with knife and straight-edge.
(iv) Joints should be butted.
(v) Fold carefully without creasing.
(vi) Smooth with felt or rubber roller.
(vii) Remove any paste from the surface immediately and polish with a soft cloth.
(viii) The mirror-like finish of these papers highlights every surface defect, therefore the surface preparation must be perfect to obtain an acceptable effect.

## 4.8.9 Cork

Thin veneer of natural cork laminated onto coloured or metallic paper backing. Holes occur naturally in the thin cork veneer, and the coloured or metallic backing shows through to add to the decorative effect.

***Size*** (i) 7.30 m long × 910 mm wide.
(ii) 10.05 m long × 530 mm wide.

***Use*** Quite hard-wearing and used in domestic and office buildings.

***Hanging***

(i) Applied with tub paste or PVA-reinforced paste.
(ii) Trim with knife and straight-edge or Ridgley track and wheel trimmer.
(iii) Dampening the edges with water before pasting will sometimes prevent excessive curling of the edges which is common to these papers.
(iv) Joints should be butted.
(v) Smooth with felt roller.

*Note:* Cork is also available in tile and sheet form. Refer to manufacturer's hanging instructions.

# 4.9 Defects of surface coverings

**Table 4.3** Surface coverings: defects due to surface or room conditions

| *Cause* | *Visible defect* |
|---|---|
| *Alkaline surfaces* which are still active and not sealed may bleach or discolour printing inks | Staining<br>Discoloration |
| *Cold damp room* causes paste to dry slowly | Staining<br>Blistering |
| *Condensation* causes slow drying of the paste, encourages mould growth or discolours the paper or printing inks | Blistering<br>Mould growth<br>Staining |
| *Damp walls* have similar effect to condensation | Blistering<br>Mould growth<br>Staining |
| *Efflorescence* prevents adhesion, darkens the paper or chemically affects the pigments in the printing inks | Peeling<br>Edge springing<br>Staining |
| *Excessive porosity* tends to absorb adhesive, leaving covering loosely adhering and difficult to slip into alignment | Edge springing<br>Peeling<br>Poor matching |
| *Exposure to excessive sunlight* may cause certain pigments to fade or discolour | Discoloration |
| *Non-porous surface* (gloss paint, tiles, plastic) which has not been lined causes the paste to dry slowly, resulting in the paper fibres stretching | Blistering<br>Staining |
| *Poorly adhering lining paper* is further weakened by both the wetting action of the paste and the weight of the covering | Blistering<br>Peeling |
| *Traces of old size and paste* left unremoved before hanging vinyl | Mould growth |
| *Uneven surface* causes paper to be moulded to the undulations, resulting in stretching | Creasing<br>Poor matching<br>Joints gapping |
| *Unsealed steel pins* on timber or wall board corrode and stain the paper | Rust spotting<br>Staining |
| *Unsealed water-soluble stains* dissolved by the paste | Staining |

**Table 4.4** Surface coverings: defects due to paste and pasting

| Cause | Visible defect |
|---|---|
| *Careless pasting* may result in paste getting onto face of paper | Staining<br>Discoloration<br>Sheen patches<br>Polishing (particularly at seams) |
| *Hot paste* penetrates paper quickly | Staining<br>Delamination<br>Tearing |
| *Incorrect paste* may cause undersoaking or oversoaking, reduce adhesion, prevent slip, make paper difficult to handle, or encourage mould | Blistering<br>Peeling<br>Edge springing<br>Discoloration<br>Poor matching<br>Mould growth |
| *Insufficient soaking* makes smoothing difficult and causes uneven tensions on paper | Blistering<br>Edge springing |
| *Misses* cause uneven swelling | Blistering<br>Overlapping<br>Poor matching |
| *Oversoaking* causes excessive expansion and difficulty of handling, and may separate duplex papers | Staining<br>Delamination<br>Blistering<br>Poor matching<br>Tearing |
| *Stale paste* becomes yellower, highly acidic and may contain bacteria. It becomes thinner as it ages | Staining<br>Darkening of metallic print<br>Mould growth<br>Peeling |
| *Thin paste* has reduced adhesive properties and high wetting power | Peeling<br>Staining<br>Tearing<br>Joint springing |
| *Uneven* application or pasting parts twice causes uneven swelling | Blistering<br>Overlapping<br>Poor matching |

**Table 4.5** Surface coverings: defects due to hanging

| *Cause* | *Visible defect* |
|---|---|
| *Blunt tools* | Tearing<br>Inaccurate finishing |
| *Careless handling* may damage the paper or transfer paste and dirt to face of paper | Staining<br>Tearing |
| *Excessive brushing, rolling or 'squeegeeing'* causes covering to stretch and may polish surface of matt papers | Poor matching<br>Creasing<br>Polishing<br>Flattening of relief |
| *Insufficient care in shading* | Variations in colour of lengths |
| *Insufficient smoothing* | Blistering<br>Joint gapping<br>Creasing |
| *Reversing alternate lengths* when not recommended by manufacturer | Apparent variation in colour and texture of adjacent lengths |
| *Use of seam roller on embossed papers* | Flattening of relief |
| *Use of seam roller too soon after hanging or without protecting paper* | Paste staining<br>Polishing |

**Table 4.6** Surface coverings: defects arising in manufacture

| *Cause* | *Visible defect* |
|---|---|
| *Inaccurate trimming* | Poor matching |
| *Poorly bonded duplex paper* | Blistering<br>Delamination |
| *Shaded papers* due to use of different tones of inks or effect of light and dirt during storage | Variations in colour of lengths<br>Darkening or lightening of edges |

**Table 4.7** Surface coverings: summary of common defects and their causes

| Cause | Blistering | Creasing | Darkening of metallic print | Delamination | Discoloration | Edge springing | Inaccurate angle cutting | Joint gapping | Flattening of relief | Mould growth[a] | Overlapping | Paste staining | Peeling | Polishing | Poor matching | Rust spotting | Shading | Sheen patches | Staining | Tearing |
|---|---|---|---|---|---|---|---|---|---|---|---|---|---|---|---|---|---|---|---|---|
| Alkaline surface | | | | | X | | | | | | | | | | | | | | X | |
| Cold, damp room | X | | | | | | | | | | | | | | | | | | X | |
| Condensation | X | | | | | | | | | X | | | | | | | | | X | |
| Damp surface | X | | | | | | | | | X | | | | | | | | | X | |
| Efflorescence | | | | | | X | | | | | | | X | | | | | | X | |
| Excessive porosity | | | | | | X | | | | | | | X | X | | | | | | |
| Excessive sunlight | | | | | X | | | | | | | | | | | | | | | |
| Non-porous, unlined surface | X | | | | | | | | | | | | | | | | | | X | |
| Poorly adhering lining | X | | | | | | | | | | | | X | | | | | | | |
| Uneven surface | | X | | | | | | X | | | | | | | X | | | | | |
| Unsealed steel pins/ screws | | | | | | | | | | | | | | | | X | | | X | |
| Unsealed water-soluble stain | | | | | | | | | | | | | | | | | | | X | |
| Careless pasting | | | | | X | | | | | | | X | | X | | | | X | X | |
| Hot paste | | | | X | | | | | | | | | | | | | | | X | X |
| Incorrect paste | X | | | | X | X | | | | X | | | X | | X | | | | | |
| Insufficient soaking | X | | | | | X | | | | | | | | | | | | | | |
| Misses in pasting | X | | | | | | | | | | X | | | | X | | | | | |
| Oversoaking | X | | | X | | | | | | | | | | | X | | | | X | |
| Stale paste | | | X | | | | | | | X | | | X | | | | | | X | |
| Thin paste | | | | | | | | X | | | | | X | | | | | | X | X |
| Uneven pasting | X | | | | | | | | | X | | | | | X | | | | | |
| Blunt tools | | | | | | | X | | | | | | | | | | | | | X |
| Careless handling | | X | | | | | | | | | | | | | | | | | X | X |
| Careless seam rolling | | | | | | | | | | | X | | | | | | | | | |
| Excessive brushing | X | X | | | | | | | X | | X | | | X | X | | | | | |
| Insufficient smoothing | X | X | | | | | | X | | | | | | | | | | | | |
| Reversing alternate lengths | | | | | | | | | | | | | | | | | X | | | |
| Rolling embossed paper | | | | | | | | | X | | | | | | | | | | | |
| Inaccurate trimming | | | | | | | | | | | | | | | X | | | | | |
| Poorly bonded duplex | | | | X | | | | | | | | | | | | | | | | |

[a]Mould growth is not always apparent on the face of surface coverings, although it may be far advanced beneath the material. Particularly on papers, it may easily be detected by staining. Advanced moulds may be detected by smell, particularly in humid conditions.

# 4.10 Glossary of terms

***Batch or shade numbers*** **(Fig. 4.10)** An identification number printed on the back of the material at the end of the roll or on a label inserted beneath the transparent cover. Indicates the code number of the colour mixed to print the paper; only coverings having the same number are likely to be exactly the same colour.

***Butt jointing*** The most common method of hanging surface coverings: the edges touch without a gap or overlap.

***Centring*** Method of setting out a wall, ceiling or panel, particularly where a large pattern is being used, to avoid an unbalanced effect. The two alternative methods are shown in Fig. 4.11.

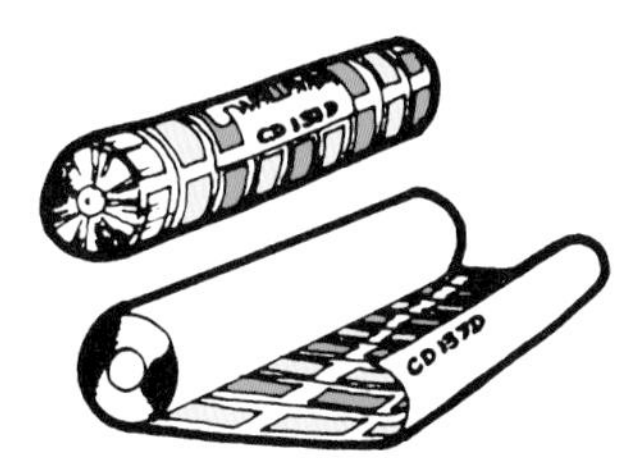

Fig. 4.10 Batch numbering

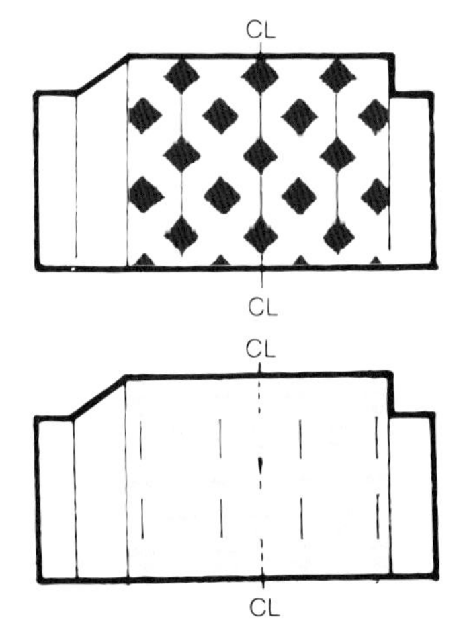


Fig. 4.11 Centring

***Concertina folding*** **(Fig. 4.12)** Method of folding long lengths of pasted covering to be applied to ceilings or horizontally to walls. Sometimes used when hanging very long lengths vertically.

***Cross-lining*** The hanging of lining paper horizontally: the normal method when subsequent covering is to be hung vertically.

***Crutch*** **(Fig. 4.13)** A support used when hanging ceiling papers to prevent the folded, pasted paper bending or creasing. May be a roll of paper, cardboard roll or a straight-edge.

***Cutting top and bottom*** **(Fig. 4.14)** Method of marking and cutting ends of lengths after application to surface to ensure accurate fitting into ceiling and skirting angles.

***Double-lining*** The hanging of two layers of lining paper, one vertically and the other horizontally. A method used to mask very irregular surfaces.

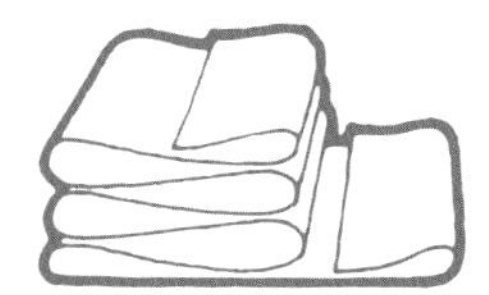

Fig. 4.12 Concertina folding

Fig. 4.13 Use of a crutch

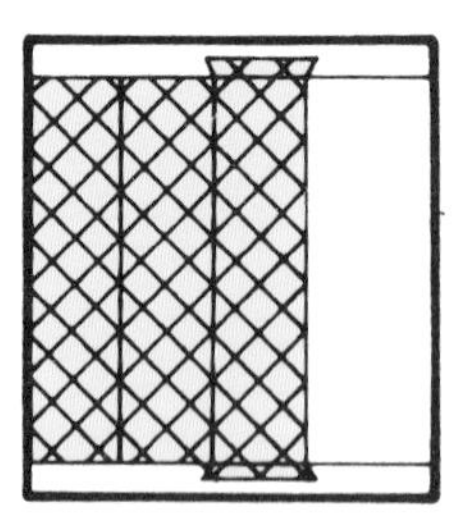

Fig. 4.14 Top and bottom cutting

Fig. 4.15 Drop pattern

Fig. 4.16 End-to-centre folding

***Drop pattern*** **(Fig. 4.15)** A pattern which does not repeat on a horizontal line from edge to edge of the material. Requires special care when cutting lengths, to ensure that matching is accurate. Also offset match.

***End-to-centre folding*** **(Fig. 4.16)** Method of pasting coverings which are to be hung vertically. Ensures that no pasted areas are exposed, makes for easy carrying and delays set of paste.

***Lapping*** The overlapping of edges of consecutive lengths of surface covering by up to 20 mm. Mainly used when applying pitch papers or metal foils, to minimise the risk of moisture penetrating through the joint.

***Length*** Part of a roll cut to fit the surface to be covered. Usually calculated as the exact distance between ceiling angle and skirting, or both wall angles when covering ceilings, plus an allowance of about 100 mm to ensure accurate marking and cutting at angles (see Cutting top and bottom).

Fig. 4.17 Random pattern

***Matching*** The method of aligning the pattern of one length of surface covering to the preceding one.

***Plumb first length*** Because not all wall angles or door and window openings are perfectly upright, it is necessary to hang the first length on each wall to a line that has been set out exactly vertical (see Plumbing).

***Plumbing*** The use of a plumb bob and line or large spirit level to produce an exact vertical line.

***Random pattern*** **(Fig. 4.17)** A design that has no apparent repeat, therefore does not require matching. Textures, stripes and jaspes come within this group. Some of these are called 'semi-plain'. Also called free match.

***Release agent*** A coating containing waxes, applied to the prepared surface to make the subsequent removal of paper easy.

***Reversing alternate lengths*** In some non-patterned or semi-plain coverings, slight variations in colour may be masked by hanging every other length in the opposite direction. This technique should be used only when recommended by the manufacturer.

***Roll or piece*** The smallest unit by which surface coverings can be purchased. Size may vary according to country of origin (see Standard size).

***Selvedge*** **(Fig. 4.18)** A narrow strip (approximately 12 mm) on each edge of a roll of surface covering, permitting accurate printing. Usually removed by the manufacturer, but may be left on some hand prints and lincrusta (see Trimming).

Fig. 4.18 Selvedge

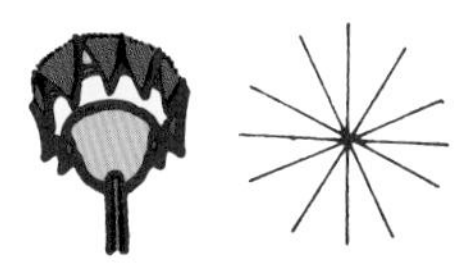

Fig. 4.19 Star cut

Fig. 4.20 Straight or set pattern

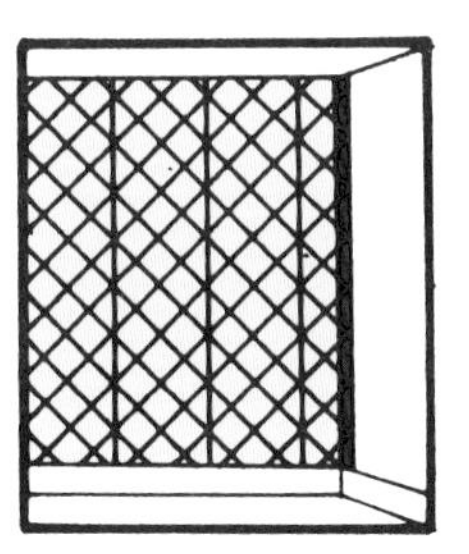

Fig. 4.21 Turning corners

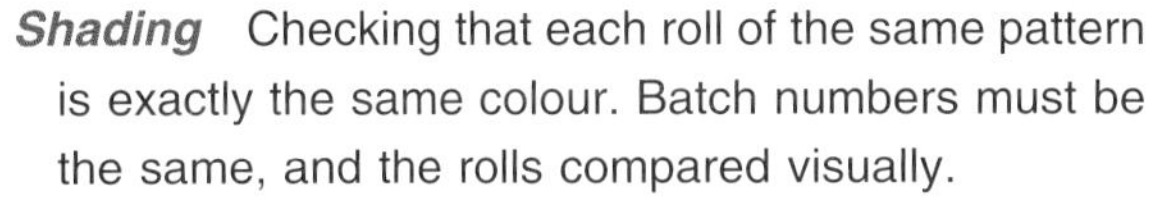

***Shading*** Checking that each roll of the same pattern is exactly the same colour. Batch numbers must be the same, and the rolls compared visually.

***Sizing*** The application of a thin coat of glue size or paste to a porous surface to ensure even absorbency, before hanging surface coverings (see 4.1).

***Slip*** The degree to which a covering can move once it is applied to a surface. This property is crucial to accurate matching.

***Smoothing*** The method of removing creases and air bubbles after the covering has been applied and of ensuring good contact with the surface.

***Snatch*** The holding of the applied covering by a too-absorbent surface, allowing no movement. The opposite to slip.

***Soaking length*** The period of time that a length of heavy surface covering requires to be left after pasting and folding before it becomes sufficiently soft and supple to be hung.

***Standard size*** English papers: 10.05 m long × 530 mm wide. A plus or minus tolerance of 5 per cent is permitted on the width, which has resulted in many standard papers being made available in 520 mm width. Foreign papers vary in both length and width; sizes must be checked before estimating.

***Star cut*** **(Fig. 4.19)** The normal method of cutting coverings to fit around obstructions on the surface which cannot be removed.

***Straight or set pattern*** **(Fig. 4.20)** A design that repeats in a horizontal line, the match being at the same level on each edge.

***Striking a line*** Method of producing on a ceiling or sloping surface, using a chalked string, a line against which to hang the first length.

***Trimming*** Removal of the selvedge by the use of either a knife and straight-edge or a hand trimmer before pasting, or with a track and wheel cutter after pasting and folding.

***Turning corners*** **(Fig. 4.21)** The lapping of wall coverings round corners. Few wall angles are perfectly true, and if the lap is more than about 15 mm, shrinkage on drying may cause blisters.

***Working from the light*** An alternative method of working when centring is not used. If lengths are hung starting by the chief source of natural light, any slight, unavoidable overlapping will not cast a shadow and will be less apparent.

# 4.11 International performance symbols

A range of internationally recognised symbols which often appear in wallpaper pattern books and on product labels (Table 4.8).

**Table 4.8** International performance symbols

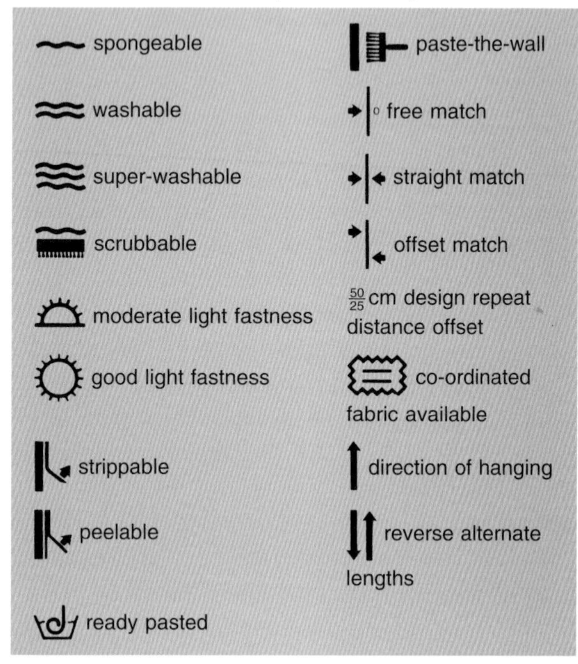

| | |
|---|---|
| spongeable | paste-the-wall |
| washable | free match |
| super-washable | straight match |
| scrubbable | offset match |
| moderate light fastness | $\frac{50}{25}$ cm design repeat distance offset |
| good light fastness | co-ordinated fabric available |
| strippable | direction of hanging |
| peelable | reverse alternate lengths |
| ready pasted | |

Reproduced by permission of the Wallcovering Manufacturers' Association.

# PART 5

# Scaffolding

**The Working at Height Regulations 2005 (Revised 2007)**

2007–2008 saw 58 fatal accidents at work and over 4000 major injuries, all due to falls from height. Of these, falls from ladders were responsible for 16 deaths and 1146 major injuries. They remain the single biggest cause of death in the workplace and one of the main causes of major injury. These regulations replace all previous Regulations regarding working at height and also implement the European Council Temporary Work at Height Directive.



*Note* In Part 5 Scaffolding, all references to regulations about the construction and safety of scaffolding are taken from the Construction (Health, Safety and Welfare) Regulations 1996 which replace the Construction (Working Places) Regulations 1996, the Construction (Health and Welfare) Regulations 1996, and the Construction (General Provisions) Regulations 1961.

# 5.1 Ladders, trestles, stepladders

## 5.1.1 Ladders

Ladders are used for work which is easily accessible and of a light nature. They are suitable for work which is too high for trestles or stepladders, or where large patent or tubular scaffolds cannot be used. Pole ladders are usually used for access on tubular scaffolds.

### Extension ladder

Constructed of wood (becoming less common due to weight and cost), aluminium alloy, steel and fibreglass.

Timber stiles should be straight-grained, free from knots and have a wire rope set into the back for strength. They should be protected by varnish rather than paint, which might hide defects. Rungs are either rectangular or round and made from hardwood such as hickory, oak or ash. Metal rods are fixed beneath the rungs at regular intervals to prevent the stiles parting (Fig. 5.1a).

The stiles of metal ladders should be constructed of rectangular box section of heat-treated aluminium alloy. The rungs are usually ribbed for a better grip (Fig. 5.1c).

Extension ladders can be of double or triple sections, hand-push or rope-operated, using various patent methods of rung locking to prevent collapse (see Fig. 5.3). They range from 3.1 m to 9.2 m (extended).

### Pole ladder (Fig. 5.1b)

Single-length ladder with stiles made from one straight tree trunk sawn in half lengthways. This ensures that both stiles have even spring. All timber should have the bark stripped from it. Construction, shape and reinforcement of rungs are similar to those of extension ladders. Lengths available are 4 m, 6 m and 8 m.

***Storage* (Fig. 5.2)**

(i) Laid flat.
(ii) Supported in at least three places along its length.
(iii) Stored under cover but not in a dry store.

***Safety precautions* (Fig. 5.3)**

(i) *Never* use ladders with missing or loose rungs or with makeshift repairs to rungs.
(ii) *Never* repair broken stiles with wooden splints nailed or lashed to them.
(iii) *Never* stand ladders on uneven, soft or loose ground.
(iv) *Never* reach out too far: it is easy to overbalance.

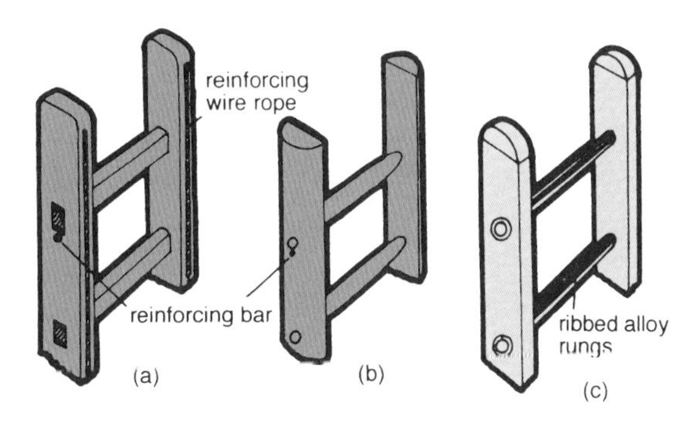


**Fig. 5.1** Types of ladder: (a) timber ladder, (b) pole ladder, (c) aluminium ladder.

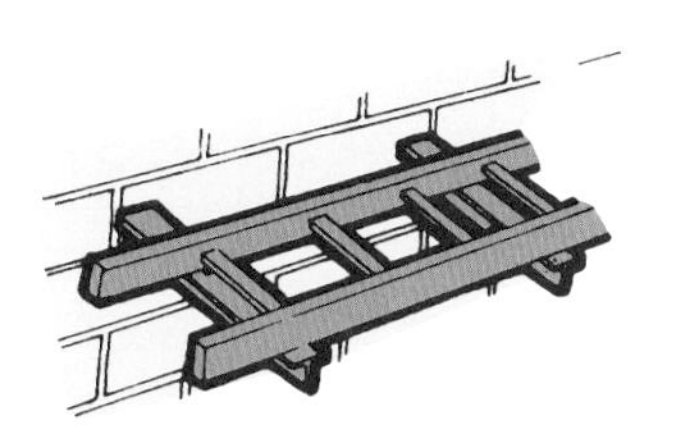

**Fig. 5.2** Ladder support

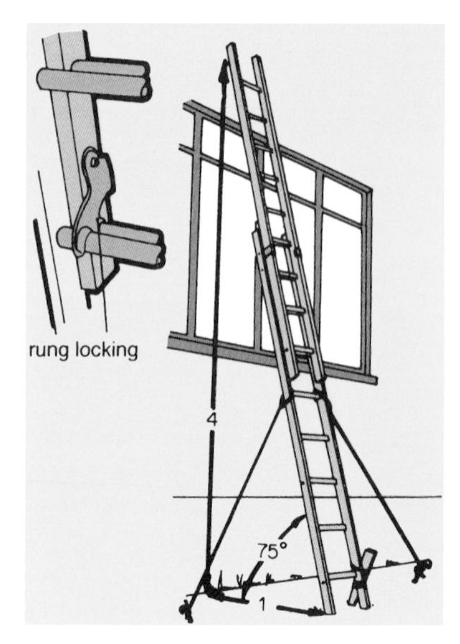


**Fig. 5.3** Securing the ladder

(v) *Never* try to gain extra height by resting the ladder on an unstable surface, or supporting it on its rungs.
(vi) *Never* climb a ladder beyond 1 m from the top.
(vii) *Always* secure the ladder
  (a) at the top, to prevent it from slipping; or, if this is not practical,
  (b) to anchorages at or near ground level; if neither of these is possible,
  (c) by a person standing on the bottom rung.
(viii) *Never* use ladders with frayed or damaged ropes.
(ix) *Always* use ladders at the correct angle of 75° or the 1 in 4 rule (for every unit out, four units up) or use the angle indicator, which is attached to the stile of some ladders.
(x) *Only* use ladders according to their Duty Rating, see 5.2.3.

## 5.1.2 Ladder attachments

**Safety feet (Fig. 5.4)**

Rubber or plastic padded metal feet which can be fitted to the bottom of ladder stiles to prevent them slipping.

**Fig. 5.4** Safety feet

**Fig. 5.5** Ladder stay

They are hinged to allow the ladder to be used at the correct angle.

**Ladder stay (Fig. 5.5)**

A metal bracket attached to the top of a ladder, used to hold the ladder away from the building for easier working and access to projecting or overhanging features such as gutters, eaves and soffits.

## 5.1.3 Roof ladder (Fig. 5.6)

Made from wood or aluminium, at least 380 mm wide with cross bars 32 mm thick fixed at 380 mm intervals. Wheels are fitted to slide up the roof and hooks to anchor the ladder over the ridge.

***Use*** (a) To gain access onto and over sloping and pitched roofs (e.g. having a pitch of more than 10°), to prevent slipping.
(b) To avoid fracturing fragile materials such as asbestos, slate or glass.

***Safety precautions*** **(Fig. 5.7)**

(i) When working on fragile roofs, two ladders should be used side by side.
(ii) Access ladders to a roof should, where possible, be fixed to a window brace or to a roof rafter.

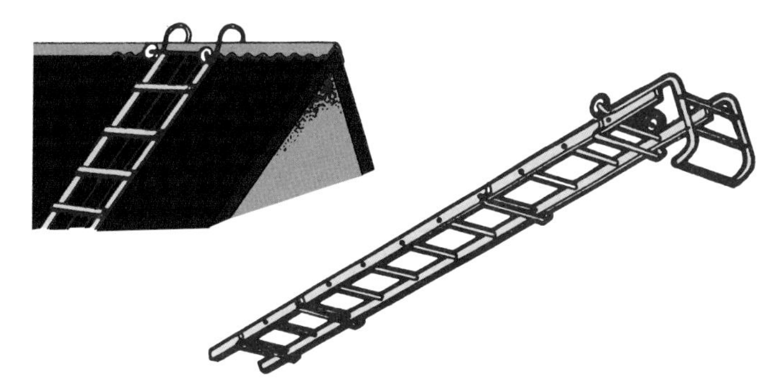

**Fig. 5.6** Roof ladder

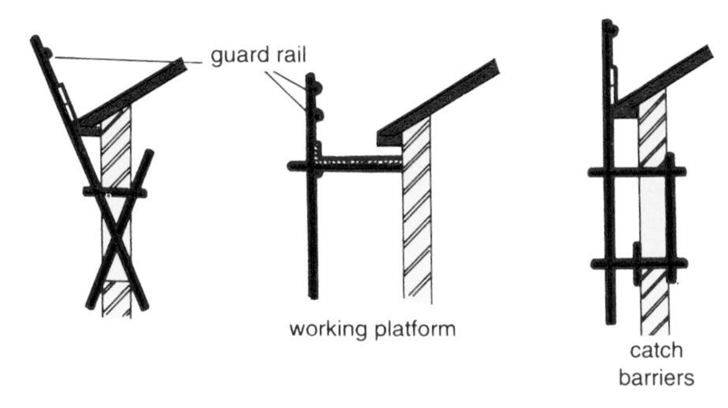


**Fig. 5.7** Safety measures for roof work

(iii) When roof slope is over 30° or is dangerously slippery, one of the following structures must be erected at the eaves:
  (a) a catch barrier or platform;
  (b) a working platform with guard rail.

(iv) A safety belt and rope fastened to a safe anchorage should be worn.

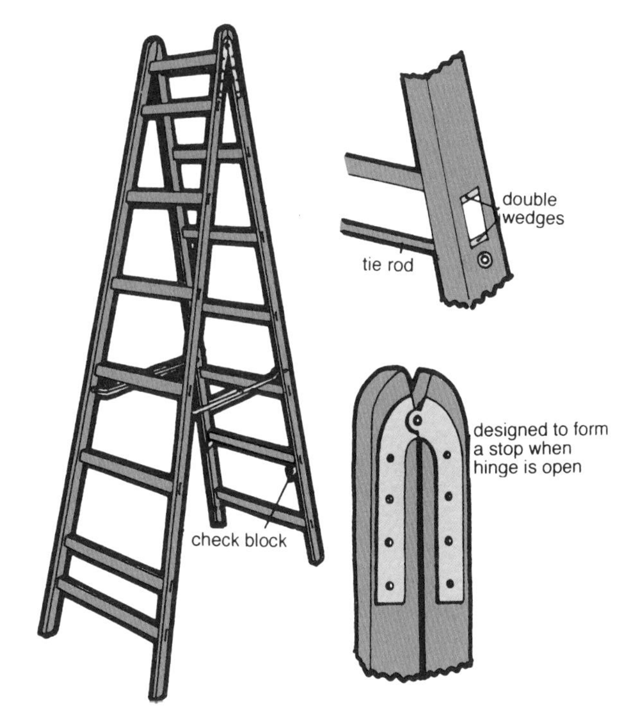


**Fig. 5.8** Folding trestles

## 5.1.4 Trestles (Fig. 5.8)

Made from rectangular sectioned softwood (becoming less common due to cost and weight), or aluminium alloy section and wide enough to take two scaffold boards or one lightweight staging. The cross members are spaced approximately 500 mm apart and staggered on each side to give a platform rise of approximately 250 mm. At least two tie rods should be fitted to both sides of timber trestles for extra strength. The timber should be protected with varnish rather than paint so that defects are not hidden from view.

They are available in a range of sizes between 1.89 m and 2.49 m high and are up to 40 per cent lighter than wood.

***Storage*** Closed, with check blocks located in the opposite stile, and standing under cover.

***Safety precautions***

(i) *Never* use trestles when only partly open.

(ii) *Never* use trestles in a closed position as ladders.

**Fig. 5.9** Swing-backed steps

(iii) Guard rails and toe boards are required for platforms over 2 m high.
(iv) *Never* use folding trestles for more than one tier.
(v) At least $\frac{1}{3}$ of any folding trestle should be above the working platform.
(vi) *Never* use trestles with missing or loose cross-members.
(vii) *Never* use trestles with loose or slack hinges.
(viii) *Always* dismantle scaffolding before moving.
(ix) Only use trestles according to their correct Duty Rating, see 5.2.3.

## 5.1.5 Stepladders (Fig. 5.9)

Stepladders are constructed of either timber, aluminium alloy or fibreglass.

Timber stepladders are made from softwood. Treads should be housed into stiles at 225 mm centres and reinforced with steel tie rods at regular intervals. Rustless hinges are bolted and screwed for extra strength. Ropes are fitted to control the opening of steps when in use. The timber should be protected by varnish rather than paint, which might hide defects.

Aluminium stepladders are lighter, and fitted with metal bars to control the opening of the steps. Rubber feet are fitted to prevent slipping.

Fibreglass stepladders are non-conductive to 30,000 volts. Allow for working heights up to 2.60 m.

Available in sizes from 5 to 10 treads high.

***Storage*** Closed, tied with their ropes and standing under cover.

***Safety precautions***

(i) *Never* use steps when only partly open.
(ii) *Never* use steps in a closed position as a ladder.
(iii) *Never* use steps with a working platform over 2 m high.
(iv) *Never* use stepladder unless the operative's knees are below the top rung.
(v) *Never* use steps with frayed or uneven-length ropes.
(vi) *Never* use steps with missing or loose treads.
(vii) *Never* use steps with loose or slack hinges.
(viii) *Never* work from the top of a stepladder unless it has been constructed as a platform with a secure hand hold.
(ix) *Always* dismantle scaffolding before moving.
(x) *Always* select steps by their Duty Rating, see 5.2.3.

## 5.1.6 Fixed-leg adjustable steel trestles

Used in conjunction with scaffold boards or lightweight staging. Each trestle accommodates four scaffold boards.

They need to be spaced in accordance with the requirements of the boards being used.

Used for lightweight building work at low level, i.e. false-ceiling erectors.

## 5.1.7 Hop-ups (Figs. 5.10 and 5.11)

Hop-ups can be hand built of wood or purchased ready made. Owing to the high cost of timber and the low cost of ready-made hop–ups, the former is now rare.

(a) The hand-built variety consists of a simple step up to a flat working platform measuring not less than 500 mm × 400 mm (see Fig. 5.10).

(b) Ready made, of aluminium similar to a two-tread stepladder and available in several sizes (see Fig. 5.11).

Height approximately 500–600 mm. Various lengths available.

***Use*** A light and versatile means of access to the top section of walls not exceeding approximately 2.5 m high, or used in conjunction with scaffold boards or lightweight staging for access to ceilings of similar height.

***Safety precautions***

(i) Ensure hand-built varieties are sturdy, remain steady in use and are protected only with varnish which will not hide any defects.

(ii) Small stepladder types must be fully opened and locked before use.

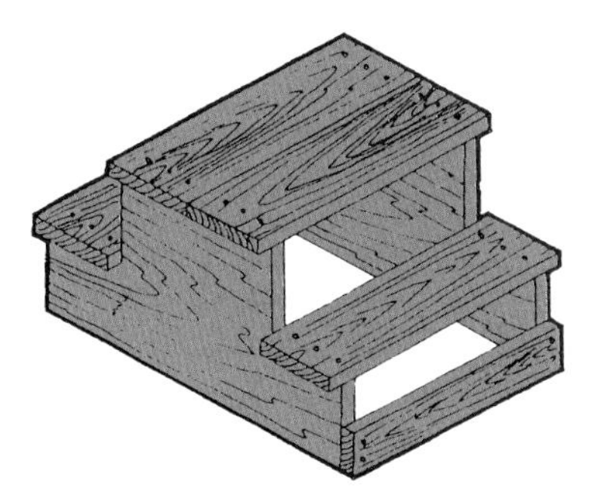

**Fig. 5.10** Hand-built hop-up

**Fig. 5.11** Ready-made hop-up

# 5.2 Boards, staging, working platforms

## 5.2.1 Scaffold boards (Figs. 5.12 and 5.13)

Made from softwood (European whitewood). The timber should be straight-grained, free from large knots and splits. Ends of boards should have the corners cut off and be protected from damage by binding with metal strips.

Boards must not be less than 200 mm wide if less than 50 mm thick. If thicker than 50 mm they may be 150 mm minimum width.

Available in a range of lengths, rarely exceeding 4 m.

***Use*** To provide a working platform on ladder, trestle and tubular scaffolds.

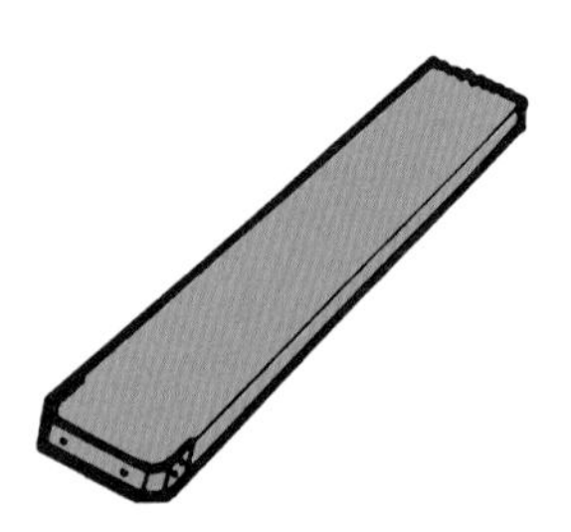

**Fig. 5.12** Scaffold board

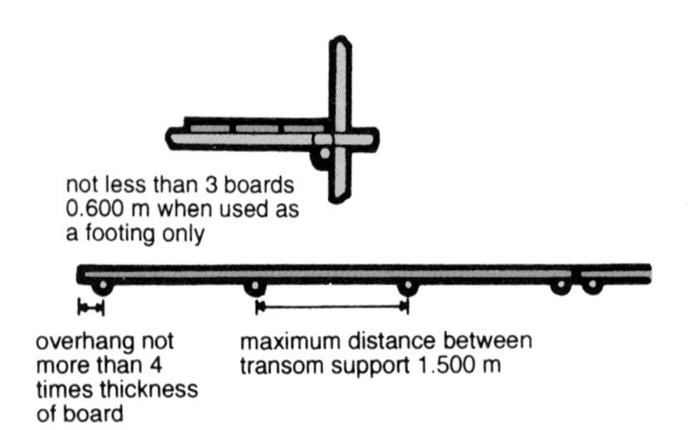


**Fig. 5.13** Scaffold board supports

***Safety precautions***

(i) The maximum distance between supports depends on the thickness of the board: 32 mm can span 1 m, 38 mm can span 1.5 m, 50 mm can span 2.6 m.

(ii) The minimum width of a working platform is 430 mm or two boards when used on trestles. On tubular scaffolds the working platform must not be less than 600 mm or 3 boards.

(iii) Maximum permissible gap between boards is 25 mm.

(iv) Maximum permissible overhang of boards on a working platform is four times the thickness of the board, and not less than 50 mm.

(v) Every board must rest securely and evenly on its support.

***Storage*** Stored under cover. Supported throughout their length on well-ventilated racks and arranged according to size.

## 5.2.2 Lightweight staging (Fig. 5.14)

Lightweight staging can be used with trestles or aluminium towers and gives a non-slip alternative to scaffold boards.

***Sizes*** 450 mm and 600 mm widths available.

Lightweight construction (aluminium stiles and timber deck).

Handrail systems available.

450 mm available in lengths from 3 m to 7.2 m.

600 mm available from 3.6 m to 6 m.

*Note:* Platforms must be a minimum of 600 mm wide. One 600 mm staging or two 450 mm stagings side by side; unless the gap is less than 600 mm wide.

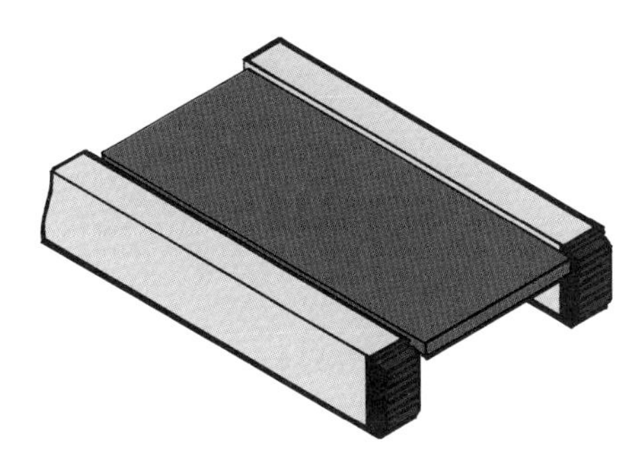

**Fig. 5.14** Lightweight staging

## 5.2.3 Classes of portable plant

All ladders, steps, trestles and stagings have a maximum vertical static load rating and should be marked accordingly along with being colour-coded for ease of identification:

| *Class* | *Maximum static vertical load* | *Use* | *Colour-coded* |
|---|---|---|---|
| 1 | 175 kg | Industrial | Blue |
| EN131 | 150 kg | Commercial | Green |
| 3 | 125 kg | Domestic | Red |

# 5.3 Scaffolding

## 5.3.1 Tubular components

Scaffolding must only be erected, modified or dismantled under the supervision of an experienced and competent person.

### Parts of a scaffold (Figs. 5.15 and 5.16)

***Standard*** Upright tubes which bear the weight of the scaffold.

***Ledger*** Horizontal tubes which extend along the whole length and act as supports for putlogs and transoms.

***Transom*** Span across ledgers to support working platform.

***Guard rail*** A safety rail running not less than 0.950 m and not more than 1.140 m above working platform. A mid-rail leaving no gap greater than 0.470 m.

***Toe board*** Usually a scaffold board on edge, running along the outside of the working platform to prevent objects falling off.

***Raker*** Inclined load-bearing tubes or buttresses to support scaffolds which cannot be otherwise tied to the building.

***Traverse or cross brace*** Triangulates the frame to give greater rigidity to the scaffold.

***Longitudinal or diagonal brace*** Serves the same purpose as cross brace.

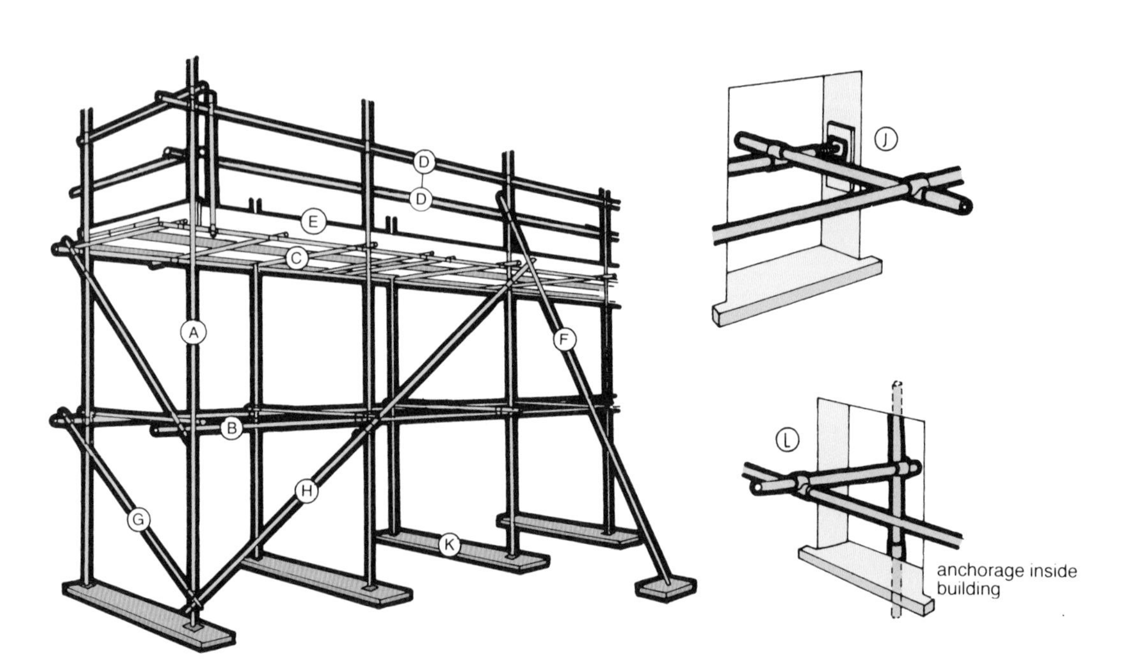


A *standard* B *ledger* C *transom* D *guard rail* E *toe board* F *raker*
G *traverse or cross brace* H *longitudinal or diagonal brace* J *reveal tie* K *sole plate* L *tie*

**Fig. 5.15** Parts of the scaffold

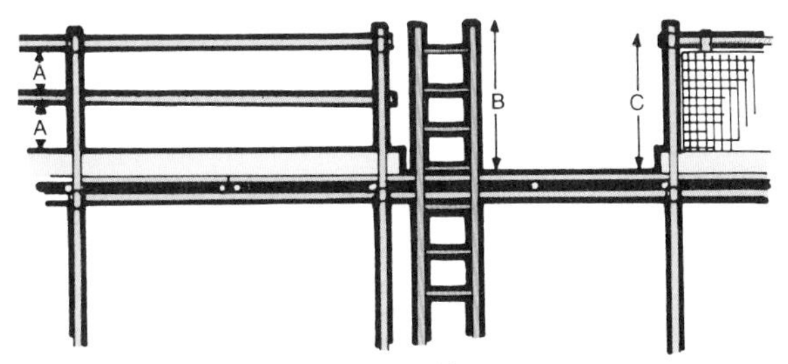


A *not more than 0.470 m if open*
B *1.000 m or over*
C *between 0.950 and 1.140 m*

**Fig. 5.16** Scaffold: safety requirements

**Fig. 5.17** Universal coupler

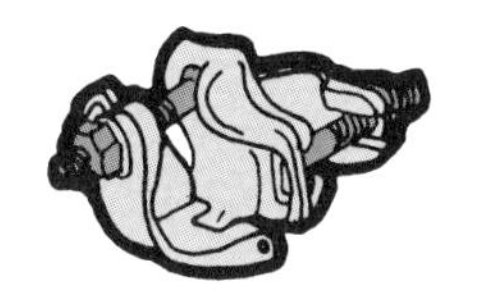

**Fig. 5.18** Double coupler

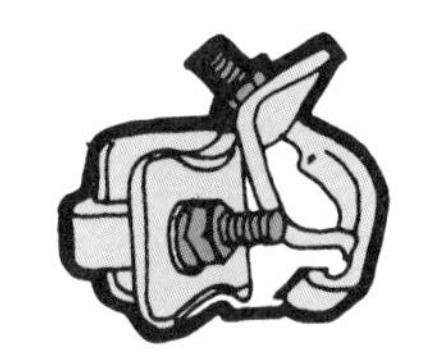

**Fig. 5.19** Swivel coupler

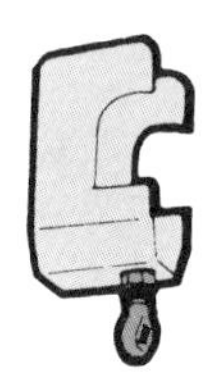

**Fig. 5.20** Putlog coupler

***Reveal tie*** Tubes jacked between opposite sides of an opening to provide an anchor to tie the scaffold to the building.

***Sole plate*** Boards or sleepers forming the main bearing for base plates to give greater distribution of load.

***Tie*** A tube used to connect the scaffold to the reveal tie or other method of tying in.

***Putlog*** Short tube with a flat end. The flat end fits into the brick joint, the open end is coupled to a ledger. It can support a working platform (see 5.3.5).

***Puncheon*** A vertical tube that does not rest on the ground. Used where a standard cannot be taken to the ground (see 5.3.8).

***Bridle*** Horizontal tube fixed between putlogs to span openings in a wall and support transoms (see 5.3.5).

## Tubes

Tubes may be of steel or of aluminium alloy. Steel tubes should be welded or seamless. If not galvanised, they should be painted or varnished to prevent corrosion. Aluminium alloy tubes need not be painted, but care should be taken to avoid contact with harmful materials, i.e. alkalis or salt water.

Steel and aluminium tubes should not be used in the same scaffold as under load they bend to different extents.

Steel and aluminium alloy fittings may be used in the same scaffold.

## Scaffold fittings

***Universal coupler* (Fig. 5.17)** A load-bearing fitting for connecting two scaffold tubes at right angles, e.g. standard to ledger.

***Double coupler* (Fig. 5.18)** A load-bearing fitting for connecting two scaffolding tubes at right angles.

***Swivel coupler* (Fig. 5.19)** A fitting used for connecting two tubes at any angle, e.g. cross brace to standard.

***Putlog coupler* (Fig. 5.20)** For securing (a) putlogs or transoms to ledgers to prevent movement; (b) guard rails to standards. Must not be used for load-bearing.

**Fig. 5.21** Sleeve coupler

**Fig. 5.22** Joint pin

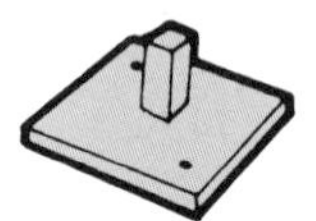

**Fig. 5.23** Base plate

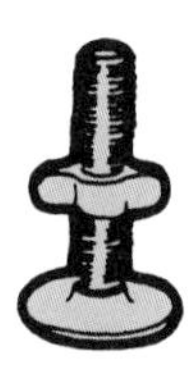

**Fig. 5.24** Adjustable base plate

**Fig. 5.25** Reveal pin

**Fig. 5.26** Putlog end

**Fig. 5.27** Toe board clip

**Fig. 5.28** Finial coupler

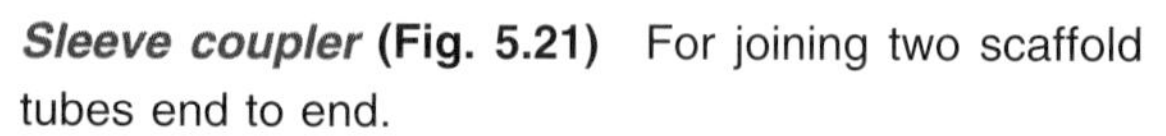

***Sleeve coupler*** **(Fig. 5.21)** For joining two scaffold tubes end to end.

***Joint pin or expanding spigot*** **(Fig. 5.22)** An alternative fitting for connecting two scaffold tubes end to end. Fitted internally, it expands to grip the wall of the tube.

***Base plate*** **(Fig. 5.23)** Placed under standards to distribute the load. It has an integral spigot and fixing holes for use with sole plates.

***Adjustable base plate*** **(Fig. 5.24)** For use on uneven or sloping ground.

***Reveal pin*** **(Fig. 5.25)** A fitting used for tightening a tube between two reveals to form a reveal tie.

***Putlog end*** **(Fig. 5.26)** For converting a scaffold tube into a putlog.

***Toe board clip*** **(Fig. 5.27)** For fixing a toe board securely to a scaffold standard on a working platform.

***Finial coupler*** **(Fig. 5.28)** For fixing a guard rail on top of a standard.

***Safety precautions*** A universal or double coupler is the only type which should be used where a load has to be transferred from one tube to another via a coupler, e.g. for joining a ledger to a standard.

When not in use, fittings should be cleaned and soaked in light oil and stored in bins.

## 5.3.2 System or unit scaffold

**(Figs. 5.29–5.31)**

A patent form of metal scaffold designed to reduce the need for loose fittings, and to speed up and simplify erection. Made of either steel (usually galvanised) or aluminium alloy, the latter being lighter and so more suitable for the painter and decorator.

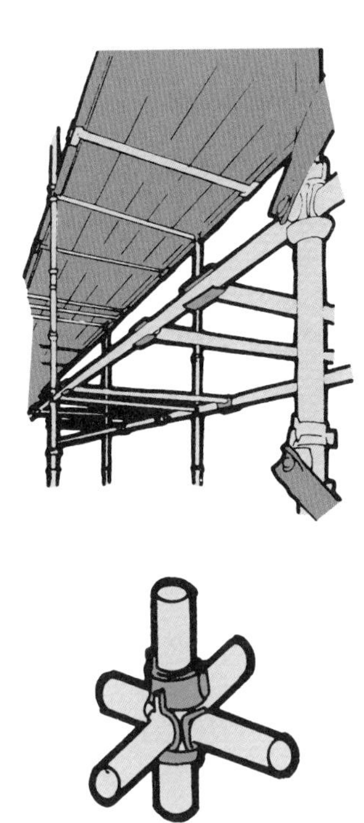

Fig. 5.29 System scaffold. Type (a)

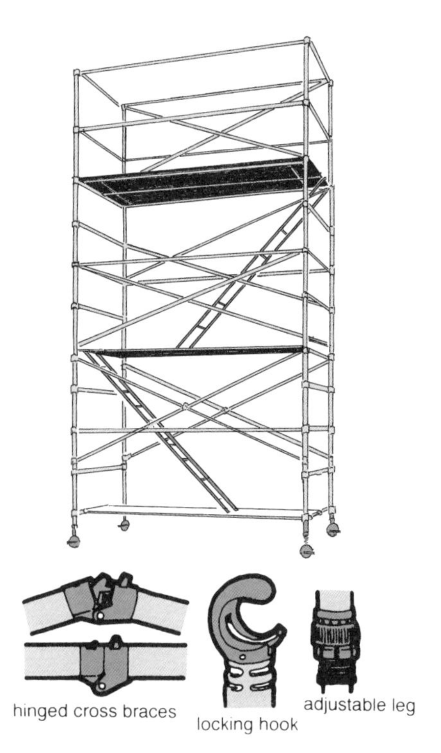


Fig. 5.31 System scaffold. Type (c)

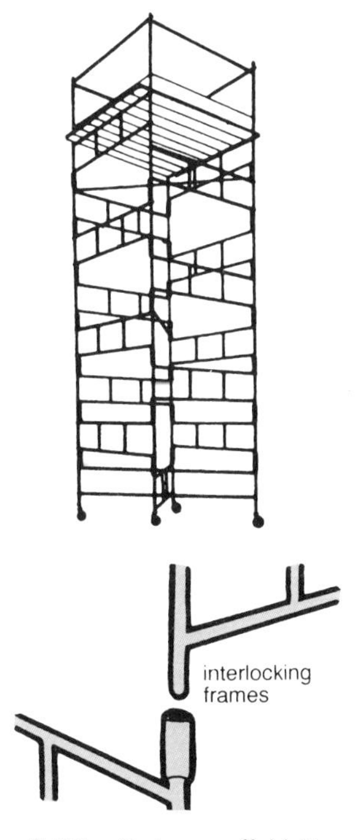


Fig. 5.30 System scaffold. Type (b)

The various systems can be divided into three groups:

(a) Standard scaffold tubing fixed by means of special brackets (Fig. 5.29).
(b) Frames varying in width but of standard depth, which slot into each other at right angles. These systems include guard rails, castors, outriggers and sometimes gantries to give access between units (Fig. 5.30).
(c) A simple tower unit in which all components are built in. The cross braces are hinged so that the tower can be erected by unfolding. It can be folded flat for storage after use. Lightweight platform units lock into the frame, giving extra rigidity to the scaffold. Access ladders are contained within the scaffold and adjustable legs take up changing floor levels (Fig. 5.31).

***Use*** Principally for erecting towers, whether used singly or in pairs to support working platforms.

***Safety precautions*** Safety lies only in carrying out the procedure laid down by the manufacturer at all times.

If instructions are not available, the following rules apply:

(i) Do not exceed manufacturer's maximum height.
(ii) Ensure that the base is firm.
(iii) Uprights must be plumb.
(iv) All joints must be securely locked in place.
(v) Only adequate and effective ties and bracing must be used.
(vi) Wheels must be secured to standards and locked when in use.
(vii) When moved, the tower should be pushed or pulled only from the base, and the working platform should be free of men and material.
(viii) Mobile towers should only be used on firm and level ground.

***Maintenance*** Ensure that sections are not bent or twisted, slots and tube ends are not blocked and welds are not fractured.

***Use*** Internally, where access is required to ceilings, walls, columns or specific areas where detailed work is required, e.g. picking-out.

Externally, where repetitive work is carried out, e.g. to rows of windows, lamp standards or balconies.

## 5.3.3 Independent scaffold (Fig. 5.32)

The scaffold stands clear of the building and is designed to carry the full weight of all load of materials and men placed upon it. It must be tied to the building for stability and to prevent movement towards or away from the building.

When used for painting or similar work, standards may be placed up to 2.700 m apart, with 1 m between the rows. This allows for a four-board-wide platform.

Ledgers are fixed 2.600 m from the ground and then at 2 m intervals. Transoms are placed 2.400 m apart except when supporting a working platform, when the distance should be 1.200 m. The maximum height for this type of scaffold is 61 m.

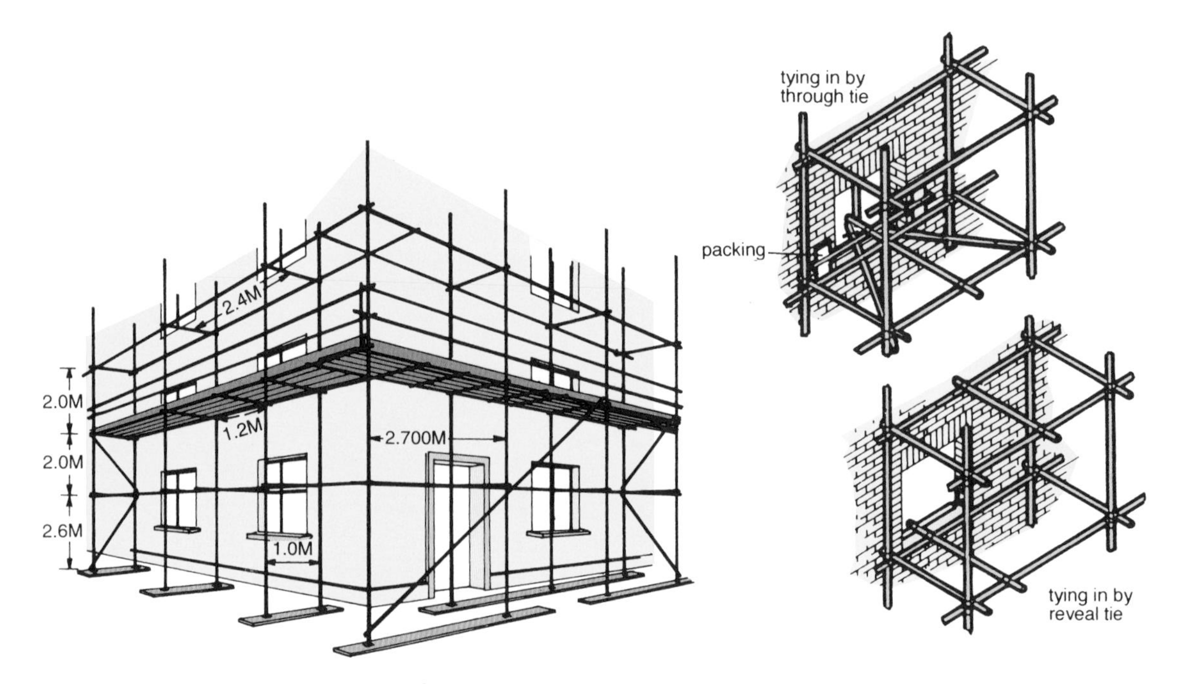


**Fig. 5.32** Independent tied scaffold

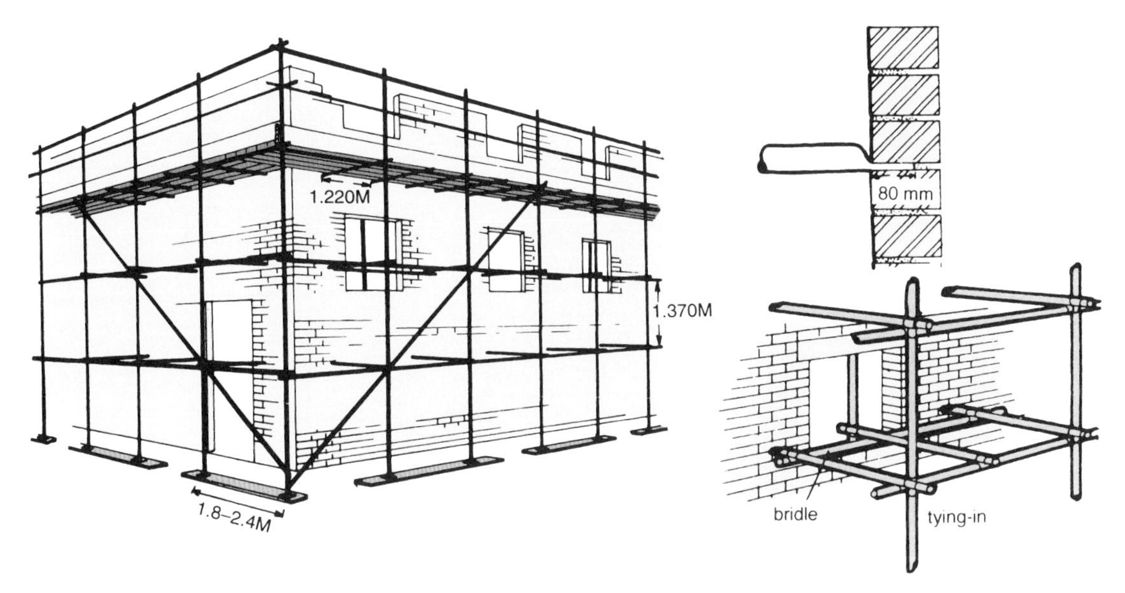


**Fig. 5.33** Putlog scaffold

***Use*** Generally used by most building trades on external surfaces of multi-storey buildings when extensive work is to be carried out.

***Safety precautions***

(i) Standards must be placed on base plates. On soft, uneven or fragile ground, sole plates must be used.
(ii) Standards must be vertical or slightly inclined towards the building.
(iii) It is *essential* that tube joints in standards and ledgers are staggered and NOT kept in line.
(iv) Boards must be evenly supported.
(v) Working platforms must have guard rails and toe boards.
(vi) Safe means of access must be provided throughout the scaffold.

***Maintenance*** Like all tubular scaffolds, they must be inspected every 7 days and immediately after a spell of bad weather.

## 5.3.4 Putlog scaffold (Fig. 5.33)

The putlog scaffold has only one row of standards and depends for stability on supporting itself by means of the wall of the building. It may be extended vertically as the building itself is built. Maximum height 46 m.

The standards are spaced 1.800–2.400 m apart, depending on the type of work. They are placed not less than 1.270 m and not more than 1.320 m away from the wall. The working platform must not exceed 5 boards wide. Ledgers are placed at vertical intervals of 1.370 m. Putlogs are fixed to ledgers at about 1.220 m intervals, and the spade end should extend into the brickwork at least 80 mm.

Ties are very important, especially in new brickwork, and raker struts also increase stability.

***Use*** Mainly where brick structures are being built or repaired. Sometimes called a single or bricklayer's scaffold. Used by painters on new building work.

***Safety precautions*** As independent scaffold.

***Maintenance*** Seven-day check, as for all types of tubular scaffold.

## 5.3.5 Birdcage scaffold (Fig. 5.34)

An interior scaffold that provides a single working platform usually filling the room or hall in which it is being used. Maximum height 61 m.

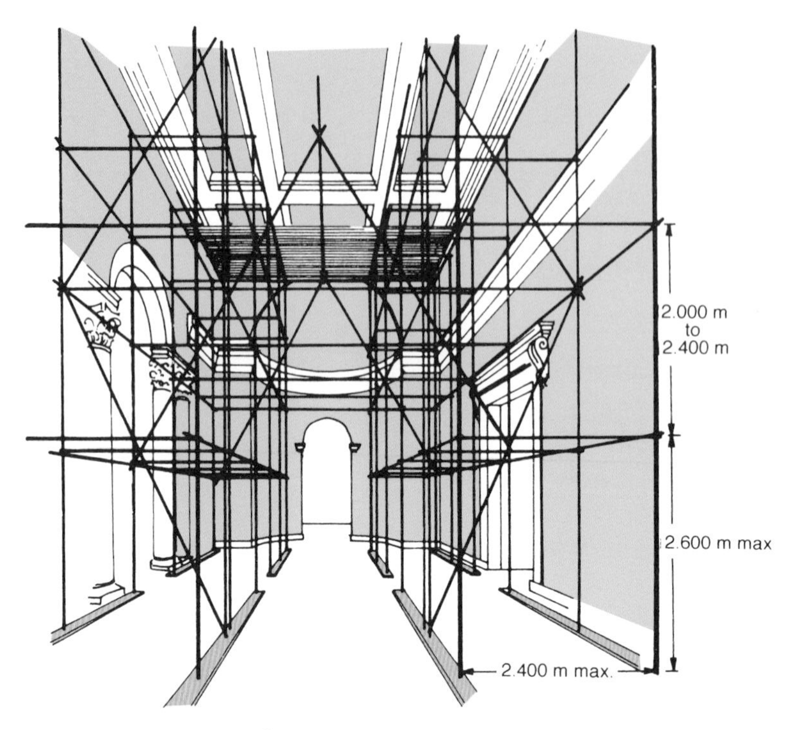


**Fig. 5.34** Birdcage scaffold

The construction is similar to that of an independent scaffold, but rows of standards cover the whole of the work area, spaced at 2.400 m intervals. Sole plates to support base plates are essential to spread the load and protect the floor. If the floor is suspended, the sole plates should run at right angles to the floor joists.

The scaffold should be tied to the surrounding walls, beams or columns. The working platform should be close-boarded and toe boards and guard rail must be provided. Access ladder must be provided.

***Use*** For access to ceilings and walls of large buildings such as cinemas, churches and ballrooms. Because of its high cost, it is used only when the work is of long duration.

***Safety precautions*** As independent scaffold, except that standards must be upright.

***Maintenance*** Seven-day check, as for all types of tubular scaffold.

## 5.3.6 Slung scaffold (Fig. 5.35)

A framework of tubular fittings, close-boarded and suspended from the ceiling of large buildings. It cannot be raised or lowered.

Wire ropes or chains must be used to secure the scaffold to structural members, e.g. rolled steel joists (RSJs). At least six of these suspension points are required for a working platform. The ropes must be equally spaced, vertical and all taut when under load. These ropes are secured to a framework of ledgers and transoms or aluminium beams by means of a round turn and two half hitches with the end tied (see 5.5.2). Ledgers should be at approximately 2.0 m centres and the transoms set on the ledgers at about 1.220 m centres.

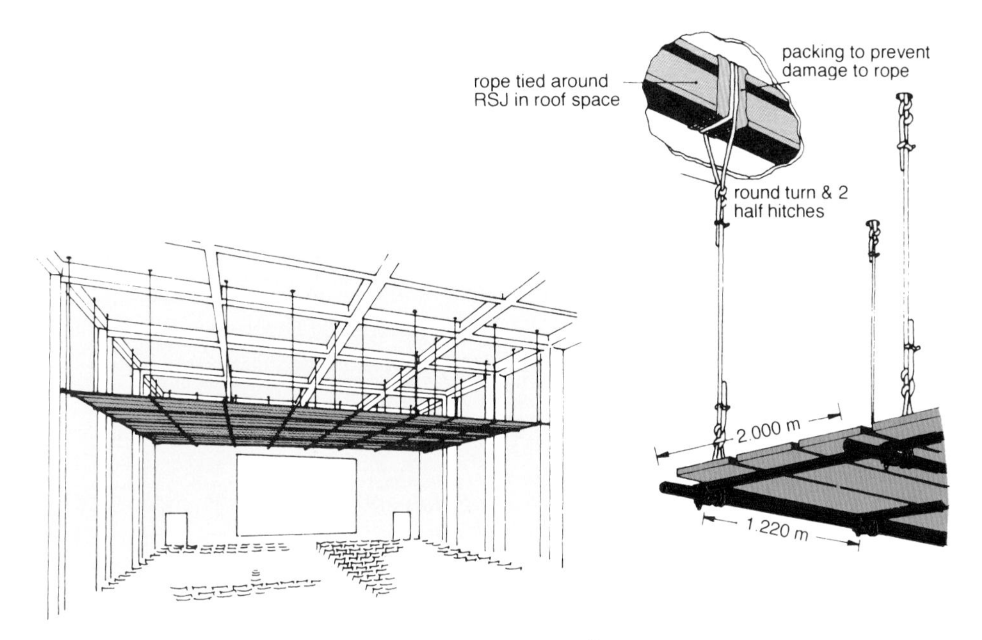


**Fig. 5.35** Slung scaffold

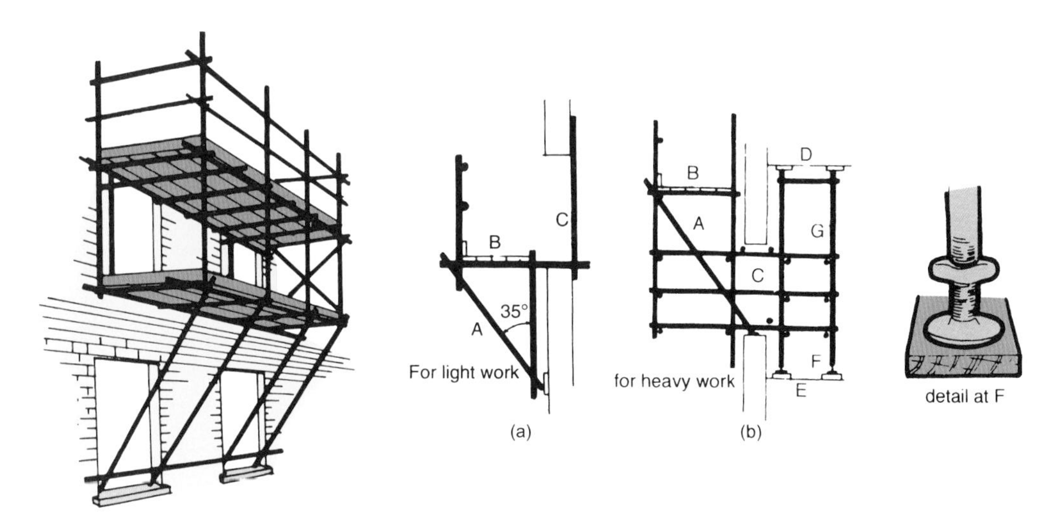


A *raker* B *working platform* C *window opening* D *ceiling* E *floor*
F *adjustable base plate* G *puncheon*

**Fig. 5.36** Truss-out scaffold

The platform must be close-boarded and fitted with toe boards and guard rails. To avoid horizontal movement, guy ropes should be attached from the platform down to convenient points above ground level.

***Use*** To provide a working platform to internal roof and ceiling areas where the floor level is not accessible or not suitable for a built scaffold, e.g. over areas of constant use such as theatres, railway platforms, work areas in factories.

***Safety precautions***

(i) Only wire ropes, chains or tubes must be used for suspension.

(ii) All ropes or chains must be taut.
(iii) The whole platform must be close-boarded.

## 5.3.7 Truss-out scaffold (Fig. 5.36)

A form of independent scaffold which is supported entirely by the building. There are two types:

(a) Horizontal cantilever tubes protrude through windows and are anchored to vertical tubes against the inside walls. The outer ends are supported by rakers bearing on the cills or ledges. Only suitable for very light work.
(b) Anchorage is provided inside the building by two rows of puncheons wedged between the floor and ceiling bearings and against sole plates, which should run at right angles to floor and ceiling joists. Rakers should not exceed 2.700 m in length without being braced. They should be fitted to every standard. Maximum height above the trusses 12 m.

This is a sturdy form of scaffold used for heavier work. Buildings must be in sound condition to take all the stresses and weights involved. All tubes and fittings should be new or as new condition.

***Use*** Where access from ground level is impossible or inadvisable, e.g. small work on the upper parts of tall buildings, or over busy roads, canals or rivers.

***Safety precautions*** As independent scaffold, but only one working platform should be used at a time.

***Maintenance*** Seven-day check, as for all types of tubular scaffold.

# 5.4 Ropes, knots

## 5.4.1 Types of rope

Three main types of rope are used by the painter: natural fibre ropes, man-made fibre ropes and wire ropes.

### Natural fibre rope

Two classes of natural fibre rope are used by the painter: manila and sisal.

***Manila*** Made from fibres of certain species of tree. The fibres are long and strong. Supplied in three grades, which are identified by a blue thread as follows:

*Superior quality* – through centre of all three strands.
*Grade 1 quality* – through centre of two strands.
*Grade 2 quality* – through centre of one strand.

***Sisal*** Made from the leaves of a cactus. Fibres shorter than and not as strong as manila ropes. Supplied in one grade only and identified by a red thread.

*Note:* If a rope has no identification it should be treated as the lowest grade.

***Size*** Diameter 18 mm and 24 mm; length 220 m.

***Use*** (i) Suspended cradles and bosun's chairs.
(ii) Guy ropes and lashing.

***Care and maintenance*** Must be kept dry and away from all acids, gases and cleaning compounds. Wet ropes must be dried naturally as heat can cause rope to become brittle. Coil and hang on wooden pegs in well-ventilated store to dry.

### Man-made rope (nylon, Terylene, polyethylene, polypropylene)

Superior to natural rope in many respects, being stronger, less liable to chemical attack, completely resistant to mildew and rot, more water-resistant. Although non-flammable they may melt at high temperatures. They are 5–6 times more expensive than natural ropes.

***Use*** Principally for safety nets, safety belts and harnesses, and cradle suspension where chemicals are being used.

***Care and maintenance*** Avoid exposure to strong sunlight, heat and chemicals. If contaminated they should be hosed down with clean water and allowed to dry.

### Wire rope

Wire strands twisted round a jute or hemp heart. Three sizes are used by painters:

6 mm for lashings only;
9 mm for lashings and cradles;
12 mm for slung scaffolds.

***Care and maintenance*** Kinked ropes or ropes with broken strands must not be used.

## 5.4.2 Knots

The knots most commonly used by the painter are the following.

***Reef knot* (Fig. 5.37)** For joining two ropes of similar thickness.

***Clove hitch* (Fig. 5.38)** For securing to scaffold tubes, outriggers and ladders.

***Bowline* (Fig. 5.39)** Used on safety lines to produce an eye.

***Round turn and two half hitches* (Fig. 5.40)** Suitable for securing ropes to tubes.

***Figure-of-eight knot* (Fig. 5.41)** A terminal knot. Useful for preventing ropes running out of pulleys and gin wheels.

***Rolling hitch* (Fig. 5.42)** Will withstand sideslip.

**Fig. 5.37** Reef knot

**Fig. 5.38** Clove hitch

**Fig. 5.39** Bowline

**Fig. 5.40** Round turn and two half hitches

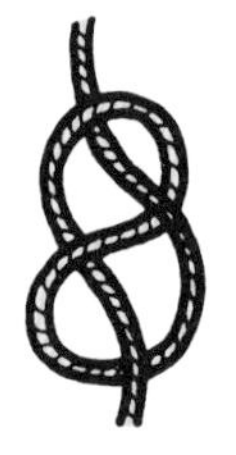

**Fig. 5.41** Figure-of-eight knot

**Fig. 5.42** Rolling hitch

**Fig. 5.43** Sheet bend

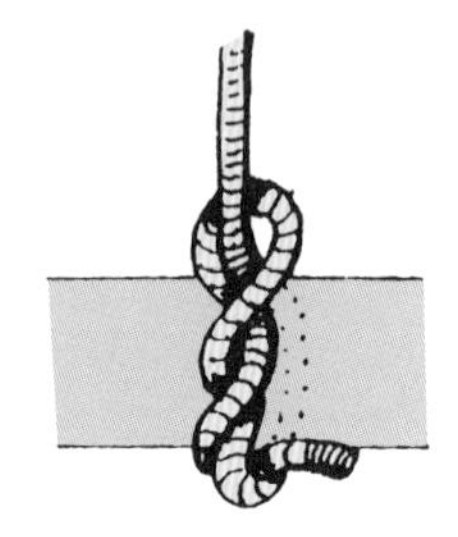

**Fig. 5.44** Timber hitch

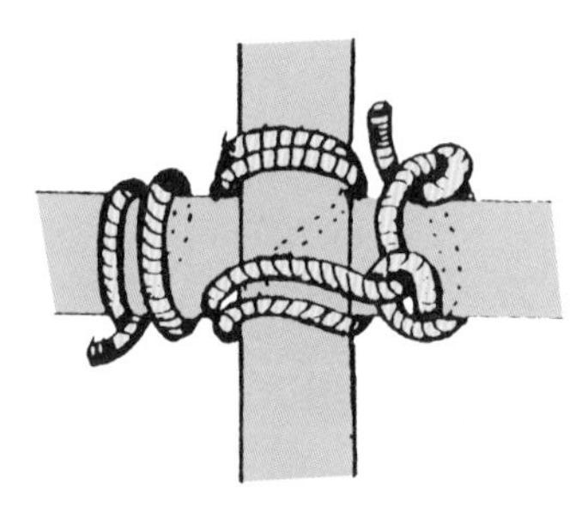

**Fig. 5.45** Square lashing and hitch

***Sheet bend* (Fig. 5.43)** For securing an end of rope to an eye or loop and for joining ropes of unequal thickness.

***Timber hitch* (Fig. 5.44)** Tightens with pressure. Used to raise and lower spars and boards.

***Square lashing and hitch* (Fig. 5.45)** For joining spars or boards at right angles. Used to secure ladder stiles to ledger.

# 5.5 Safety equipment

## 5.5.1 Safety belts, harnesses

**Safety belt (Fig. 5.46)**

A belt or harness strapped to the body and fixed to an anchored safety line.

(a) Pole belt, used for work on poles and similar work.
(b) General-purpose belt used by painters, consisting of a body belt or harness with shoulder straps. Made of man-made fibre and leather. Attached by means of a 'D' hook to a length of nylon rope which is spliced to a safety clip or hook. The length of nylon safety line should not exceed 1.800 m.
(c) Rescue harness.

***Use*** When no other means of safety can be provided and a fall would almost certainly prove to be fatal, e.g. under high bridges, under the roofs of large railway stations, on the edge of high buildings.

***Suspension points*** Suspension points must be strong as they take all the weight and strain. It is sometimes recommended that guidelines of steel rope be used, and the belts attached to them.

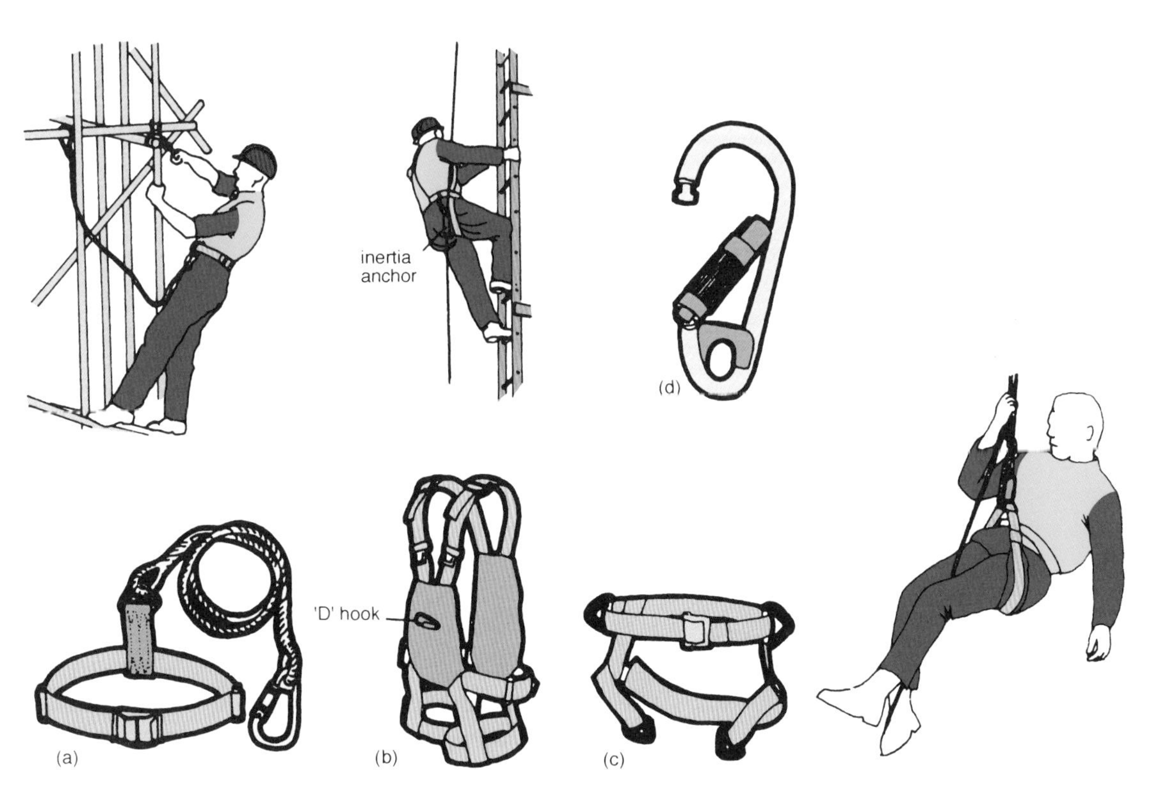


**Fig. 5.46** Safety belts and harnesses: (a) and (b) general purpose belts, (c) safety harness, (d) safety clip

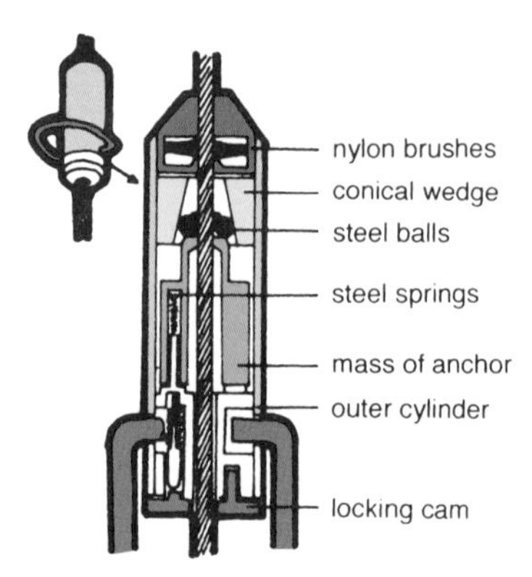


**Fig. 5.47** Inertia-operated anchor

***Inspection and maintenance*** Belts and harnesses should be inspected regularly by a competent person and recorded in a register. Man-made fibres should be kept clean by washing in water. Leather belts should be cleaned and dressed at least every three months.

### Mobile anchor – inertia-operated (Fig. 5.47)

A device attached to a safety line which allows free movement but locks when sudden movement or heavy weight is applied to it. Inside the device, at the top end, are three steel balls held in a cage which is part of the mass floating on steel springs. In this position the walls are free. With a downward fall, the mass remains stationary and the balls are forced by a conical wedge against the cable, causing it to lock on the line.

***Use*** When the regulation-length safety line is too short or restrictive for the type of work.

***Safety precautions***

(i) After a fall, the cable and equipment should be checked for damage.

(ii) The anchor should be attached as close to the user as possible – never more than 600 mm.

## 5.5.2 Safety nets and sheets

### Safety nets (Figs. 5.48 and 5.49)

These are required by The Construction (Design & Management) Regulations 2007 when it is impractical to erect a standard working platform. They are intended to catch (a) falling personnel, and (b) debris, to protect those working below.

*Personnel nets* normally have a mesh size of 100 mm, the mesh being either square or diamond shaped.

*Debris or protection nets* have a much finer mesh, average size 16 mm, and are normally used in conjunction with personnel nets to prevent injury or damage to property from objects falling from above.

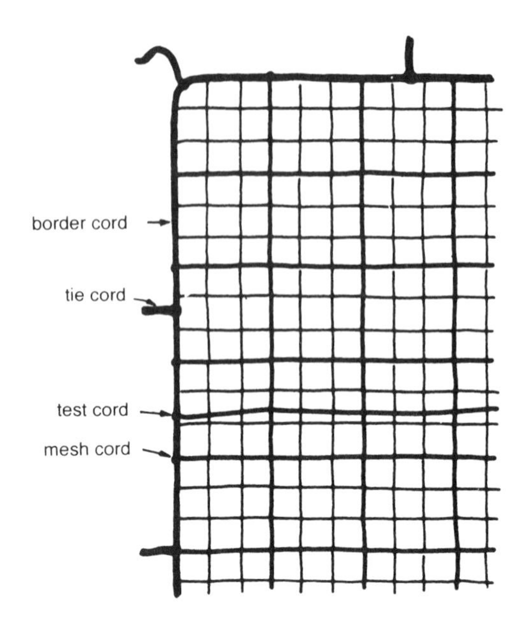


**Fig. 5.48** Safety net

**Fig. 5.49** Safety net in position

Nets are designed by the manufacturer for the specific work to be carried out. Whether made of man-made or natural fibre, they must be rot-proof. Terylene or nylon are the most commonly used materials. They consist of the following parts:

*Mesh cord:* the cord from which the mesh is constructed.

*Border cord:* the cord surrounding the mesh which determines the size of the net.

*Tie cord:* secures the border cord to its supports.

*Test cord:* loosely threaded through the net and tested every 3 months to ensure that the material is not deteriorating.

*Note:* Once a safety net has been deployed, i.e. someone has fallen into it, it should be destroyed.

## Safety sheets

Heavy sheeting fixed under working platforms to protect people and equipment underneath from paint spots and fine debris. Made of sound, rot-proof material such as tarpaulin, edges double-hemmed and double-stitched, all corners double thickness. All support eyelets should be brass-ringed or buttonhole-stitched. Support ropes should be passed through eyelets and spliced.

***Use*** When working (a) over public and pedestrian areas; (b) in cinemas and factories or similar situations where people and equipment must be protected. They can be used alone or in conjunction with safety nets.

# 5.6 Statutory obligations

Before any type of scaffold is erected which may obstruct a public highway, an application for a licence must be made to the Highways Department of the Local Authority concerned.

The proposed site of the scaffold would be inspected by the Authority and, if appropriate, a licence issued subject to certain duties and obligations by the person applying for the licence.

The duties of the person to whom the licence is issued can be enforced by the Authority, which has powers under the Highways Act 1980 to issue fines and penalties.

A public highway includes pavements as well as roads, and these regulations could therefore affect the use of tower scaffolds, trestles and ladders if erected in such a way as to cause any obstruction.

All Local Authorities interpret the Highways Act in a different manner, but a summary of some of the duties are as follows:

(i) To act in accordance with existing statutory regulations related to the work in progress, i.e. Construction (Design & Management) Regulations 2007 or Workplace (Health, Safety & Welfare) Regulations 1992.

(ii) To display the licence in a conspicuous place on the scaffolding.

(iii) To ensure that the scaffolding is adequately lit during the hours of darkness.

(iv) To paint the uprights white between the ground level and 2–3 metres above.

(v) To erect and maintain safety signs warning the public of any potential hazards.

(vi) To protect the public from hazards by the use of suitable barriers.

(vii) To ensure adequate and safe passage around any obstruction.

(viii) To erect and maintain traffic signs if the safe passage includes part of the road.

# PART 6

# Safety, Health and Welfare

# 6.1 Health and safety at work

## 6.1.1 Health and Safety at Work Act

Since the 1920s, Government statutes (laws) have existed which stated safe working conditions for building workers in general and, in some cases, painters and decorators specifically.

Because these statutes tend to take many years to pass through Parliament, and because technology in the industry is developing so rapidly, the specific regulations which have been introduced are often out of date or redundant in a short time. This situation prompted the introduction in 1974 of the Health and Safety at Work Act (HSWA) which made it illegal for any worksite to be unsafe, and for any material or equipment to have unknown hazards. The Act does not identify any specific site conditions, equipment or materials but directs the onus for site working conditions onto:

(i) the employer;
(ii) the employee;
(iii) the manufacturer.

Briefly, the responsibilities for each are:

***The employer***

(i) Provide safe workplaces, machines and methods of work.
(ii) Ensure that machines and materials are used, stored and transported safely.
(iii) Provide information, training and supervision to ensure the health and safety of the workers.
(iv) Provide safe entrances and exits to workplaces.
(v) Provide good, well-maintained welfare facilities.
(vi) Publish a written safety policy.
(vii) Ensure that members of the public are not exposed to risks.

***The employee***

(i) Take care of own and workmates' health and safety.
(ii) Follow the health-and-safety regulations which apply to the workplace.
(iii) Cooperate with the employer in making areas safe.
(iv) Not misuse anything provided to make work safe.

***The manufacturer***

(i) Ensure that products are without risks to health.
(ii) Ensure that products have been fully tested.
(iii) Provide adequate information on the safe use of the products.

HSWA is enforced by the Health and Safety Executive (HSE). They provide Health and Safety Inspectors (HSI) who have the power to enter workplaces and either advise on conditions or practices and issue Improvement Notices, or to condemn hazardous conditions and practices by issuing Prohibition Notices. HSIs can also prosecute offenders, whether they be employer, employee or manufacturer.

The following sections identify specific situations which may prove hazardous to the painter and decorator. Although specific statute regulations are quoted, the HSWA encompasses them all, and applies equally to those which are not quoted.

## 6.1.2 Health-and-safety regulations

As a result of the Health and Safety at Work Act 1974, statutory regulations are being continually updated and developed to identify the many sources of hazards in the construction industry and to give safety directions and advice to employers, employees and manufacturers/suppliers. A list of the most recent and/or important regulations for the construction worker is as follows:

- Control of Asbestos at Work Regulations 2002
- Construction, Design and Management Regulations 2007
- Construction (Head Protection) Regulations 1989
- Control of Lead at Work Regulations 2002
- Control of Substances Hazardous to Health Regulations 2004
- Electricity at Work Regulations 1989
- Health and Safety (First Aid) Regulations 1981
- Dangerous Substances & Explosive Atmospheres Regulations 2002
- Management of Health and Safety at Work Regulations 1999
- Manual Handling Operations Regulations 1992
- Noise at Work Regulations 2005
- Personal Protective Equipment at Work Regulations 1992
- Provision and Use of Work Equipment Regulations 1998
- Workplace (Health, Safety and Welfare) Regulations 1992

To follow is a brief, introductory guide to some of the above regulations.

*Note:* It must be stressed that the following is only a brief, introductory guide to the regulations, and fuller details should be obtained from the actual regulations, approved Codes of Practice or guidance notes issued by the Health and Safety Executive.

Many of these new regulations may overlap with existing requirements where they are still in force. Compliance with the new regulations should normally be sufficient. However, where other regulations are still in force they should be considered together with the new.

## Construction, Design and Management Regulations 2007

The Construction & Design Management (CDM) Regulations apply to any construction work, which is notifiable to the Health and Safety Executive. That is, projects for non-domestic clients lasting longer than 30 days or involving more than 5 workers. This triggers the appointment of the 'Principal Contractor' and the 'CDM coordinator'. The Health and Safety Plan and Health and Safety File compiled by the CDM coordinator in the early stages and further down the line by the principal contractor, help manage H&S from start to finish. They provide a combination of instructions and also provide a record for the future, so that anyone coming onto the site later or returning for maintenance work knows what H&S issues were considered and how they were dealt with.

***The Health and Safety Plan***

The Health and Safety Plan provides the health and safety focus for the construction phase of a project.

This is compiled by the CDM coordinator with principal contractor involvement later on and should include:

(i) A general description of the work.
(ii) Details of timings within the project.
(iii) Details of risks to workers.
(iv) Information required by principal contractor to demonstrate competence or adequacy of resources.
(v) Arrangements for the health and safety of all who may be affected by the construction work.
(vi) Arrangements for the management of health and safety of construction work and monitoring of compliance with health and safety law.
(vii) Information about welfare arrangements.

## The Health and Safety File

This is a record of information for the client/end user, which tells those who might be responsible for the structure in the future, the risks that have to be managed during maintenance, repair or renovation.

The planning supervisor has to ensure that the file is prepared as the project progresses, and that it is given to the client/end user when the project is complete. The client has to make it available to those who will work on any future design, building, maintenance or demolition of the structure.

***The client***

The client should be satisfied that only competent people are appointed as CDM coordinator, principal contractor, designers and contractors. Duties on clients do not apply to domestic householders when they have construction work carried out.

***The designer***

The designer should ensure that structures are designed to avoid, or where this is not possible, to minimise, risk to health and safety while they are being built and maintained. Design includes the preparation of specifications as well as drawings.

***The CDM coordinator***

The CDM coordinator has overall responsibility for co-ordinating the health and safety aspects of the design and planning phase and for the early stages of the health and safety plan and safety file which need to be developed.

***The principal contractor***

The principal contractor should take account of health and safety issues when preparing tenders. They are also responsible for developing the health and safety plan and coordinating all activities to ensure they comply with health and safety legislation.

***Contractors and the self-employed***

All should cooperate with the principal contractor and provide information on the health and safety risks created by their work and how they will be controlled.

## Construction (Head Protection) Regulations 1989

These regulations require the wearing of suitable head protection on all building operations and construction sites, unless there is no risk of head injury other than to the person falling. Sikhs wearing turbans are exempt from the regulations.

*The employer must:*

(i) Assess the suitability of any head protection that is to be used.
(ii) Bring the rules to the attention of all concerned in writing.
(iii) Provide safety helmets that are suitable for the job.
(iv) Make sure that head protection is worn when instructed.
(v) Replace equipment if damaged, exposed to chemicals, or at intervals recommended by the manufacturer.

*The employee must:*

(i) Wear the safety helmets provided.
(ii) Take care of the equipment.
(iii) Report damage or loss of the helmet.

## Control of Substances Hazardous to Health Regulations 2004

These regulations, sometimes referred to as the COSHH Regulations, are aimed at the protection of employees and others from the effects of working with substances which are hazardous to health. COSHH does not apply to work with asbestos, lead or radioactive materials which are covered by their own specific legislation.

*The employer must:*

(i) Know what substances are being used.
(ii) Assess the hazards to health that the substances can cause.
(iii) Eliminate or control the hazards identified.
(iv) Give information, instructions and training to employees.
(v) Monitor the effectiveness of any controls.
(vi) Keep records.

*The employee must:*

(i) Make use of any control measures and equipment provided.
(ii) Comply with any arrangements the employer has put into effect.

## Electricity at Work Regulations 1989

These regulations require precautions to be taken against the risk of death or personal injury from electricity and create duties for employers, the self-employed and employees in all aspects of electrical work.

The regulations require that no one shall do any work where technical knowledge or experience is necessary to prevent danger, or injury, unless they have the knowledge or experience necessary, or are under proper supervision as may be appropriate, having regard to the nature of the work.

Having appropriate technical knowledge means that the person must be competent. This means having:

(i) Adequate knowledge of electricity.
(ii) Good experience of electrical work.

(iii) Practical experience of the system.
(iv) Knowledge of hazards and precautions that need to be taken.
(v) Ability to recognise unsafe situations.

Information, instructions and training must be given. Written safe systems of work should form part of a 'permit to work' procedure.

### Health and Safety (First Aid) Regulations 1981

These require that employers and self-employed persons provide adequate and appropriate first-aid facilities which should be based on an assessment of the need for first-aid depending on the nature of the work and not just on the number of people on site.

*The employer must:*

(i) Appoint a sufficient number of suitable and trained personnel to give first-aid to employees.
(ii) Appoint other persons who, in the absence of the trained first-aider, will be capable of taking charge in an emergency and looking after first-aid equipment.
(iii) Provide and maintain suitable first-aid boxes in places that can be easily reached by employees.
(iv) Display notices giving the identity of first-aiders and location of the first-aid equipment.
(v) Keep an accident-report book which details all accidents.
(vi) Investigate the circumstances of all accidents.
(vii) Report notifiable events and diseases to the Health and Safety Executive.

### Management of Health and Safety at Work Regulations 1999

These require that in any company, large or small, suitable assessment of the risks to health and safety of employees and others is carried out. The regulations apply to companies with as few as five or more employees. If these regulations overlap with any existing regulations, to adopt the more specific regulations will normally be sufficient to comply with the Management of Health and Safety at Work Regulations. For example, the COSHH require that risk assessment takes place.

The regulations are about:

- Identifying the roles and responsibilities of all involved in the management of health and safety.
- Setting up safety standards to be achieved.
- Creation of procedures to be followed and ensuring that they are carried out.

*The employer must:*

(i) Assess the risks to the health and safety of employees to which they are exposed whilst at work.
(ii) Make suitable arrangements for the effective planning, organisation, control, monitoring and review of preventative and protective measures.
(iii) Provide health surveillance for risks identified in the assessment.
(iv) Appoint suitably competent persons to undertake any measures taken for health and safety.
(v) Instigate procedures for serious and imminent danger areas.
(vi) Provide information to employees.

*The employee must:*

(i) Use all work items correctly.
(ii) Cooperate with the employer to comply with the regulations.
(iii) Notify any shortcomings in the health and safety arrangements and take remedial action as may be needed.

### Manual Handling Operations Regulations 1992

These regulations deal with safe manual handling. Where the assessment required by the Management of Health and Safety at Work Regulations (see above) indicates risks to employees from manual handling of loads, the employer must take steps to:

- Avoid manual handling as far as is reasonably practicable.
- Assess any manual-handling operations that cannot be avoided.
- Reduce the risk of injury as far as is reasonably practicable.

*The employer must:*

(i) Identify all manual-handling tasks which might involve risk of injury.
(ii) Avoid such tasks where reasonably practicable.
(iii) Where not reasonably practicable, the employer must:
 (a) Carry out a suitable and sufficient assessment of each manual-handling task.
 (b) Determine and implement measures to reduce risk of injury to lowest reasonably practicable level.
(iv) Provide employees with necessary training and information concerning weight of loads, etc.
(v) Review assessments when no longer valid or changes in handling operations have happened.

*The employee must:*

(i) Use appropriate equipment in accordance with training and instructions.
(ii) Follow systems of work laid down by the employer.

### Noise at Work Regulations 2005

These regulations set noise-exposure levels and place duties on both employer and employee.

*The employer must:*

(i) Reduce noise to the lowest reasonably practicable level.
(ii) Have a noise assessment carried out by a competent person when exposure levels require it.
(iii) The level at which employers must provide hearing protection and hearing-protection zones is now 85 decibels (daily or weekly average exposure) and the level at which employers must assess the risk to workers' health and provide them with information and training is now 80 decibels.
(iv) Ensure proper use and maintenance of equipment when it is required.

*The employee must:*

(i) Properly use any ear protectors when exposure is at, or above, 85 decibels.
(ii) At all times, use any protective equipment provided and report any defects to the employer.

### Personal Protective Equipment (PPE) at Work Regulations 2002

These deal with the use of personal protective equipment in the workplace which includes protective clothing such as aprons, gloves, footwear, helmets, etc., as well as protective equipment such as eye protectors, respirators and safety harnesses. Ensure any PPE is 'CE' marked and complies with the requirements of the Personal Protective Equipment Regulations 2002. The CE marking signifies that the PPE satisfies certain basic safety requirements and in some cases will have been tested and certified by an independent body.

The regulations cover:

- Head protection.
- Eye protection.
- Foot protection.
- Hand and arm protection.
- Protective clothing for the body.

*The employer must:*

(i) Ensure suitable personal protective equipment is provided to employees.
(ii) Ensure it is appropriate to the risks involved.
(iii) Take account of the requirements and state of health of the employee.
(iv) Ensure it is capable of fitting the wearer correctly.
(v) Provide proper maintenance and repair facilities and training.
(vi) Ensure that the personal protective equipment is effective as far as is reasonably practicable.
(vii) Ensure that the personal protective equipment complies with other statutes, regulations and standards.

*The employee must:*

(i) Wear the equipment when provided.
(ii) Take care of the equipment.
(iii) Report damage or loss.

### Provision and Use of Work Equipment Regulations 1998 (PUWER)

These regulations, known as the PUWER regulations, set the standard for the provision and use of work equipment. The primary objective is to provide safe equipment and to ensure its safe use so as not to risk the health and safety of the user and others.

Generally, any equipment, which is used by employees at work, is covered; similarly, if you allow employees to provide their own equipment, it is also covered by PUWER and you will need to make sure it complies. They also cover equipment such as:

- Chain saws
- Abrasive wheels
- Disc cutting tools
- Cartridge operating tools
- Woodworking machinery
- Ladders

*The employer must:*

(i) Ensure that equipment is constructed or adapted for the purpose for which it is used or provided.
(ii) Consider working conditions and risks to health and safety to persons where the equipment is to be used.
(iii) Ensure that the equipment is used only for operations, and is under conditions, where it is suitable.
(iv) Provide information, instructions and training.

### Workplace (Health, Safety and Welfare) Regulations 1992

These regulations apply to all workplaces, for example to shops, offices, schools, hospitals, hotels and places of entertainment. Although they do not cover construction sites, they do cover temporary worksites where much of the maintenance and refurbishment parts of construction activity takes place.

The regulations are extensive and include requirements for:

- Maintenance of workplace and equipment.
- Ventilation.
- Temperature in indoor workplaces.
- Lighting.
- Cleanliness and waste material.
- Room dimensions and space.
- Workstations and seating.
- Condition of floors and traffic routes.
- Fall or falling objects.
- Windows, skylights and ventilators.
- Sanitary conveniences.
- Washing facilities.
- Drinking water.
- Accommodation for clothing.
- Facilities for changing clothes.
- Facilities for rest and eating meals.

# 6.2 Fire precautions

Painters use many flammable materials and flame- or spark-producing apparatus. They therefore need to take precautions in the use and storage of them to prevent fire. It is essential to read manufacturers' instructions before using any material which may constitute a fire risk.

## 6.2.1 Common causes of fire

### Highly flammable liquids (low flash point)

Usually petroleum spirit or mixtures of cellulose solution having a flash point below 32°C. Most of the following have a flash point much lower than this:

petrol
cellulose thinner
cellulose paint
certain adhesives
white spirit
spirit varnish
polyurethane thinner
methylated spirit
chlorinated rubber thinner

The storage and use of these materials are controlled by The Dangerous Substances and Explosive Atmospheres Regulations 2002 (see Table 6.2).

***Safety precautions***

(i) Do not handle near a naked flame, unprotected electrical equipment or source of spark.
(ii) Take from store only the amount required for immediate use.
(iii) Do not store in open or plastic containers.
(iv) Replace screw tops as quickly as possible.
(v) Return empty containers to store.

### Other flammable liquids

Although not as volatile as petroleum spirits, these materials may produce an explosive vapour:

paraffin (kerosene)
oil paint
creosote
bitumen paint

***Safety precautions*** The use and storage of these materials are not governed by such strict statutory regulations as apply to highly flammable liquids, but if the same rules are observed, there will be less fire risk.

### Gas stored in cylinders

(a) Butane, propane, acetylene
(b) Oxygen

The storage and use of these gases are controlled by The Dangerous Substances and Explosive Atmospheres Regulations 2002.

***Safety precautions***

(a) (i) Avoid sparks or naked lights, which are dangerous where 'free' gases are present.
(ii) Do not use a hose which leaks or which can be easily creased.
(iii) Ensure that connections to hoses, cylinder and burners are correct, tight-fitting and not cross-threaded.
(iv) Keep flame away from cylinder and hoses.
(v) Keep cylinder upright and do not knock violently.
(vi) If acetylene cylinder becomes heated accidentally, shut valve, detach fittings and take into open air. Immerse in or apply water to cool; open valve till empty.
(vii) Always ensure that acetylene-regulator controls are turned off. Never open valves with the regulator on.

(b) The above rules also apply to oxygen. In addition, special precautions are necessary because, while oxygen does not burn, it supports and accelerates combustion. Oil and oil-impregnated clothing can be ignited by a spark and will burn fiercely in the presence of a concentrated supply of oxygen.
(i) Keep fittings free from grease and oil.
(ii) Do not use with greasy hands or clothing.

(iii) Keep oxygen cylinders away from other cylinders.

***Safe storage***

(i) When not in use, store in fire-resisting building or in a compound away from fire risk.
(ii) Store empty and full cylinders separately.
(iii) Mark empty cylinders clearly.
(iv) Close all valves tightly.

### Flame-producing apparatus

Blowlamps, gas torches and flame-cleaning torches may be dangerous unless used with great care in the vicinity of combustible materials.

***Safety precautions***

(i) Remove all litter and rubbish from area.
(ii) Protect combustible materials (e.g. timber floors) with metal or fireproof sheeting.
(iii) Never leave lamps or torches burning when not in use.
(iv) Inspect area carefully after use to make sure that no smouldering material remains.
(v) Remove dark goggles regularly to inspect area to make sure that no surface is smouldering. (See also 1.1.4.)

### Spark-producing apparatus

When working in an atmosphere which may be laden with flammable vapour or gas (e.g. when spraying oil or cellulose paint, in spray booths, in or near empty or full spirit/gas storage tanks), great care must be taken to avoid causing a spark which could ignite it.

***Safety precautions***

(i) Use spark-proof electric motors.
(ii) Use phosphor-bronze or berylium copper tools, as steel may cause a spark when scraped on metal or stone.
(iii) Avoid clothing of man-made fibre, which can cause a spark due to static electricity.
(iv) Hoses for airless spray and abrasive cleaning must be earthed to avoid static electricity.

### Spontaneous combustion

Some materials used by the painter do not require a spark or flame for ignition. Oil-soaked rag, if screwed up and left in a heap or a bin, and paint-soaked foam rubber, if placed in an enclosed bin, may smoulder and burn.

## 6.2.2 Fire-fighting (see Table 6.1)

A fire requires:

(a) *fuel* – combustible material which may be solid (e.g. wood), liquid (e.g. oil) or gas (e.g. petroleum vapour);
(b) *oxygen* – the flame requires oxygen to maintain it;
(c) *heat* – most solids and liquids require to be heated before they emit flammable vapours.

If any *one* of these factors is removed, burning will cease. The three types of fire most commonly met by the painter are fought in different ways:

(a) Solid fuel fires (wood, paper, cloth or rubbish): controlled by cooling with water.
(b) Liquid or gas fires (paint, oil or solvents): controlled by blanketing, using foam, powder or gas, to cut off the oxygen supply.
(c) Fires in electrical equipment (motors, wiring or switches): controlled by blanketing using non-conducting materials.

## 6.2.3 Treatment for burning clothing

(i) Lay the person down to keep flames away from face.
(ii) Quench the flames with water or other non-flammable liquid.
(iii) Wrap the person tightly in a rug, blanket or coat until the flames are smothered (nylon or other synthetic material must not be used).

*If alone*, roll on the floor and smother flames with the nearest available wrap; do not run about. (See 6.8.2.)

On all burns – do not use lotions, ointments and creams. Do not use adhesive dressings. Do not break blisters.

**Table 6.1** Fire extinguishers

| *Type* | *Average range* | *Colour of banding* | *Application* | *Prohibited use* |
|---|---|---|---|---|
| Water | 4 m | Red | Wood<br>Paper<br>Rags<br>Blowlamps | Electrical fires<br>Oil<br>Solvents |
| AFFF Foam (Aqueous Film Forming Foam) | 4 m | Cream | Liquids<br>Oil<br>Paint | Electrical fires |
| Use of 'Halon' now banned (since December 2003), under European Ozone Depleting Substances Regulations (ODS) | | | | |
| Vapour-forming gas (CO2) | 1 m | Black | Electrical fires<br>Highly flammable liquids | Material containing own oxygen supply, e.g. cellulose |
| Dry powder | 2 m | Blue | Highly flammable liquids<br>Electrical fires<br>Petrol or diesel engines | |
| Sand | | | Small isolated fires | |
| Glass-fibre blanket | | | Burning clothes<br>Small isolated fires | |
| Aerosol (many types available) | Short | | Small fires | |

As a result of merging European standards the system of colour coding extinguishers has changed. All new extinguishers are coloured red. The contents are indicated by a coloured label or band around the shell of the extinguisher.

# 6.3 Storage of materials

Any materials and tools kept in stock represent a large amount of money. Careless storage can prove to be expensive and, in some instances, dangerous.

***Temperature*** Wherever practical, the temperature of a storeroom should be kept constant. The ideal general temperature is 15°C, but because of the risk of evaporation and the resultant dangerous vapour, solvents should be stored at a lower temperature.

***Dampness*** Many materials and brushes deteriorate when subjected to damp. A heated storeroom as described above rarely offers a degree of dampness that would affect materials.

***Lighting*** Adequate natural and artificial lighting is essential for easy identification of stock and to avoid hazards. Care must be taken to ensure that all electrical wiring and fittings used for lighting are in good condition, otherwise they may become fire risks.

***Fire precautions*** Very few painters' materials do not present a fire risk, therefore every precaution must be taken to keep this to a minimum by:

(i) avoiding naked lights;
(ii) banning smoking;
(iii) ensuring that correct types of fire extinguisher are available and in good condition.

For storage of materials that come within the scope of The Dangerous Substances and Explosive Atmospheres Regulations 2002, check with the regulations.

**Table 6.2** Storage of materials

| ***Material*** | ***Type of storage*** | ***Special considerations*** |
|---|---|---|
| Oil paint<br>Alkyd paint<br>Polyurethane paint<br>Oil varnish<br>Polyurethane varnish<br>Scumble<br>Oil filler<br>Spirit varnish<br>Putty<br>Bitumen | On shelves<br>Clearly marked<br>Arranged with new stock at back to avoid deterioration of old stock | Keep lids tightly closed to avoid evaporation and skinning<br>Even temperature will ensure a reliable consistency<br>Keep heavy containers at low levels to avoid difficult lifting<br>Tins of heavily pigmented materials should be inverted at regular intervals to prevent settlement |

**Table 6.2** Storage of materials (cont'd)

| *Material* | *Type of storage* | *Special considerations* |
|---|---|---|
| Emulsion paint<br>Emulsion varnish<br>Acrylic paint<br>Prepared paste<br>Multi-colour paint | On shelves<br>Clearly marked<br>New stock placed behind existing stock | Protect from frost<br>Many water-carried paints have limited shelf life – it is essential that all stock is used within the time limit |
| Whiting<br>Dry pigment<br>Plaster<br>Size<br>Paste powder<br>Powder filler | Small items on shelves<br>Larger items on platforms at ground level<br>Loose materials in closed bins | Protect from moisture<br>Plasters have a limited shelf life and may 'air set' if humidity or condensation is high |
| Spirit paint remover | On shelves | Temperatures in excess of 15°C may cause spirits to expand and blow out containers<br>Avoid naked flames |
| Wallpaper<br>Lining paper | In racks with ends protected<br>Stacked by pattern and clearly marked | Keep wrapped to protect from dust and direct sunlight which may cause fading |
| Abrasive paper | In packets or sleeves for easy identification and to keep sheets flat | Avoid excessive heat which causes brittleness<br>Avoid dampness which weakens glasspapers and garnet papers |
| Glass | In vertical racks | Requires dry storage to prevent sheets sticking together<br>Dirty storage conditions can discolour glass (see 9.2.2) |
| Dust sheets | Folded<br>On shelves | Keep clean and dry to avoid mildew |
| Brushes | Suspended or laid flat in cupboards<br>New brushes should be kept wrapped | Protect against moth attack by using insecticide<br>Keep dry to avoid mildew |
| Rollers | Suspend in cupboards | Lambswool and mohair should be treated as brushes |
| Steel tools<br>Sprayguns<br>Paraffin<br>Creosote | Suspended or laid flat in cupboards<br>(a) In large drums on trestles, tapped and locked<br>(b) In 5- or 20-litre screw-topped cans at low level | Protect from corrosion by oiling or using anti-rust paper wrapping<br>Keep tightly stoppered<br>Store outside in enclosed area, or in a building separate from main building (see also 6.1) |

**Table 6.2** Storage of materials (cont'd)

| *Material* | *Type of storage* | *Special considerations* |
|---|---|---|
| Liquefied gas<br>Compressed gas<br>Petrol<br>Cellulose paint<br>Cellulose thinner<br>Chlorinated rubber paint and thinner<br>Methylated spirit<br>Polyurethane thinner<br>Adhesives containing petroleum spirit<br>White spirit | (a) In the open, protected from ice, snow and direct sunlight<br>(b) In a special store, constructed as follows:<br>With sloping floor, so that spilt solvent does not remain under its container<br>Of brick, blocks, concrete or other fireproof material<br>With roof of easily shattered material (asbestos) to minimise effect of explosion<br>With doors 50 mm thick, opening outwards<br>With all glass wired, and not less than 6 mm thick<br>With concrete floors<br>With storage area surrounded by sill deep enough to contain contents of largest container<br>Lit with flameproof lights switched from outside<br>Unheated<br>Naturally ventilated at both high and low levels<br>Prohibited to naked lights or spark-production objects<br>Equipped with two or more exits<br>Prominently marked with red and white squares and a sign stating: 'Highly flammable' | Gas stored in cylinders controlled by The Dangerous Substances and Explosive Atmospheres Regulations 2002<br>Note: These regulations apply only when more than 50 litres of material are stored<br>Permission for storage must be obtained from the District Inspector of Factories |

# 6.4 Lead paint

## 6.4.1 Control of Lead at Work Regulations 2002 ('CLAW')

These regulations supersede the previous lead regulations, dated 1998. Their aim is to protect workers and others likely to be affected by their work, such as building occupants, who may breathe in or swallow any lead-containing material (LCM). Lead is toxic. Inside the body it harms the brain and nervous system and causes learning and behaviour problems in young children and reproductive problems in adults. It speeds up the onset of age-related diseases and is linked to raised blood pressure and cancer. The body cannot tell the difference between lead and the calcium and iron it needs to build bones and form blood.

CLAW compliance is necessary when there is, or is likely to be, a 'significant' exposure risk. This is defined as 'where there is a substantial risk of an employee ingesting (swallowing) lead' from regular hand-to-mouth contact. The biggest danger to the painter is when old surface coatings are being prepared for redecoration. An amount of lead no bigger than a crystal of sugar, if ingested, could put a few small children in hospital.

Paints used up into the 1980s contained lead pigments. These were banned for consumer use in 1988. About 75% of UK homes and schools are old enough to contain layers of lead-based paintwork. When lead dust and fumes are released, from abrasion or 'burning off', they can be inhaled or swallowed. This is a 'significant' exposure risk. Even when all the lead paint has been stripped from timber its surface can still contain amounts of lead that would become a 'significant' exposure risk during any further rubbing down.

CLAW places equal responsibilities on employers and employees to work and behave on-site in a manner that eliminates or minimises and contains lead exposure risks. Briefly, the regulations cover the following points and apply to any site on which workers may be exposed to lead:

(i) The employer must provide similar protection to both his own employees and other people who may be affected by their work. This means that in occupied buildings like offices, schools and factories, every precaution must be taken to prevent lead dust being allowed to travel freely through the site.

(ii) Employers must inform employees of the risks to which they may be exposed. They must also tell them how to protect themselves against the possible risks.

(iii) Employers must ensure that no methods of preparation are carried out which will endanger the health of the employees. Specifying wet abrasion reduces the risk of breathing in lead dust but does not stop wet lead residues being swallowed as a result of hand-to-mouth contact (smokers and nail-biters please note!).

(iv) Even if lead dust can be prevented, employees must be provided with adequate respiratory protective equipment (RPE), rated for 'Toxic Dust' (min. FFP3/P3).

(v) Employers must provide protective clothing to all employees.

(vi) Washing facilities, and places to store protective clothing and personal clothing not worn during working hours, must be provided.

(vii) Smoking, eating and drinking are not permitted except in a place provided which is free from any danger of lead contamination.

(viii) Every care must be taken to prevent spreading the risk of lead exposure to other places. This requires proper containment or isolation of the work areas, regular cleaning using a combination of high efficiency (HEPA) vacuuming and wet mopping; and use of protective clothing and properly fitted RPE. Clothes and boots worn on the site should be kept on the site.

(ix) When there is or is likely to be a 'significant' exposure risk, as defined by the regulations (which is almost always the case where lead-based paint is being disturbed), some of the most important compulsory requirements include:

a) *Consideration of the actual amount of lead involved.* Traditional laboratory analysis is now being challenged by cutting edge on-site non-destructive testing with handheld XRF-i (X-Ray Fluorescence 'isotope') devices.

b) *Checking blood lead levels (BLLs).* These can be confirmed by analysing blood samples. Laboratory analysis of saliva samples can provide an effective initial 'screen' for BLLs. If a certain blood lead level is exceeded an employee must be re-tested, put under medical surveillance and strict records kept of the results.

c) *Preliminary air monitoring to determine airborne lead dust levels.* These readings are important for helping an approved doctor to decide how frequently a person's BLLs should be checked. This could be as frequently as once a fortnight, for example, if shot-blasting lead paint on a bridge. They are also used to confirm the appropriate selection of respiratory protective equipment.

*Note:* Paints containing lead shall not be applied by spray in the interior of buildings. All of the above will be addressed by the completion of a risk assessment of the job in hand.

## 6.4.2 Removal of paint systems containing lead

All work involving the disturbance of coatings containing lead must be treated as a likely 'significant' exposure risk. Even if the airborne lead dust limit for 'significant' exposure is not reached a 'significant' exposure risk can, nevertheless, arise from the ingestion risk. Where old coatings have to be removed:

(i) Using a simple chemical test kit (available from some trade outlets) to identify lead in paint is acceptable for DIY, but does not satisfy the CLAW requirement for employers and the self-employed to 'consider the amount' of lead involved.

(ii) The Health and Safety Executive (HSE) now recommends lead surveys wherever painted surfaces are likely to be disturbed. Although many paint manufacturers advise special care on 'pre-1960s paintwork', HSE advice applies no such date limit.

(iii) If lead is present, wet methods such as wet sanding and chemical stripping are preferred although power sanding with HEPA-filtered dust extraction can be effective.

(iv) The use of blow-torches and/or heat guns for 'burning off' lead paint is best avoided because it is difficult to supervise their controlled use below 450° C. At higher temperatures toxic fumes can be released which are easily inhaled and rapidly absorbed into the lungs and the bloodstream. An effective 'low heat' alternative (<170° C) is available which uses infrared heat. And because there is no forced airflow, as from a heat gun, there is no risk of spreading dust contamination around.

*Note:* Most naked flame torches exceed 450° C. All softened paint must be scraped directly into a suitable container and disposed of in accordance with the Hazardous Waste (England and Wales) Regulations, 2005.

(v) Whichever paint preparation/removal methods are used, thorough and effective containment and use of polythene floor protection will minimise clean-up times.

(vi) As with asbestos, there are established 'clearance' standards where lead dust contamination is involved. Dust wipe sampling of floors and window sills by specialists can provide independent verification that work areas have been cleaned to a standard considered 'safe' for re-occupation.

# 6.5 Hygiene

## 6.5.1 General principles of hygiene

It is an essential feature of good painting practice to work only in clean conditions. Such principles must also be applied personally. Most of the following points refer to site conditions required by the Construction, Design & Management Regulations 2007.

***Hand-washing*** must be carried out regularly, especially before meals. The regulations state that washing facilities must be provided on any site where people are employed for more than 4 hours.

The swallowing of old paint-film dust, paint particles, fungicide washes and pastes and many industrial dusts can have serious effects on the body, particularly the bowels.

***Overalls or protective clothing*** must be worn for the entire period of work. The regulations require that accommodation for personal clothing must be provided on every site and this should be used to store outdoor clothing while overalls are worn on site.

Overalls should be washed regularly, at least once a week.

Wearing of overalls off the site is anti-social: it spreads the dirt and debris of the site and subjects other people to discomfort, irritation and the risk of disease.

Site facilities to comply with the regulations must include:

(i) a place that offers protection during bad weather, in which warming and drying facilities are provided;
(ii) a messroom in which there are arrangements for boiling water and heating food;
(iii) sufficient drinking-water points that are easily accessible.

***Sanitary conveniences*** must be provided which are reasonably accessible to all workplaces. They must be covered, partitioned to give privacy, have a door with a fastener, be lit and well-ventilated.

## 6.5.2 Industrial dermatitis (occupational skin disease)

Dermatitis affects people in different ways: in a mild form it may result in slight irritation or redness of the skin; in more severe dermatitis the skin may harden, blister or crack, becoming exposed to a variety of infections. Warts and ulcers may result from severe skin damage. The disease does not necessarily show itself immediately: lack of care in handling harmful materials when young may result in skin disease later in life.

Ways in which the skin may be affected are shown in Table 6.3.

### Prevention

The hands are the principal part of the body to be affected, and there are various ways by which they can be protected.

***Industrial gloves*** Impervious gloves made from rubber, polyvinyl chloride (PVC) or an impregnated material. Available in a variety of thicknesses and sizes. May cause severe perspiration which can have a detrimental effect on some skins, but this may be avoided by wearing thin cotton inner gloves. Gloves can be worn when working with any of the chemically active materials listed in Table 6.3.

***Barrier cream*** Only effective for short periods. Rubbed into every part of the hands *before* commencing work. There are two principal types:

**Table 6.3** Causes of industrial dermatitis

| *Cause* | *Agent* |
|---|---|
| Chemical action<br>(a) Irritation | Glass fibre<br>Cement<br>Epoxy, polyurethane, polyester, formaldehyde resins and hardeners<br>Tar, pitch, bitumen<br>Fungicide, insecticide<br>Acid<br>Alkali<br>Certain dyes<br>Certain glues |
| (b) De-fatting: dissolving of the protective fat or oil in the skin, leaving it open to infection | White spirit<br>Paraffin<br>Cellulose thinner<br>Polyurethane, epoxy and chlorinated rubber paint thinners<br>Methylated spirit<br>Detergent<br>Petrol<br>Liquid paint remover |
| Obstruction of skin pores | Dirt<br>Oil<br>Grease<br>(*Note:* Also causes pimples and boils) |
| Mechanical damage: piercing of outer skin | Grit and dirt |
| Softening of the skin, with reduced protection against infection | Immersion in oils or watery solutions |
| Bacterial infection | Air-carried organisms trapped in infected skin |

(a) *Water-soluble,* which protects hands against attack by solvents and solvent-carried paints and resins. After work the hands are washed in water to remove all traces of cream and paint.

(b) *Water-repellent,* which is less common but essential when using water-based materials such as detergent or cement. Removed with skin cleanser, *not* solvent.

Skin cleansers are special blends of solvent oils, antiseptics and emulsifiers which remove paint, resins and barrier cream from hands without dissolving the natural oils in the skin. Available in liquid, jelly or paste form.

*Note:* The use of any solvent or detergent to clean hands is a dangerous practice which may cause dermatitis.

# 6.6 Fumes, dust, noise, vibration

## 6.6.1 Fumes

Asphyxia is one of the most common causes of unconsciousness. It occurs when a person is unable to breathe sufficient oxygen.

Asphyxia presents a treble danger:

(a) Becoming unconscious while on a scaffold can prove fatal.
(b) If the brain is deprived of oxygen for over 4 minutes it can be permanently damaged.
(c) Prolonged oxygen starvation causes death.

### Solvent fumes

The space occupied by the vapour of a solvent is many thousand times that of the liquid. As spirit-carried paint dries, its solvent evaporates and, if unable to disperse, gradually replaces the air content of the space in which the work is being carried out. In a short time, the painter is breathing spirit fumes which do not contain the oxygen that the body needs.

Materials that produce this kind of vapour are:

oil paint
alkyd paint
polyurethane, epoxy and chlorinated
rubber paint
cellulose paint
bitumen paint
spirit paint remover
oil and spirit varnish

When spraying solvent-thinned materials in enclosed spaces, the fumes can build up into an explosive mixture. Special extraction fans and trunking may be necessary if open doors and windows may release the fumes into the building and cause further hazards.

### Gas

Unbreathable atmospheres can be produced by other forms of gas.

*Carbon monoxide* Exhaust gas from petrol or diesel engines progressively replaces oxygen in the blood, with lethal results.

*Town gas* Has similar effect to carbon monoxide.
*Natural gas*
*Propane*
*Butane*

Non-poisonous, but if breathed in for sufficient length of time, asphyxia will occur.

*Smoke* from any burning material gradually replaces oxygen (smoke is the cause of death more often than the actual fire).

*Isocyanate* A chemical used in polyurethane coatings and foams whose vapour can cause irreversible damage to lungs.

### Chlorinated hydrocarbon solvents

As well as having the properties described for solvents above, these materials present another danger: in vapour form they decompose if exposed to a naked light or flame and form an irritant gas.

Chlorinated hydrocarbon solvents are used in:

spirit paint remover
some cellulose paints
some polyurethane and epoxy paints

### Paint-burning

Any form of combustion requires oxygen, therefore prolonged burning-off in a confined space will quickly use up the atmospheric oxygen. Some paints, e.g. polyurethane, also produce poisonous gases when burning (see 6.4.2).

***Safety precautions***

(i) The Construction (Design & Management) Regulations 2007 require that when operatives are working in the presence of injurious fumes either:
  (a) adequate ventilation must be provided, or
  (b) suitable respirators must be supplied.

(ii) When working inside a building with any of the solvents or paints listed above, ensure ample ventilation by opening windows and doors.

(iii) If ventilation is not possible, work in short spells only, going into the open air regularly to ensure adequate oxygen intake.

(iv) In very confined spaces such as storage tanks, pipelines, tunnels, breathing apparatus may be necessary: the use of this equipment requires special training. (See Table 6.4.)

(v) Do not smoke, use burning equipment, or allow any naked flame in the area, particularly when using chlorinated hydrocarbon solvents.

(vi) Ensure that petrol and diesel engines do not emit their exhaust into the interior of a building (this is also a requirement of the regulations quoted above).

(vii) Check that no hose connections, gas bottle or gas tap are leaking.

### Note

(i) Face masks and respirators are of no protection when working in low-oxygen atmospheres. They only clean the air being breathed; they do not replace the oxygen. Only breathing apparatus is effective under these conditions.

(ii) Many of the gases and vapours mentioned are heavier than air, therefore working at a low level in a room will prove more dangerous than in a standing or elevated position.

## 6.6.2 Dust

Unrestricted inhalation of dust can cause considerable damage to the respiratory organs. The damage is apparent as:

running nose and eyes
sore throat
headache and dizziness
inflammation of the lungs, which can lead to bronchitis

Dust and grit can also severely damage the eyes unless protective equipment is worn.

### Asbestos

The Control of Asbestos at Work Regulations 2006 control the use of asbestos products because serious lung disease (asbestosis) can be caused by inhaling the dust. The painter is generally concerned with asbestos cement whose effect is less severe than that of asbestos lagging, asbestos insulation boards or sprayed asbestos. Precautions should, however, be taken when producing any asbestos dust.

### Stone, brick and concrete

Cleaning or abrading by mechanical tools, or grit-blasting, may release siliceous dust which, if inhaled, can cause severe damage to the lungs (silicosis). All these dusts can severely damage the eyes.

### Timber

Excessive inhalation of any timber dust will cause respiratory damage but certain hardwoods, such as teak, obeche or iroko, and impregnated timbers, contain elements which can cause greater damage. Protection is essential when mechanically abrading timber in confined spaces or indoors.

### Plastics

Plastic dust can cause lung damage. Glass-fibre dusts also severely irritate the skin and damage internal organs if inhaled.

***Safety precautions***

(i) The Construction (Design & Management) Regulations 2007 require suitable respirators to be provided where operatives are required to work in dust-laden atmospheres.

(ii) The Personal Protective Equipment at Work Regulations 2002 require that:
  (a) Suitable eye protectors or shields are provided where processes are carried out which could damage operatives' eyes.

**Table 6.4** Protective equipment

| | *Equipment* | *Hazard* | *Notes* |
|---|---|---|---|
| Respiration: prevention of inhalation of dust, fumes, spray, gas | *Mask* (Fig. 6.1)<br>Disposable paper mask<br>Disposable cotton filter pad | Large particles<br>Non-toxic dust, e.g. paint spray, timber dust | Not safe for use in oxygen-deficient atmosphere |
| | *Dust respirator* (Fig. 6.2)<br>Facepiece with replacement cotton/ felt or impregnated wool filter | Fine or toxic dust, e.g. glass-fibre, cement | Ditto |
| | *Light fume respirator* (Fig. 6.3)<br>Facepiece with replaceable activated carbon filter | Fine dust and non-toxic fumes, e.g. spray | Ditto |
| | *Breathing apparatus* (Fig. 6.4)<br>Orinasal facepiece or full-face mask or hood connected to a filtered air supply by a flexible air hose | Asphyxia in oxygen-deficient atmosphere, e.g. unventilated or very confined areas in which solvent-carried materials are used | Recommended for spraying two-pack epoxy and polyurethane coatings in enclosed spaces<br>Destroys sense of smell<br>Particularly dangerous when working with solvent-thinned materials as explosive atmospheres may build up |
| | *Blast cleaning helmet* (Fig. 6.5)<br>Glass-fibre helmet with two-ply plastic window (outer can be replaced when badly abraded) and attached canvas cape covering shoulders, arms and chest. Can have air supply | Abrasive action of grit<br>Breathing in highly concentrated dust atmosphere | |
| Eyes | *Impact and chemical goggles* (Fig. 6.6)<br>(a) Clear or tinted acetate or polycarbonate lenses in tight-fitting PVC frames<br>(b) Rubber or PVC frames with toughened-glass lenses | Flying particles<br>Liquid splashes<br>Dust | Insufficient protection against glare or oxyacetylene flame |
| | *Welders' goggles* (Fig. 6.7)<br>As above, with special blue or green filter glass or plastic lenses | Glare and flying particles from oxyacetylene torches | Allow only poor vision for normal use |
| Head | *Safety helmet or hard hat* (Fig. 6.8)<br>Glass-fibre or plastic, with soft, padded headband and harness | Falling objects<br>Collision with protrusions, scaffold poles or fittings | |

**Table 6.4** Protective equipment (cont'd)

| | *Equipment* | *Hazard* | *Notes* |
|---|---|---|---|
| Hands | *Industrial gloves*<br>Leather, cotton twill<br>Asbestos cloth<br>Rubber, PVC, neoprene or rubberised fabric | Abrasion, sharp edges<br>Extreme heat<br>Solvents, detergents | |
| | *Industrial gauntlets*<br>Rubber, PVC or neoprene<br>Leather or rubber | Caustic soda<br>Flying debris when grit-blasting | |
| Ears | *Earmuffs* (Fig. 6.9)<br>Pair of padded plastic cups, either solid or liquid fillings, connected by simple head frame. Reduce high-frequency noise but allow hearing of loud voices | Harmful noises at very high level, e.g. explosions, blasting, jet engines | Should be issued only after sound intensity has been measured and correct type of protection specified |
| | *Ear plugs or defenders*<br>Plastic or rubber plugs that fit into the ears and muffle noise<br>Various sizes available | As above | For use when earmuffs are not available |
| | *Waxed cotton wool or glass fibre*<br>Moulded and fitted personally<br>They loosen with jaw movement, are less hygienic than plugs and more likely to cause infection | As above | For use when other forms of protection are not available |
| Feet | *Safety shoes or boots*<br>With steel toe caps<br>With steel sole lining<br>(Many incorporate both features) | Dropping of heavy objects<br>Stepping on sharp protrusions or objects | |

**Fig. 6.1** Mask

**Fig. 6.3** Light-fume respirator

**Fig. 6.2** Dust respirator

**Fig. 6.4** Breathing apparatus

**Fig. 6.5** Blast-cleaning helmet

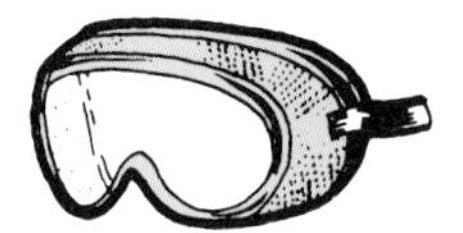

**Fig. 6.6** Goggles

**Fig. 6.7** Welders' goggles

**Fig. 6.8** Safety helmet

**Fig. 6.9** Earmuffs

(b) All persons provided with eye protectors must wear them while the conditions for which they are supplied exist.

(iii) Wherever possible, use a wet-abrading process so that dust is not released.

(iv) Always wear a face mask or respirator when dust cannot be avoided. Breathing apparatus is necessary when asbestos dust is released.

(v) Remove excessive dust with an industrial vacuum cleaner, preferably one with High Efficiency Particle Air (HEPA) filter, rather than by brushing or sweeping.

## 6.6.3 Noise

Prolonged or high-level noise can cause tiredness, nausea or temporary or permanent deafness. Many mechanical tools used on construction sites can produce these symptoms.

***Safety precautions*** The Noise at Work Regulations 2005 suggest that, as a rough guide, you will need to do something if any of the following apply:

(a) Is the noise obtrusive, as from a crowded street, vacuum cleaner or busy restaurant for most of the working day?

(b) Do you need to shout in order to be heard over a distance of approximately 2 m for at least part of the day?

(c) Do employees use noisy power tools or machinery for more than half an hour each day?

(d) Do you work in a noisy industry, i.e. construction?

(e) Is there impact noise, i.e. pneumatic impact tools or cartridge-operated tools?

*Note:* dB(A) is a decibel scale which measures noise intensity. An increase of 3 dB(A) represents a doubling of sound intensity.

## 6.6.4 Vibration

Vibration produced by percussion hammers and needle guns used for long periods can cause operatives to lose all sensation in their hands: the condition is called 'white finger' or 'dead hand' (medical name, 'Raynaud's phenomenon'). There is no medical remedy for this disease; once established, attacks may occur at any time, most often when not using vibrating tools and when the hands are cold.

Apart from the danger of dropped tools, it may make it possible for a man to injure his hand without feeling it.

***Safety precautions***

(i) Restrict work to short spells.
(ii) Wear gloves: keep hands warm.
(iii) Grip tool as lightly as possible.
(iv) Once symptoms are suspected, stop work immediately and seek medical advice.

## 6.6.5 Radiation

Painters working in nuclear power stations can be exposed to ionising radiation.

***Safety precautions*** A health register is required to be kept, in which exposure is recorded until the permissible maximum is reached or until a period of 30 years has elapsed. Records must be passed to any new employer.

# 6.7 Lifting

During the normal course of their work, the painter is required to lift heavy objects. If this lifting is carried out incorrectly, it can cause injury. The Manual Handling Operations Regulations 1992 state that a person must not lift or carry any load which may cause an injury. Bad handling does not necessarily cause immediately noticeable injury, but may be responsible for the later development of rheumatism, fibrositis or slipped disc. Rupture or hernia is another serious injury which can be caused by incorrect lifting.

Correct handling and lifting also reduce effort and fatigue.

The *basic principles of handling and lifting* are as follows (see Fig. 6.10):

**Fig. 6.10** Lifting

***Keep back straight*** If the back is bent while lifting, the spinal column may be damaged. The safe method is to keep the back straight throughout the lift.

***Bend the knees*** Crouch to the object with knees bent and back straight.

***Tuck chin in*** With the head raised and the chin tucked in, a straight back is maintained.

***Position of feet*** One foot should be placed behind and the other alongside the object, pointing in the direction of movement after the lift.

***Firm grip*** Finger-tip grip is unsafe. The roots of the fingers and palm of the hand must be fully used to ensure a safe grip.

***Arms close to the body*** If arms are spread out, the weight is transferred to fingers, arms and back, which could result in injury. The body takes the weight if the arms are kept close while lifting and carrying.

***Smooth action*** Push off with rear foot, straighten the legs and move off in the required direction in one smooth movement.

*Note:* Awkward-shaped objects require special care. When more than one person is lifting, appoint a leader to ensure that the lifting action is carried out together. If lifting gear is available, always use it.

# 6.8 First Aid

Safety rules and regulations are made to prevent accidents, but because the nature of the work and the materials used by the painter are potentially hazardous, precautions must be taken to deal effectively and efficiently with an accident should it occur.

## 6.8.1 First-aid requirements

The Health and Safety (First Aid) Regulations 1981 lay down the following for the provision of first-aid boxes and suitable first-aider support.

The regulations require that employers and self-employed persons provide adequate and appropriate first-aid facilities which should be based on an assessment of the need for first-aid depending on the nature of the work and *not just on the number of people on site.*

### First-aid boxes and kits

On small sites, where no special or unusual hazards exist and normal emergency services are readily available, the appointment of sufficient suitable persons trained in first-aid and the provision of first-aid boxes, with guidance and the presence of first-aid notices, may be adequate. Sites which present a high risk from hazards should be provided with a first-aid room under the responsibility of a suitable person with a suitable person available at all times. A suitable person is a first-aider who holds a current first-aid certificate issued by an organisation whose training and qualifications are approved by the Health and Safety Executive. Practising registered medical practitioners and nurses are considered as suitable persons.

First-aid boxes and travelling first-aid kits should contain a sufficient quantity of suitable first-aid materials *and nothing else*. In most cases these will be:

- 1 guidance card
- 20 individually wrapped sterile adhesive dressings
- 2 sterile eye pads
- 6 individually wrapped triangular bandages
- 6 safety pins
- 6 medium-sized individually wrapped sterile unmedicated wound dressings (approximately 100 mm × 80 mm)
- 2 large individually wrapped sterile unmedicated wound dressings (approxmately 130 mm × 90 mm)
- A pair of disposable gloves

Where mains tap water is not available for eye irrigation, sterile water or sterile normal saline (0.9%) in sealed, disposable containers should be provided. *Eye baths/cups/refillable containers should not be used for eye irrigation.*

## 6.8.2 First-aid treatment

No matter how good the first-aid provisions on a site, there may always occur a situation when immediate, positive action has to be taken, before trained help can arrive, to save the life or reduce unnecessary suffering of a workmate after an accident.

The following notes indicate first-aid treatment that an *untrained* person may be able to give in an *emergency* while awaiting the qualified first-aider or determining the extent of the injury. The treatment is not meant to be applied as an alternative to obtaining medical attention.

### Burns and scalds

Caused by:

*Dry heat* Fire, flame, hot objects, sun
*Electricity* Electric current
*Friction* Revolving wheel, rope or wire

*Corrosive chemicals* Acids: sulphuric, hydrochloric, hydrofluoric, carbolic. Alkali: caustic, ammonia, lime.

*Scalds* Moist heat: boiling water, steam, hot oil or tar

***Treatment***

(i) Reduce local effect of heat under slow-running water, or by immersing in cool water until pain has ceased (may be 10 minutes).

(ii) Remove constricting objects, e.g. rings, belts, boots.

(iii) Remove soaked clothing but not burnt clothing, which is sterilised.

If the effect is severe:

(iv) Lay patient down.

(v) Cover burn with dressing or cling film.

(vi) Arrange immediate removal to hospital.

### Eye burn

Caused by corrosive chemicals.

***Treatment***

(i) Hold head under gently running tap; or plunge head into bucket or basin of water, instructing patient to blink continuously; or sit or lay patient with head tilted and flush copiously with cold or tepid water, milk or sterile water from an approved Eye Station if available.

(ii) Apply dressing or clean cloth.

(iii) Arrange immediate removal to hospital.

### Minor wounds

***Slight bleeding*** Wash, if possible with running water; apply dressing with pad and bandage.

***Severe bleeding*** Stop bleeding by applying finger pressure to wound, either direct or over a dressing if immediately available; lay patient down and make comfortable; arrange for hospital treatment.

***Bleeding from head*** Apply large dressing and fix with bandage; *do not press wound* in case skull is fractured.

***Nose bleeding*** Support patient in sitting position with head slightly forward; ensure that the patient is breathing through mouth; instruct patient to pinch the soft part of the nose firmly for about 10 minutes.

***Discharge from ear*** Place and secure pad over ear; lay patient down with head and shoulders slightly raised; obtain medical advice.

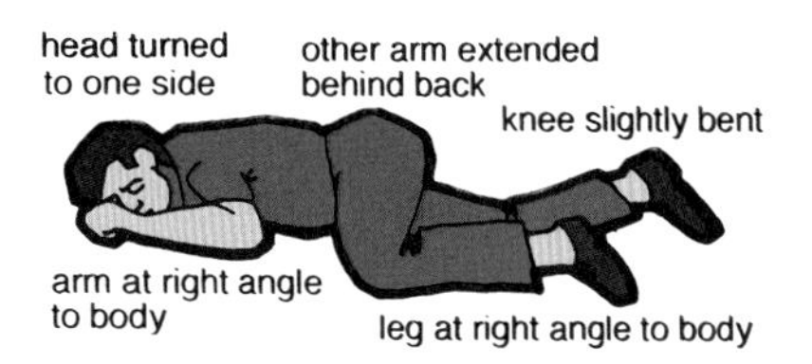


**Fig. 6.11** Recovery position

***Bruising*** Apply ice bag or damp material to reduce swelling and relieve pain.

### Fractures

***Limb fracture*** Steady and support injured part to prevent further damage; immobilise and support; arrange immediate removal to hospital.

***Skull fracture*** Symptoms: blood or fluid discharge from ear, vomiting, bloodshot eyes.

***Treatment*** Place patient in recovery position with discharging ear downwards (Fig. 6.11); apply sterile dressing to an ear discharge; keep watch on breathing; start artificial respiration if breathing stops; arrange urgent medical help.

***Jaw fracture*** Ensure airways not obstructed by tongue or blood; support jaw by hand or bandage; if severely injured, place in recovery position; obtain urgent removal to hospital.

***Severe back or spinal injury*** Do not move patient – warn him to lie still; cover with blanket or coat; arrange urgent removal to hospital.

### Eye damage

All eye injuries are potentially serious and should receive immediate medical aid.

***Injury to eyeball*** Foreign matter on eyeball or pupil, cause of damage unseen: close the eyelid; cover with large soft pad; arrange removal to hospital.

Foreign matter which is visible and *not* on pupil or adherent to eyeball: pull down lower lid or raise upper lid; remove matter with corner of clean handkerchief or small piece of damp cotton wool.

### Unconsciousness due to asphyxia (See 6.6)

Caused by:

swelling of throat due to scalding or swallowing corrosives

drowning

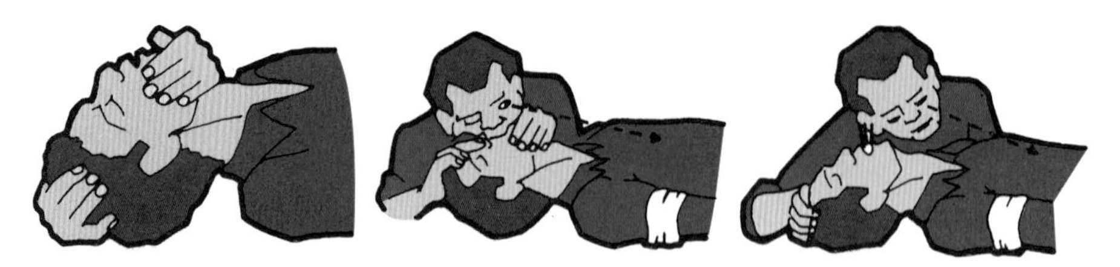

**Fig. 6.12** Mouth-to-mouth resuscitation

poisoning by industrial gases (phosgene, solvent vapour, gas, carbon monoxide)

breathing smoke-laden or oxygen-deficient atmospheres.

***Treatment***

(i) In the case of the last two causes, remove patient into open air or away from the affected area immediately.

(ii) Remove any constricting garment, e.g. tie, buttoned collar, belt, waistband.

(iii) Clear mouth of blood, vomit or swollen or swallowed tongue.

(iv) If breathing is laboured or has stopped, apply respiratory resuscitation.

## Severe electric shock

Caused by contact with low voltage (domestic supply) electrical current.

***Treatment***

(i) Break the electrical contact by switching off current, removing the plug or wrenching the cable free. If these methods are not possible or cannot be carried out quickly, stand on dry, insulating material (plastic, rubber) and, using dry wood, folded newspaper or rubber, push the patient's limbs away to break the contact.

*Do not touch the patient with your hands*

(ii) Treat burns as recommended earlier.

(iii) If patient is unconscious, apply respiratory resuscitation immediately.

## Methods of resuscitation

***Mouth-to-mouth*** **(Fig. 6.12)** Used if mouth or face is not damaged and there is no vomiting:

(i) Pinch patient's nostrils together.

(ii) Seal lips around mouth.

(iii) Blow into lungs until chest rises.

(iv) Remove mouth until chest falls.

(v) Repeat at natural rate of breathing.

# PART 7

# Specification

# 7.1 Components of buildings – exterior

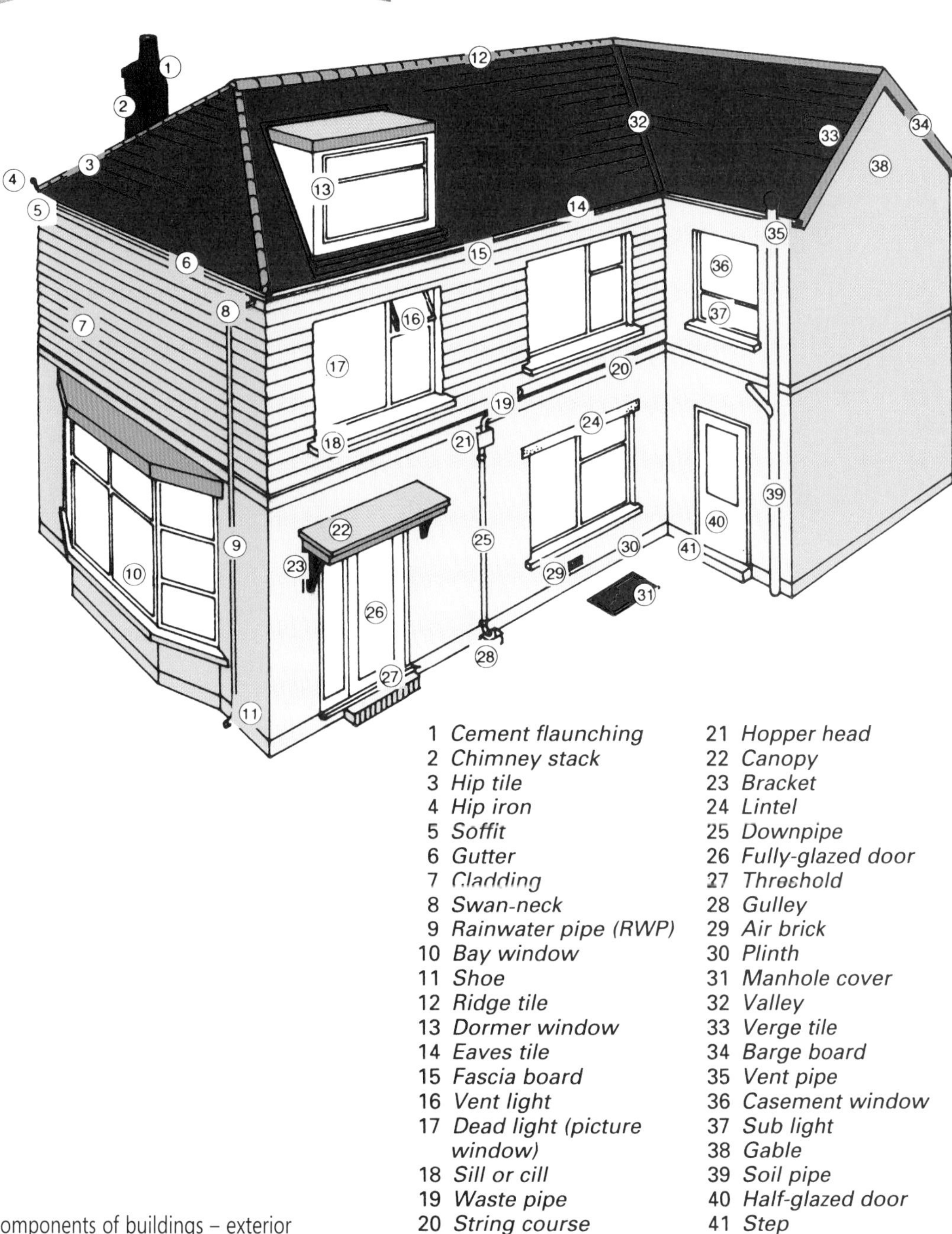


**Fig. 7.1** Components of buildings – exterior

# 7.2 Components of buildings – interior

1 *Ceiling rose*
2 *Cove cornice*
3 *Frieze*
4 *Picture rail*
5 *Wall filling*
6 *Niche*
7 *External mitre*
8 *External angle*
9 *Chimney breast*
10 *Mantle shelf*
11 *Return angle*
12 *Fire surround*
13 *Plinth block*
14 *Hearth*
15 *Internal mitre*
16 *Internal angle*
17 *Arch*
18 *Soffit*
19 *Architrave*
20 *Reveal*
21 *Sash window*
22 *Dado*
23 *Reeded column*

**Fig. 7.2** Components of buildings – interior

1 *Ceiling bed*
2 *Pelmet*
3 *Bay window*
4 *Glazing bar*
5 *Apron panel*
6 *Skirting board*
7 *Cornice*
8 *Capital*
9 *Fanlight*
10 *Transom*
11 *Architrave*
12 *Pilaster (fluted or plain)*
13 *Wall panel*
14 *Panel moulding*
15 *Panelled door*
16 *Dado or chair rail*
17 *Dado panel*
18 *Plinth block*
19 *Margin*

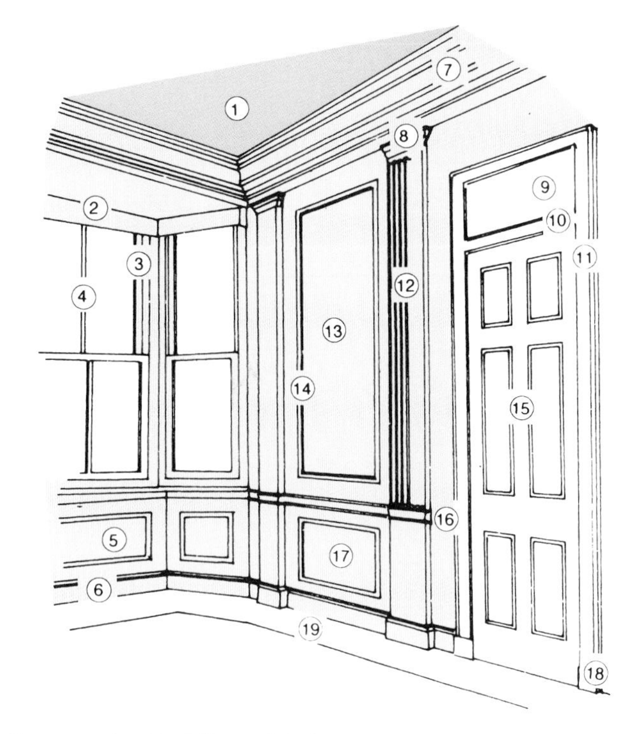


**Fig. 7.3** Components of buildings – interior

# 7.3 Components of buildings – staircases

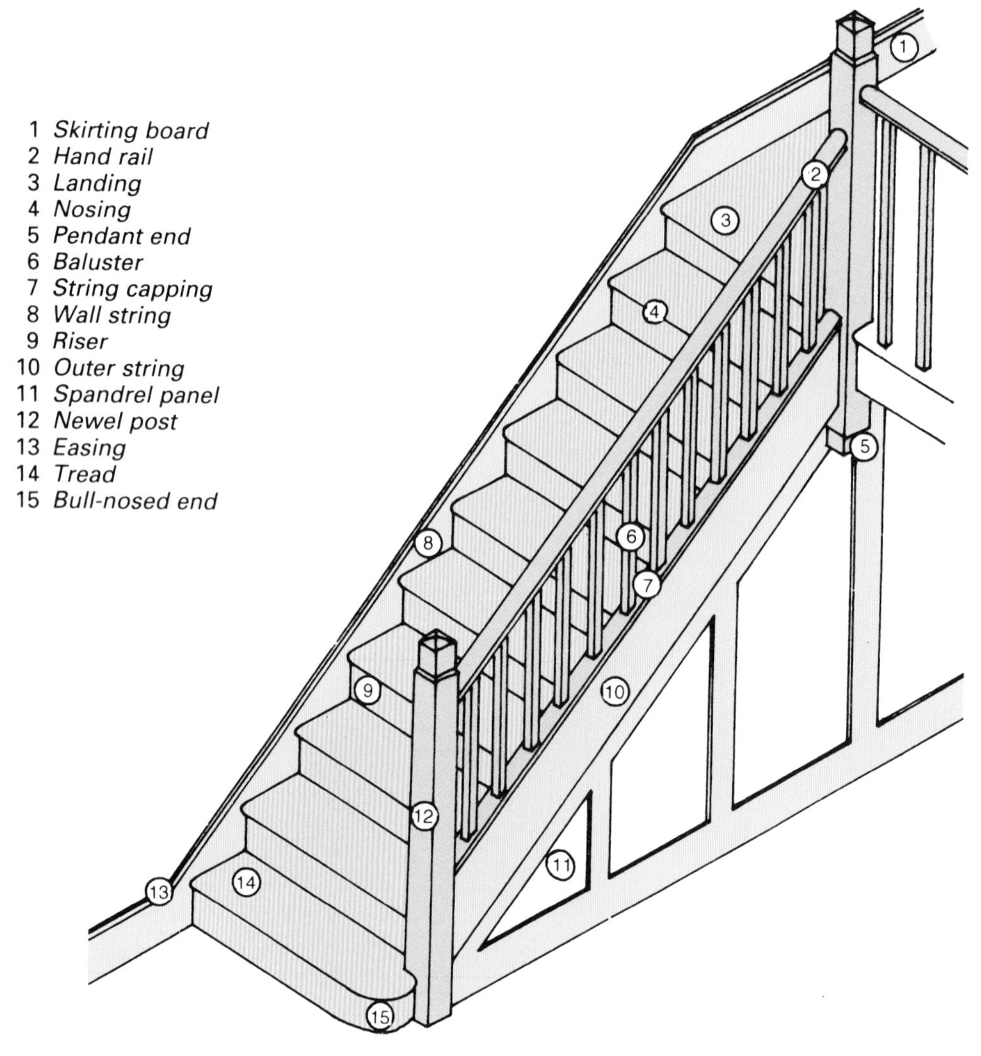


**Fig. 7.4** Components of staircases

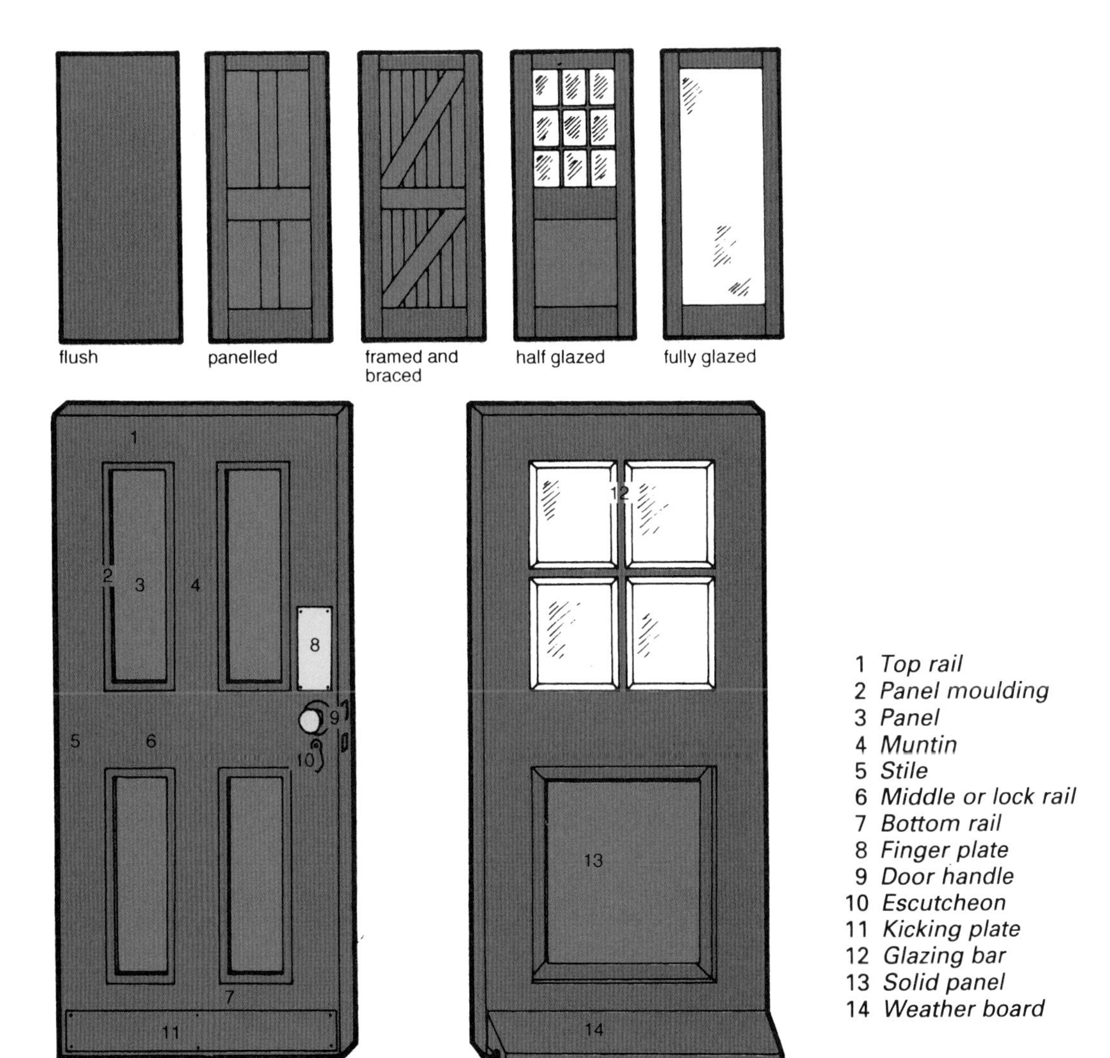
flush
panelled
framed and braced
half glazed
fully glazed
1
2
3
4
5
6
7
8
9
10
11
12
13
14
1 Top rail
2 Panel moulding
3 Panel
4 Muntin
5 Stile
6 Middle or lock rail
7 Bottom rail
8 Finger plate
9 Door handle
10 Escutcheon
11 Kicking plate
12 Glazing bar
13 Solid panel
14 Weather board

**Fig. 7.5** Components of doors

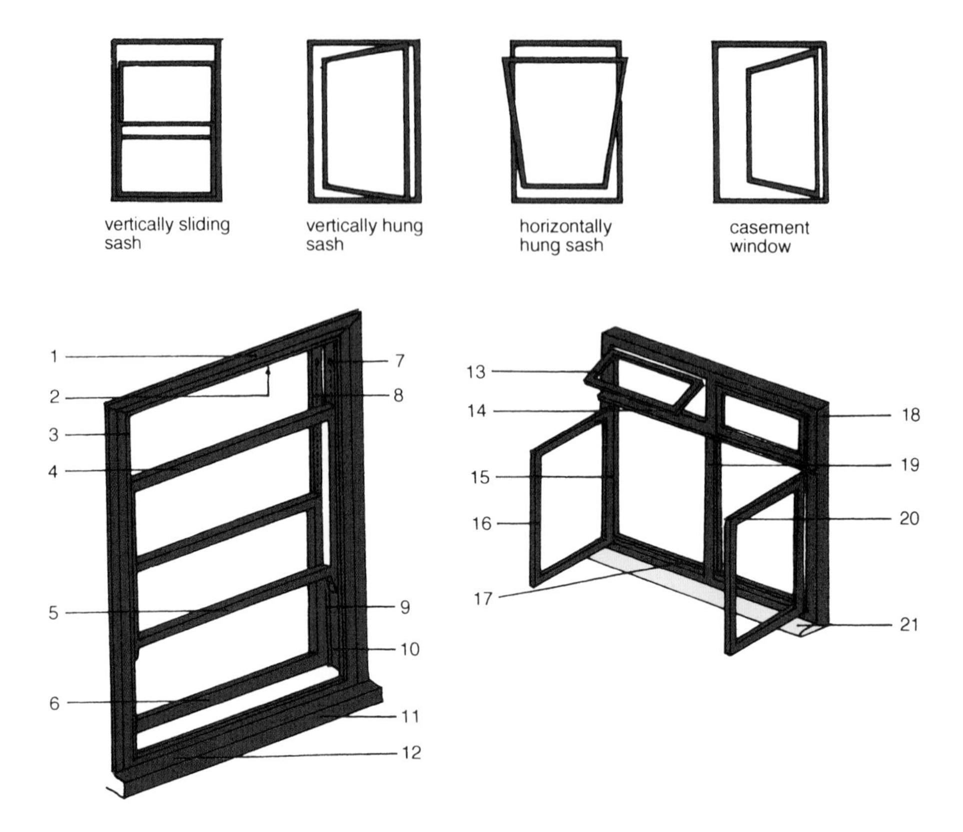


1 *Architrave*
2 *Soffit*
3 *Stop bead*
4 *Top (outer) sash*
5 *Meeting rail*
6 *Bottom (inner) sash*
7 *Pulley*
8 *Sash cord*
9 *Parting bead*
10 *Runner*
11 *External sill*
12 *Weathering*
13 *Ventlight*
14 *Transom*
15 *Hanging stile*
16 *Meeting stile*
17 *Rebates*
18 *Window frame*
19 *Mullion*
20 *Casement*
21 *Window board*

**Fig. 7.6** Components of windows

# 7.5 Job planning

## 7.5.1 Interior painting

A typical method of organising the decoration of the room illustrated in Fig. 7.2 would be as follows:

### Sequence of operations before painting

Remove carpets, pictures, mirrors and light fittings.

Remove small and light furniture.

Place heavy and bulky furniture in the centre of the room.

Cover floor and furniture with dust sheets.

Remove curtains and pelmets.

Remove all door and window furniture and fittings.

### Sequence of operations for preparation

All processes should be carried out by starting at the highest point, and working downwards.

(i) Wash, abrade and make good surfaces in the following order:
 Ceiling bed and cornice
 Frieze, wall-filling and dado
 Windows; replace defective glass and putties, and clean and straighten existing putties by scraping paint from glass
 Doors and other woodwork.

(ii) Remove all rubbish from site.

(iii) Spot-prime with suitable primer all bare patches exposed during the preparatory process.

### Sequence of operations for painting

Working from top:

Ceiling bed and cornice
Frieze, wall-filling and dado
Windows
Doors
Other woodwork, completing with the skirting board.

### Precautions when working

Always wear protective clothing.
Always use suitable plant and scaffolding.
*Never* stand on shelves, window boards or furniture.
*Never* rest stepladders on glass or glazing bars.

### On completion

Clean all glass.

Clean and replace door and window furniture.

Replace carpets, curtains, pelmets and other furniture.

## 7.5.2 Exterior painting

To avoid working over clean, finished work, all preparation and painting should be started at the highest level. A recommended method of working on the house illustrated in Fig. 7.1 is as follows.

### Sequence of operations before painting

Remove or tuck away curtains, especially when burning off.

Lay protective sheets to protect paving and stonework.

Cover plants and shrubs near the building.

### Sequence of operations for preparation

(i) Wash, abrade and make good surfaces in the following order:
 Dormer window: also replace broken glass and defective putties
 Barge boards
 Gutters: also clean insides and repair defective joints
 Fascia boards

Soffits

First floor windows: also replace broken glass and defective putties

Cladding

Ground floor window and sills: also replace broken glass and defective putties

Canopy

Brickwork or rendering

Rainwater, waste and soil pipes

Doors and other woodwork

(ii) Remove all rubbish from site.

(iii) Spot-prime all bare patches exposed during preparatory process.

### Sequence of operations for painting

Working from the top, paint in the same order as for preparation.

### Precautions when working

*Always* stand ladders on scaffold boards to prevent damage to lawns.

*Always* pad tops of ladders to prevent damage to freshly painted surfaces.

*Always* use a kettle hook when working on ladders (see 1.1.2).

*Never* rest ladders on glass, glazing bars or gutters.

*Never* drag ladders down walls or across gutters.

### On completion

Clean up paint spots.

Clean and replace door furniture and fittings.

Clean all glass.

## 7.5.3 Wall and ceiling hanging

### Sequence of operations before hanging surface coverings

Prepare as for interior painting.

Undercoat as for interior painting, omitting surfaces to be papered.

Size and cross-line if necessary (see 4.1).

Complete painting as for interior painting, taking paint 6 mm onto lining paper at cornice, architraves and skirting.

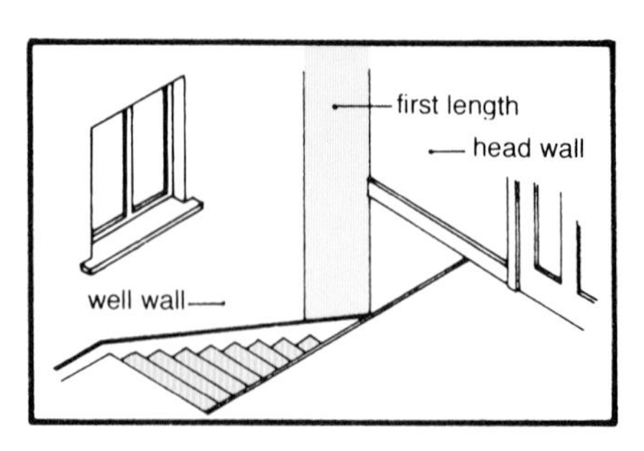


**Fig. 7.7** Papering a staircase well

### Sequence of operations for papering

(i) *Walls*

Either (a) start at the window and work away from it in both directions or (b) work either side of centre line (see 4.10, Centring).

If chimney breast or other focal point exists, strike a centre line and work from either side.

If possible, remove or loosen switches and fittings as they are reached and fit paper behind before refixing.

*Note:* Turn off power before loosening switches.

Do not fit foil wall vinyls behind switches.

Finish in a dark or unobtrusive corner where the loss of matching pattern will not be noticed.

(ii) *Ceilings*

Work from the light.

Hang paper across the room, working away from the window.

(iii) *Staircases* (Fig. 7.7)

Hang first length on well wall, at junction with head wall or apron lining.

Work away from the first length in both directions, finishing in a dark corner.

### Precautions when working

Match and cut all lengths before pasting and hanging (with the exception of staircase work).

Fold up waste material and place under paste bench or in suitable container.

Remove paste from paintwork while still wet.

### On completion

As for interior painting.

## 7.5.4 Construction documents

The planning and running of jobs of all sizes require organisation and records. A few of the most important documents used for this purpose are as follows:

### Architects' drawings

A range of scale, working drawings which form the primary means of communication in the construction industry. Consist of six main types:

***Block plan*** Shows the proposed building site in relation to the surrounding houses and roads, etc. Usually to a scale of 1:2500 and 1:1250.

***Site plan*** Shows the position of the proposed building and the general layout of adjacent roads, services, trees and drainage, etc. Usually to a scale of 1:500 and 1:200.

***Location drawings*** Show plans and elevations of the various areas within a building and the location of the principal elements of the building. Usually to a scale of 1:200, 1:100 and 1:50.

***Range drawings*** Show the sizes of a standard range of building components such as kitchen units and windows, etc. Usually to a scale of 1:100, 1:50 and 1:20.

***Detail drawings*** Show all the information required to manufacture particular components such as windows and doors, etc. Usually to a scale of 1:10, 1:5 and 1:1.

***Assembly drawings*** Show details of the junctions between the various elements and components of a building such as between door frame and wall, etc. Usually to a scale of 1:20, 1:10 and 1:5.

### Bill of quantities

A book of calculated quantities of materials, labour and other items for a proposed job. Compiled by the quantity surveyor and based on the architect's drawings, specifications and schedules.

### Delivery notes

Paper record which accompanies all deliveries of materials and plant to the site. Someone is required to carefully check the items delivered compared to the details on the delivery note and sign that they have been delivered. Any items not delivered should be noted on the form. All notes should be sent to the employer to confirm that goods have been received before the employer pays invoices.

### Method statement

A method statement is a method of control that is used after a risk assessment of a process has been carried out. The method statement is used to control the operation and to ensure that all concerned are aware of the hazards related to the work and the safety precautions to be taken.

In addition, construction method statements are used on site. For a particular section of similar work, they detail its boundaries, materials and plant requirements and temporary works' signs.

### Progress charts

A visual chart on which a proposed job can be planned from start to finish ensuring that appropriate deliveries of materials and supply of labour can be organised at the right time in the job. The two main types used are the bar chart and the Gantt chart.

### Requisition/order forms

Form on which to order plant and materials for a particular job. The order-form number and delivery note for the goods delivered should match, as should the invoice sent to the employer for payment.

### Risk assessment

The Management of Health and Safety at Work Regulations 1999 require employers to assess the health and safety risks involved in all activities. Risk assessment is now a common requirement of all Health and Safety legislation, the emphasis being on preventing accidents and work-related ill health, rather than just reacting to incidents and making improvements after the event.

### Schedules

These are used to record repetitive information (size, colour, position, etc.) for a range of similar components such as paint finishes, doors, windows, etc. There can also be a schedule which lists all the drawings related to a job.

### Schedule of rates

A list of prices per $m^2$ for specified tasks. They include all materials, labour, overheads and profit for the work described.

### Specification

Contains precise descriptions of materials to be used and processes to be carried out which cannot be shown on the architect's drawings. Any variations to the specification can only be authorised by the agent. Also contains other information such as requirements for site clearance, making good on completion and who is responsible for approving finished work, etc.

### Timesheets

An individual daily and weekly record of work done and time taken. Usually completed by the individual operative and verified by the supervisor.

### Variation order

A form which authorises variations to the specification. They must be agreed and signed by the agent so that they can legitimately be used to change the final sum of the contract.

# 7.6 Choice of method of paint application

As paint technology has advanced, so alternative methods of application have been devised to cope with the new materials produced. The conventional brush has been superseded for some purposes by methods of application designed to overcome the problems involved in applying rapid-drying, high-viscosity and textured coatings, while at the same time producing a high-quality, even coating. In some cases, an alternative method may simply be more productive than the brush.

**Table 7.1** Comparison of methods of paint application

| *Method* | *Advantages* | *Disadvantages* | *Common use* |
|---|---|---|---|
| Brush | Smooth finish<br>Ability to cut-in<br>Works paint into crevices to gain adhesion<br>Essential for priming<br>Ability to reach otherwise inaccessible places<br>Very little splashing<br>Economical with material | Slow and laborious on large or textured areas<br>Unsuitable for rapid-drying paints<br>Animal-hair fillings affected by caustic and liquid paint removers | All materials having a sufficiently long initial set<br>Typical use –<br>25 mm: cutting-in glazing bars<br>50 mm: panelled door<br>75 mm: flush door<br>100 mm: ceilings and walls<br>150 mm: large areas |
| Paint roller | 2–3 times faster than the brush on suitable surfaces<br>Applies thicker coating than the brush<br>No brush marks<br>Can be used fixed to extension pole | Splashes more than the brush<br>Loose fibres may spoil the finish<br>Inaccurate for cutting-in<br>Cleaning can be time-consuming<br>Paint containers cumbersome on scaffolding<br>Textured finish not always acceptable | All materials having a sufficiently long initial set<br>Large areas with the minimum of cutting-in<br>Relief and textured areas<br>Flat oil paints, avoiding the need to stipple<br>Chain-link fencing and railings<br>Acoustic tiles |
| Conventional spray | 4–5 times faster than the brush on suitable surfaces<br>No brush marks<br>Can reach almost inaccessible places<br>High quality of finish<br>Easy control of spray pattern | Produces overspray (paint in the air); adjacent work not required to be painted must be protected from overspray by masking<br>Respirators and extractor fans may be required for inside use<br>Uneconomical with paint and cleaning solvents<br>Cleaning can be time-consuming<br>Paints must be thinned to pass through the gun<br>The preparation and maintenance of the equipment makes its use uneconomic for small jobs | Large, unbroken areas<br>Textured areas<br>Rapid-drying paints<br>Awkward, irregular shapes<br>All paints that can be satisfactorily thinned and will flow<br>Multi-colour paints |

**Table 7.1** Comparison of methods of paint application (cont'd)

| Method | Advantages | Disadvantages | Common use |
|---|---|---|---|
| Airless spray | Up to 30 times faster than the brush on suitable surfaces<br>No brush marks<br>Very little or no overspray<br>Large spray pattern<br>Ability to cut-in within 15 mm<br>Paint-thinning unnecessary<br>Applies a thick coating with the minimum of strokes<br>Paint flows out better than conventional spray<br>No contamination from air-line impurities | Equipment initially expensive<br>Cleaning can be time-consuming<br>In spite of precautions, guns tend to block up, although reversible tips simplify the clearing of blocked fluid-tips<br>High velocity of fluid jet presents a hazard at close quarters, especially when fluid-tip is removed<br>Fluid-tip must be changed to alter the spray pattern | Very large areas: gas holders, ships, aircraft hangars<br>Thick or thin paints<br>Heavily textured surfaces<br>Awkward and irregular shapes |

# 7.7 Costing

## 7.7.1 General procedure for large contracts

(i) Architect or designer prepares a specification and drawings of the proposed job based on the client's requirements.
(ii) Quantity surveyor prepares a bill of quantities based on the specification and drawings.
(iii) Bill of quantities and drawings are sent out to decorating contractors with an invitation to submit a tender.
(iv) Decorating contractor costs the job, prepares the estimate from the bill of quantities and submits the tender.
(v) The accepted estimate becomes the price which the client will pay at the satisfactory completion of the job (plus or minus certain extras).

## 7.7.2 General procedure for small (jobbing) contracts

(i) Decorating contractor is approached by the prospective client.
(ii) Decorating contractor obtains the details of the job from the client or the client's agent.
(iii) Decorating contractor takes all necessary site measurements of the job.
(iv) Decorating contractor prepares a specification of the work to be done and an estimate of its cost which is submitted to the client for approval.

## 7.7.3 Information for costing small painting jobs

A method of taking site measurements and preparing the specification is as follows.

**Job dimensions**

Measuring-up on site should be carried out accurately with the use of 2-metre rod or stick, measuring tape or ordinary folding rule. Dimensions are then recorded in a 'Dim' book, or on paper ruled in a similar way. Job details can be recorded at the same time as dimensions, e.g.

| | | | |
|---|---|---|---|
| 6/ | 0.750 m<br>2.000 m | 9 $m^2$ | |
| | | | 2 oil paint on flush doors |
| 2/ | 2.000 m<br>2.000 m | 8 $m^2$ | |
| | | | 2 oil paint on casement windows plus 20% cutting-in |

*Column 1* Quantity of items measured (6 doors, 2 windows).
*Column 2* Actual dimensions of the item, recorded from the top: *length* followed by *width* followed by *height*.
*Column 3* Actual total area of the items measured.
*Column 4* Description of the item and brief specification of the work to be done, often abbreviated, e.g.

| Component | Existing condition | Work to be carried out |
|---|---|---|
| Ceiling and cornice | Papered and painted in sound condition | 2 coats white emulsion |
| Walls | Papered in sound condition | Strip, cross-line, hang supplied paper |

G & V: grain and varnish
K.P.S: knot, prime and stop
E.G: eaves gutter
R.W.P: rainwater pipe
2: two coats

Note the additional percentages to cover time-consuming work, e.g.

Cutting-in windows: 10–50 per cent
Painting pipes: 10–50 per cent
Painting corrugations: 30–50 per cent
Painting gutters: 10–50 per cent

## Job details and specification

Notes should be made on the condition of the job, e.g. Is burning-off necessary? What is condition of wall beneath old wallpaper? Client's requirements of work should be recorded accurately room by room in the sequence of operations for internal painting (see 7.5).

*Preparing the specification* As the specification is a binding contract, it must be written in a simple and clear manner so that there is no confusion as to what the client will be paying for, or how the work is to be done. The full preparatory process should be described, and also the material and colour to be used and the method of application.

Supplied specifications are usually laid out in the following order:

(i) *Job headings:* brief description of the work, client's name and address, site address, architect.
(ii) *Drawings:* copies of all necessary plans and elevations to assist in defining means of access.
(iii) *Preliminaries:* a list of general clauses which would otherwise appear many times in the specification, covering, for example, British Standards, quality of workmanship and materials, construction regulations.
(iv) *Clauses:* separate, explicit instructions and details of work to be carried out. Each page numbered and the clause lettered, e.g. Clause 2D (clause D on page 2).
(v) *Appendices:* other relevant information, e.g. position of electrical power supply, water supply and rest rooms.

## Costing the job

The price of a job consists of:

(i) the total cost of labour for the job;
(ii) the total cost of materials for the job;
(iii) an allowance for overheads for the period of the job;
(iv) an allowance for profit.

The following variables affect the costs involved, and they are used in different ways depending on the size and complexity of the job, and on the administrative organisation of the contractor.

## Labour hourly rate

The total cost of labour per employee hour. It is calculated not only on the employee's weekly wage, but also includes numerous other expenses such as travelling expenses, overall allowance, public and annual holiday payments, redundancy contributions, National Health and Insurance contributions.

The weekly total of all these expenses is divided by the standard number of hours worked per week to calculate the labour hourly rate.

## Material cost

Material costs are supplied by the distributor and expressed as cost per litre. They may vary from month to month and current pricelists should always be consulted.

## Spreading capacities of materials

These are supplied by the manufacturer and expressed as square metres per litre ($m^2$ per l) and are calculated on normal brush applications to surfaces of average absorption and texture. Spreading capacity can vary by up to 25 per cent either way, depending on method of application, texture, temperature and absorption. For spreading capacities of specific materials, see Part Two: Surface Coatings.

### Labour constant

The amount of work an employee can be expected to do in 1 hour, expressed as square metres per hour. It is dependent on the condition and type of surface, the material being used, the process involved and, most importantly, the method of application:

| *Process or material* | *$m^2$ per hour* |
|---|---|
| Wash, abrade, make good walls | 3–5 |
| Gloss painting by brush | 7 |

### Cost of overheads

This will vary from one contractor to another depending on the size of the firm and the quality of the building and administration. Overheads cover such items as insurances, non-productive staff salaries, rent, rates, telephone, power, depreciation of tools, plant and equipment, stationery and advertising. Overheads, when coupled with profit, may be expressed as a percentage of the prime cost.

### Profit

This will vary depending on the firm's overheads and estimated profit. It is expressed as a percentage typically ranging from 25 to 100 per cent which is added to the prime cost. Calculated to cover the cost of overheads and leave a net profit for the contractor.

## 7.7.4 Glossary of terms

***Agent*** Person acting on behalf of the client.

***Architect*** Designer of buildings. Specifies the materials to be used and the method of use. Often acts as agent on small sites.

***Bill of quantities*** A book of calculated quantities of materials for a proposed job, made up by a quantity surveyor, based on the architect's specification.

***Client*** Person or persons requiring and paying for work to be done. Usually the owner of the property.

***Daywork rate*** A unit rate per hour to cover labour and overhead costs. Often used as the basis for a provisional sum, or for work which is too complex to quote accurately.

***Dim book*** Small pocketbook ruled in a special way in which to record dimensions when measuring work.

***Estimate*** An estimate of the cost of the job prepared by the contractor. It is not a binding contract, and allows for unforeseen additions which could not be seen under normal circumstances, such as the surface condition under existing wallpaper.

***Estimated contract*** A quotation which makes allowances in the final price for rising labour and material costs. The period of the contract is agreed between the client and the contractor.

***Estimated profit*** The net profit that a contractor expects to make on an estimated job.

***Fixed-price contract*** An estimate which cannot under any circumstances be altered by subsequent rising costs of labour or materials.

***Gross pay*** Employee's pay before deductions such as income tax, insurance contributions and superannuation.

***Gross profit*** Profit made from a contract before the deductions to cover overheads.

***Hourly rate*** For costing purposes, the employee's hourly paid rate plus additions to cover the cost of overheads such as travelling, insurance and holidays.

***Labour constant*** The estimated and averaged amount of work one painter can do in one hour ($m^2$ per hour).

***Net pay*** Employee's pay after deductions of income tax, insurance contributions and superannuation. A painter's 'take home' pay.

***Net profit*** Profit made from a contract after the deductions to cover overheads. Employer pays income tax on this amount.

***Overheads*** The cost of running a business, such as non-productive staff salaries, rent and transport.

***Prime cost*** The actual cost of labour and materials before the addition of overheads and profit.

***Provisional sum*** A quotation for a separate quantity of work. Supplied for parts of a job which are too complex to accurately estimate time or cost. Often quoted on a daywork-rate basis.

***Quotation*** A binding contract prepared by the contractor which shows the client the exact amount to be paid for a stipulated quantity, and quality of work. It can only be amended for additional work ordered by the client at an agreed rate.

***Schedule of rates*** A list of prices per $m^2$ for specified tasks. Includes all overheads and profit. Formulated from current material and labour costs.

***Specification*** A binding contract which defines in detail the work to be carried out, and also colours and types of material to be used on a contract.

***Tender*** An estimate submitted to a client. In an *open tender*, all contractors are invited; in a *closed tender*, only specified contractors are invited.

# 7.8 Calculations

## 7.8.1 International system of units (SI)

A metric system of measurement in which the principal units are usually divided or multiplied by one thousand to obtain the smaller or larger units. The prefix *milli* means a one-thousandth part. The prefix *kilo* means one-thousand times.

### Correct writing of numbers

(i) Omit the comma and leave one space when writing numbers exceeding ten thousand, e.g. 32 000, 170 455, 11 060 (*but* 1000, 3240).

(ii) Arrange large numbers in groups of three, e.g. 302 270 355.

(iii) Express decimal fractions to three decimal places, e.g. 2.350 m, 0.750 m, 9.075 m.

(iv) Express fractions of metres either as decimals or as whole numbers of millimetres, e.g. 0.750 m = 750 mm, 0.065 m = 65 mm. (Also, £0.75 = 75p.)

(v) Give dimensions in order of length × width × height, e.g. a room measures 4.400 m × 3.050 m × 2.995 m.

### Correct writing of symbols

(i) In giving dimensions, do not mix units, e.g. use 2.100 m × 0.750 m or 2100 mm × 750 mm (*not* 2.100 m × 750 mm).

(ii) Leave a space between numbers and symbol, e.g. 5.300 m.

(iii) Do not add a comma or a full stop after a symbol unless it is required for normal punctuation.

## 7.8.2 Calculating superficial area (area super) (Table 7.3)

The surface area of a shape is calculated in squared units which represent the number of squares contained on the surface of the shape, e.g. a wall measuring 5.000 m × 3.000 m has a surface area of 5 × 3 = 15 m$^2$ (Fig. 7.8).

## 7.8.3 Levers and beams

The main application of the principle of levers to painting and decorating is in the calculation of the counterweights required to secure safely a cradle or safety chair (see 5.4.1).

A further application is in the calculation of the loads carried by trestles (Fig. 7.9).

$$\text{Load } L = \frac{W \times A}{T} \quad \text{and} \quad \text{load } R = \frac{W \times B}{T}$$

where $T$ = length of board between supports
$B$ = distance $WL$
$A$ = distance $WR$
$L$ = left load
$R$ = right load
$W$ = total imposed weight

In the example shown,

$$L = \frac{76 \times 1}{4} = \frac{76}{4} = 19 \text{ kg}$$

**Table 7.2** Principal units used in painting and decorating

| *Quantity measured* | *Unit* | *Symbol* | *Smaller unit* | *Symbol* | *Larger unit* | *Symbol* |
|---|---|---|---|---|---|---|
| Length or linear measurement | metre | m | millimetre (m ÷ 1000) | mm | kilometre (m × 1000) | km |
| Area or superficial measurement | square metre | $m^2$ | square millimetre ($m^2$ ÷ 1,000,000) | $mm^2$ | hectare ($m^2$ × 10 000) | ha |
| Volume | cubic metre | $m^3$ | cubic millimetre | $mm^3$ | | |
| Capacity | litre | l | millilitre (l ÷ 1000) | ml | | |
| Weight | kilogram | kg | gram (kg ÷ 1000) | g | tonne (kg × 1000) | t |
| Pressure | bar | bar | millibar (bar ÷ 1000) | mbar | | |
| Volume rate of flow | litres per second | l/s | | | | |
| Temperature | degree Celsius | °C | | | | |
| Precision measurement | Micrometre | µm | | | millimetre (µm × 1000) | mm |

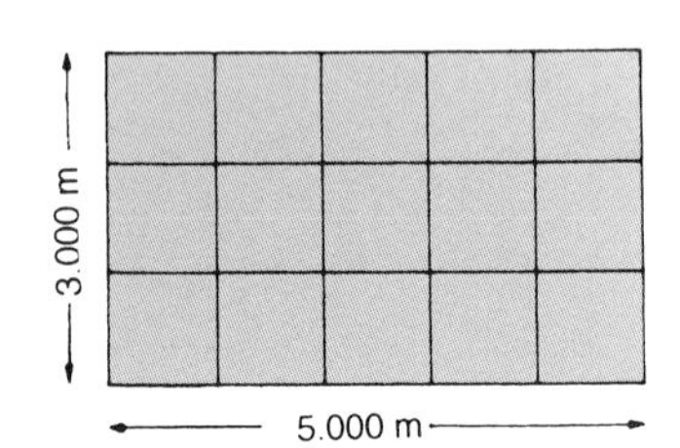


**Fig. 7.8** Calculation of surface area

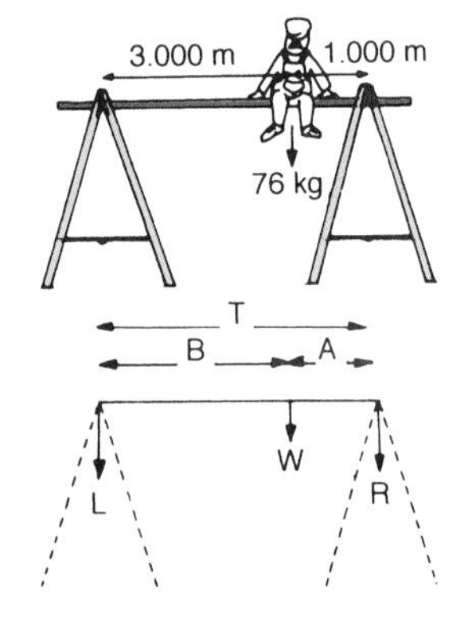


**Fig. 7.9** Calculation of load

**Table 7.3** Calculating the surface areas of common shapes

| ***Shape*** | ***Appearance*** | ***Formula for calculating the area*** |
|---|---|---|
| Square | | Multiply length × height (l × h) |
| Rectangle | | Multiply length × height (l × h) |
| Triangle | | Multiply half length × perpendicular height ($\frac{1}{2}$l × ph) or $\frac{l \times ph}{2}$ |
| Circle | | Multiply π (3.142) × radius × radius (π $r^2$)<br>*Note:* The circumference is calculated by multiplying π × diameter (π d) |
| Cylinder | | Multiply π (3.142) × diameter × length (π dl) or circumference × length |
| Sphere | | Multiply 4 × π × radius × radius (4π $r^2$)<br>*Note:* A hemisphere or dome is 2π $r^2$ |
| Walls of a room | | Multiply the distance around the room (girth) × height of the room (girth × h) |

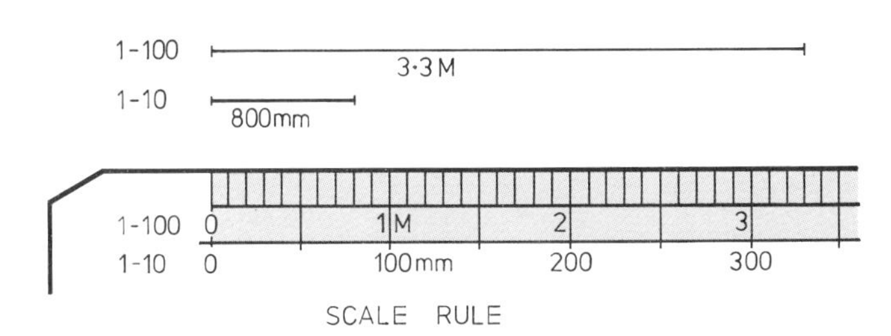


**Fig. 7.10** Scale rule

As the total load $W = 76$ kg and load $L = 19$ kg, load $R = 76 - 19 = 57$ kg. It will be seen that the support nearer the load bears the greater part of the weight.

## 7.8.4 Scales

Scales are a way of using ratio to relate measurements on drawings to the real dimensions of actual spaces. Examples of metric scales include 1:1, 1:2, 1:5, 1:10, 1:20, 1:50, 1:100, 1:200, 1:500.

Scales are not only used for making large areas smaller, but can be used to enlarge very small things in order to be able to view them more easily, in which case they are indicated in reverse, i.e. 5:1.

### Scale rules (Fig. 7.10)

Scale rules are used for preparing and reading scale drawings.

# 7.9 British standards

The letters EN and/or ISO will accompany some British Standards (BS). These mean the standard was developed as a European (EN) or International (ISO) standard and then adopted by the UK as a British Standard. These may be referred to when writing or interpreting specifications. They also contain specifications for products to support safety legislation.

## 7.9.1 British standards

| | |
|---|---|
| BS EN 131:2007 | *Ladders* |
| BS 144:1997 | *Wood preservation using coal tar creosotes* |
| BS EN 235:2002 | *Wallcoverings – Vocabulary & Symbols* |
| BS 245:1976 | *Specification for mineral solvents for paints and other purposes* |
| BS EN 266:1992 | *Specification for textile wall-coverings* |
| BS EN 354:2002 | *Personal protective equipment against falls from heights* |
| BS 381c:1996 | *Specification for colours for identification, coding and special purposes* |
| BS 476 (Parts 1–33) | *Fire tests on building materials and structures* |
| BS 952 (Parts 1–2) | *Glass for glazing* |
| BS EN 971-1:1996 | *Glossary of paint and related terms* |
| BS ISO 1000:1992 | *Specification for SI units and recommendations for the use of their multiples and of certain other units* |
| BS 1070:1993 | *Specification for black paint (tar-based)* |
| BS 1129:1990 | *Specification for portable timber ladders, steps, trestles and lightweight staging* |
| BS 1139 (Parts 1–5) | *Metal scaffolding* |
| BS 1192 (Parts 1–5) | *Construction drawing practice* |
| BS EN 1263-1997 | *Specification for industrial safety nets* |
| BS 1336:1971 | *Specification for knotting* |
| BS 1710:1984 | *Specification for identification of pipelines and services* |
| BS 2037:1994 | *Specification for portable aluminium ladders, steps, trestles and lightweight staging* |
| BS 2482:2009 | *Specification for timber scaffold boards* |
| BS 2830:1994 | *Specification for suspended access equipment for use in the building, engineering, construction, steeplejack and cleaning industries* |
| BS 2992:1970 | *Specification for painters' and decorators' brushes* |
| BS 3046:1981 | *Specification for adhesives for hanging flexible wall-coverings* |
| BS 3416:1991 | *Specification for bitumen-based coatings for cold application* |

| | |
|---|---|
| BS 3761:1995 | *Specification for solvent-based paint remover* |
| BS EN ISO 4618:2006 | *Paints and varnishes; terms and definitions for coating materials* |
| BS 4652:1995 | *Specification for zinc-rich priming paint (organic media)* |
| BS 4756:1998 | *Specification for ready-mixed aluminium priming paints for woodwork* |
| BS 4764:1986 | *Specification for powder cement paints* |
| BS 4800:1989 | *Schedule of paint colours for building purposes* |
| BS 5252:1976 | *Framework for colour co-ordination for building purposes* |
| BS 5499 (Parts 1–3) | *Fire safety signs, notices and graphic symbols* |
| BS 5499–5:2002 | *Safety signs and colours* |
| BS 5589:1989 | *Code of practice for preservation of timber* |
| BS 5974:2010 | *Code of practice for temporarily installed suspended scaffolds and access equipment* |
| BS 6100 (Parts 1–6) | *Glossary of building and civil engineering terms* |
| BS 6150:2006 | *Code of practice for painting of buildings* |
| BS 6262:2005 | *Code of practice for glazing in buildings* |
| BS 7028:1999 | *Guide for selection, use and maintenance of eye protection for industrial and other users* |
| BS 7079 (Parts A–F) | *Preparation of steel substrates before application of paints and related products* |
| BS 7664:2000 | *Specification for undercoats and finishing paints* |
| BS 7719:1994 | *Specification for water-borne emulsion paints for interior use* |
| BS 7956:2000 | *Primers for woodwork* |
| BS 7956:2000 | *Specification for semi-transparent priming stains (stain basecoats) for joinery* |
| BS 7956:2000 | *Specification for solvent-borne priming paints for woodwork* |
| BS 7956:2000 | *Specification for water-borne priming paints for woodwork* |
| BS EN 12811:2003 | *Code of practice for access and working scaffolds and special scaffold structures in steel* |
| BS EN ISO 12944-2 (Parts 1–2):1998 | *Code of practice for protective coating of iron and steel structures against corrosion* |
| BS EN 13279 | *Specification for gypsum building plasters* |
| BS EN 23270:1991 | *Methods of test for paints* |
| BS EN 50144:1999 | *Hand-held electric motor-operated tools* |

# PART 8

# Colour

# 8.1 Colour and light

## 8.1.1 The spectrum

The source of all colour is light. Without light, colour does not exist. The purest form of light originates from the sun and is known as 'white light'. It is received by us in rays which are not straight but formed of waves of different lengths of frequencies. When light strikes a surface, waves of certain lengths are absorbed and others are reflected. It is the reflected waves which are picked up by our eyes and registered by the brain as colours.

**The spectrum (Fig. 8.1)**
When light is directed through a transparent material such as glass its speed is slowed down. If at the same time the light rays can be 'bent', e.g. through a glass prism, when the light emerges from the other side it will have been divided into individual colour waves. This can easily be seen if rays are projected onto a white screen. This coloured reflection will show a distinct range of colours always in the same order: red, orange, yellow, green, blue and violet, known as the spectrum. A similar effect is produced in nature when sunlight is broken up by raindrops and a rainbow is produced.

**Seeing colour (Fig. 8.2)**
An object which is seen by a person with normal vision as red is absorbing all the light of other wavelengths and reflecting only the waves of 'red' length; the sensation caused by these waves meeting the eye is registered as red by the brain.

A white surface reflects all waves, whereas a black surface absorbs all, reflecting none. If the light source is not white the surface will appear a different colour: e.g. if red light is directed onto a green surface (green=blue+yellow), only a neutral or 'grey' colour

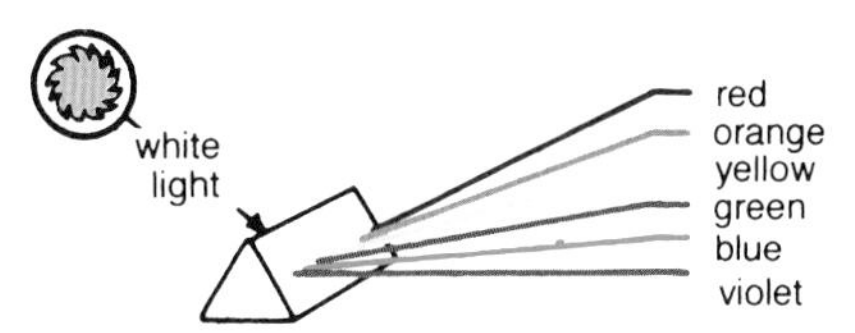


**Fig. 8.1** The spectrum

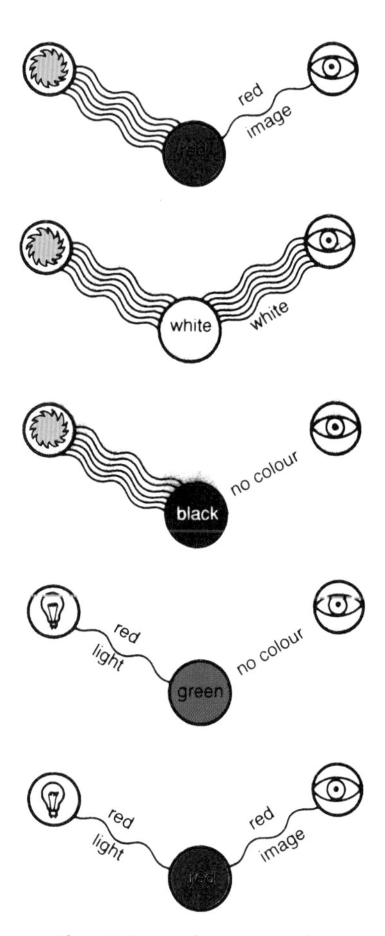


**Fig. 8.2** Colour perception

will be seen because the light contains no waves of 'green' length to be reflected.

## 8.1.2 Effect of artificial light on colour

Colours seen in 'white' or natural light may appear different when seen in artificial light. This happens because the spectrum of artificial lights differs from that of white light; therefore, the reflective rays picked up by our eyes will be different.

For this reason it is essential to work only in the main light source of a particular situation when selecting or mixing colours. Colours selected in natural light may look entirely different under artificial light.

## 8.1.3 Types of artificial light

There are many forms of artificial light, but they can be grouped under three main sources which have common spectra.

***Tungsten*** Also known as filament lamps or normal domestic bulbs. The glass bulb contains a gas and the filament inside is made of thin tungsten wire which gives off a bright yellow/red glow when electricity passes through it. The light given off contains very few blue and violet rays.

***Fluorescent*** Also called tube or strip lights. The tube contains low-pressure mercury vapour which produces ultra-violet light when electricity passes through it. The inside of the tube is coated with fluorescent powders which glow when in contact with the ultra-violet light. The colour of the light depends on the type of powder in the tube. Most types of domestic tubes give out more blue/violet rays than red/yellow.

***Discharge*** Similar to fluorescent tubes but without a coating of fluorescent powder. The vapour in the tube is at a much higher pressure and the colour of light emitted depends on the type of vapour. Sodium vapour gives off only yellow/amber light rays. Mercury vapour gives out violet, blue and green rays but very few red. Discharge lamps are used mainly for industrial and street lighting.

## 8.1.4 Metameric colour effect

Some dyes used in fabrics and paints may appear completely different when seen under various light sources. This is caused by the chemical properties of the dye which react to the various lights in different ways. This is another reason for only working in the relevant light source when selecting and mixing colours.

## 8.1.5 Disability Discrimination Act 1995 (DDA)

Since October 2004 companies and organisations that provide services to the public have been required by the above Act to ensure that those services are reasonably accessible to disabled people.

Within this framework, colour can play an important role.

(i) Colour can be used creatively to help some customers with visual impairments who may see 'blocks' of colour more easily than detail. Compared to a monotone colour scheme, they will find it easier to move around independently if floors, walls, ceilings and doors are distinguished from each other using contrasting colours or light and dark tones of the same colour. A practical tip for assessing colour contrast is to take a black-and-white photograph and see how easy it is to distinguish between different coloured surfaces.

(ii) Signage is read more easily if the lettering is well contrasted with the background (i.e. with strong contrast of light and dark between lettering and background).

(iii) Toilet areas – customers with visual impairments may find 'all-white' areas difficult to orientate themselves in and to identify fittings such as basins and the toilet itself. Making fixtures and fittings stand out more easily through the use of contrasting colour makes toilet compartments and washroom areas much easier to use.

(Above obtained from the Disability Rights Commission – DRC.)

# 8.2 Colour organisation

Many methods of putting the known facts about a colour into a practical and usable order have been prepared. The two most useful to the painter and decorator are the pigment-mixture colour circle (sometimes referred to as the Brewster theory) and the Munsell scale.

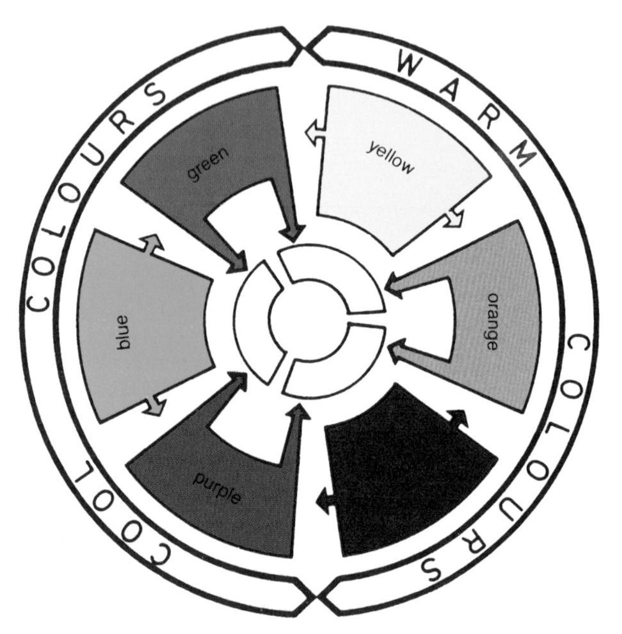


Fig. 8.3 The colour circle

## 8.2.1 Pigment-mixture colour circle (Fig. 8.3)

When the spectrum colours are arranged in a circle, they are easier to use and to relate to each other. The pigment-mixture circle consists of *three primary colours* and *three secondary colours*. The primary colours are yellow, red and blue; these are the basis of all other colours and cannot be produced by mixing other colours. The secondary colours are orange, purple and green and are produced by mixing together two primary colours. When the three primary colours are mixed together, neutral grey is the result.

*Tertiary colours* are mixtures of the secondary colours. Orange and purple; purple and green; and green and orange.

*Note:* When colours are mixed together, the resulting colour is not as bright or pure as the original colours. Therefore, secondaries are less bright than primaries, and tertiaries are less bright than secondaries.

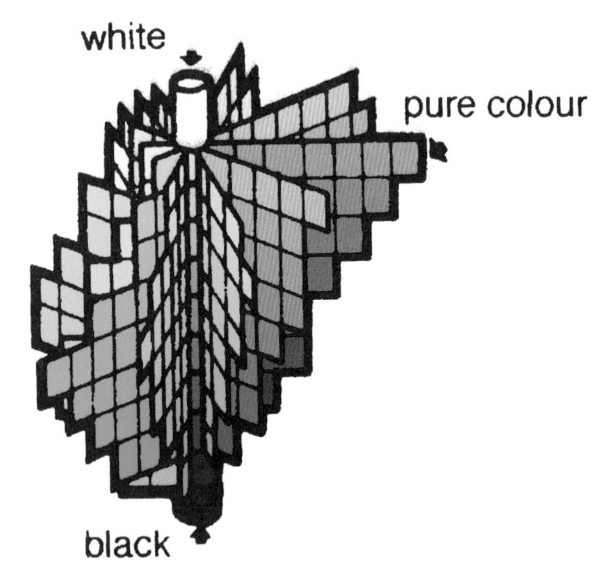


Fig. 8.4 The Munsell-scale atlas

## 8.2.2 Munsell scale (Fig. 8.4)

This is an international system which permits an accurate verbal description of colour. The method can be used:

(a) to describe a colour without using colour names: 'light orange' or 'off-white' may suggest different colours to different people, whereas a Munsell notation defines one exact colour;
(b) to compare and select colours accurately: the hundreds of colours defined by the Munsell scale allow the correct colour for a particular purpose to be chosen easily.

The full range of colours is illustrated in an atlas which can be used to select and specify colours.

The Munsell scale of colours has been used to prepare BS 4800: 1989 and BS 5252: 1976 which give standard colours for decorative paints. A list of approximate Munsell references for each of the colours in the range is published by the British Standards Institution.

The system is based on the three properties of *hue, value* and *chroma*.

### Hue (Fig. 8.5)

Hue denotes the basic colour, distinguishing yellow from red, blue from green, etc. Munsell uses 10 principal hues:

| | |
|---|---|
| yellow (Y) | purple-blue (PB) |
| yellow-red (YR) | blue (B) |
| red (R) | blue-green (BG) |
| red-purple (RP) | green (G) |
| purple (P) | green-yellow (GY) |

Each principal hue is subdivided into 10 sections, providing a full circle of 100 hues. The true hue is always prefixed by a 5, e.g. 5Y represents pure yellow. 7.5Y is two and a half divisions away from pure yellow in the blue direction, therefore slightly greenish; 2.5Y is the other side of pure yellow and is slightly reddish.

### Value (Fig. 8.6)

Value measures the lightness and darkness of a colour. The scale is based on a vertical pole divided

**Fig. 8.5** Hue

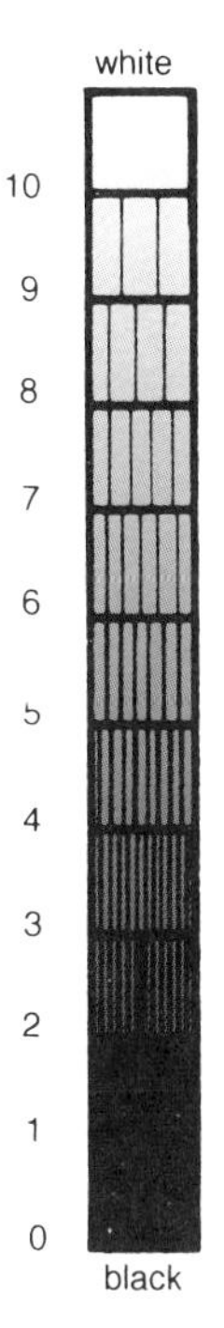


**Fig. 8.6** Value

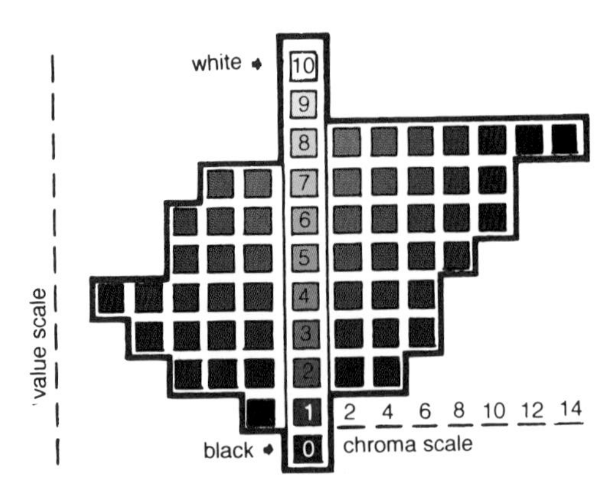


**Fig. 8.7** Chroma

into 11 divisions. White is 10 and black is 0 and the remaining 9 divisions represent a scale of greys from light (9) to dark grey (1). The value scale is shown by a number between 1 and 9 written to the right of the hue, e.g. pure yellow will be written as 5Y8 because the tone of pure yellow is equivalent to 8 on the value scale.

## Chroma (Fig. 8.7)

Chroma means the greyness of a colour. It is measured on a horizontal scale from neutral grey (0) to the pure hue which may be 10, 12 or 14 depending on the strength of the colour, e.g. pure PB is chroma 10 whereas pure Y is chroma 14. The chroma range number is written after the value scale and separated from it by an oblique stroke, e.g. pure yellow is written as 5Y8/14, whereas this colour darkened with black and reduced in strength by the addition of grey may be described as 5Y6/6.

# 8.3 Colour standards

## 8.3.1 BS 1710: 1984 Specification for identification of pipelines and services

This standard specifies a colour-coding system for use on pipes and conduits in land installations and on board ships. Each colour identifies a particular liquid, gas or electrical service. The system ensures that even in a complex pipework system, in the case of fire, other emergency or maintenance work, the contents of every pipe can be easily identified and traced to source.

Where it is sufficient to identify the general nature of the contents, basic identification colours (see Table 8.1) are used. Where more precise information is required, code indications are added according to regulations.

Basic identification colours can be applied either by painting the whole length of the pipe in the specified colour, or by placing 150 mm wide bands of paint or coloured adhesive tape at all junctions, on both sides of valves, service appliances and wall penetrations or wherever identification is necessary.

## 8.3.2 BS 4800: 1989 Schedule of paint colours for building purposes

The standard range contains 100 colours chosen from a basic range of 237 colours established in BS 5252: 1976 *Framework for colour co-ordination for building purposes*. The range is divided into a Basic selection of 32 colours, including black and white, and two Supplementary selections relating to oil-based finishes, category G, and emulsions, category M. The Basic selection comprises those colours which should be available from stock. Each colour is identified by a code consisting of three parts.

### Hue

The first part signifies hue or colour and consists of an even number of two numerals, e.g. 04. Twelve main hues are used, numbered:

**Table 8.1** Basic identification colours for pipelines

| *Pipe content* | *Colour name* | *Approx BS 4800 colour reference* |
|---|---|---|
| Water | Green | 12 D 45 |
| Steam | Silver-grey | 10 A 03 |
| Mineral, vegetable and animal oils: combustible liquids | Brown | 06 C 39 |
| Gases and liquid gases (except air) | Yellow ochre | 08 C 35 |
| Acids and alkalis | Violet | 22 C 37 |
| Air | Light blue | 20 E 51 |
| Other fluids including drainage | Black | 00 E 53 |
| Electrical services | Orange | 06 E 51 |
| Fresh water | Auxiliary blue | 18 E 53 |

02 red purples
04 reds
06 yellow reds
08 yellow reds
10 yellows
12 yellow greens
14 greens
16 green blues
18 blues
20 purple blues
22 violets
24 purples

plus greys which are all 00.

### Greyness

The second part signifies greyness, i.e. the difference in the apparent amount of greyness in one colour compared with another. Five grades are used, each defined by a single letter. There are four steps of diminishing greyness from A maximum to D minimum. Beyond this, colours are pure or free from greyness, graded as E: e.g. pure yellows are prefixed 10 E.

### Weight

When coded only on the above two grades, it was found that the yellow hues (yellow-red and yellow-green) at minimum greyness looked heavy in comparison with the other colours. This was overcome by raising the value of the yellowish colours. The result was more uniform in weight. The weight is given in pairs of numbers from 01 to 56.

Groups of colours within each of the five greyness ranges are graduated from high value to low value. Each of these graduations is numbered:

A greyness – 01 to 13
B greyness – 15 to 29
C greyness – 31 to 40
D greyness – 43 to 45
E greyness – 49 to 56

For example, the lighter yellow is 10 E 49 while the deeper pure yellow is 10 E 53.

## 8.3.3 BS 5499-5: 2002 Graphical symbols & signs incl. fire safety and signs with specific safety meanings

### Supersedes BS 5378: 1995 (part 1) Safety signs & colours

This standard specifies an international system that gives safety information with the minimum use of words. It defines safety colours and safety signs (see Tables 8.2 and 8.3) for giving messages for use in:

(i) Prevention of accidents.
(ii) Identification of health hazards.
(iii) Meeting emergencies.

*Note:* All safety signs must conform to these standards.

### Definitions

***Safety colour*** A colour which has a special safety meaning or purpose.
***Safety symbol*** A graphic symbol used in a safety sign.
***Safety sign*** A sign which states a safety message by combining a safety colour with a safety symbol or both on a geometrical shape.
***Supplementary sign*** A sign with text only.

## 8.3.4 NCS – Natural Colour System®©: an international colour standard

The six colours White, Black, Yellow, Red, Blue and Green (Plate 19) correspond to the perception of colour in the brain.

NCS colour notations are based on how much a given colour seems to resemble these six elementary colours.

**S 1050-Y90R**

In the NCS notation 1050-Y90R, for example, 1050 describes the *nuance*, i.e. the degree of resemblance

**Table 8.2** Safety signs

| *Safety sign* | *Layout of signs* | *Example of use* |
|---|---|---|
| | *Prohibition signs*<br>White background with red circular band and cross bar.<br>Safety symbol in black, placed centrally. Must not obliterate the cross bar.<br>Sign should be at least 35 per cent red. | smoking prohibited |
| | *Warning signs*<br>Yellow background with triangular band in black.<br>Safety symbol or text in black, placed centrally.<br>A black and yellow identification sign can be used where there is a risk of hazards.<br>Sign should be at least 50 per cent yellow. | caution risk of fire<br>risk of hazard |
| | *Mandatory signs*<br>Blue circle.<br>Safety symbol or text white, placed centrally.<br>Sign should be at least 50 per cent blue. | respiratory protection must be worn |
| | *Emergency signs*<br>Green square or rectangle.<br>Safety symbol or text in white, placed centrally.<br>Sign should be at least 50 per cent green. | first aid |
| | *Supplementary signs*<br>White square or rectangle. Text in black.<br>Alternatively, the background can be the same safety colour as the safety sign it is supplementing with the text in the appropriate contrasting colour. | FIRE EXIT |

**Table 8.3** Safety colours

| Colour | Meaning or objective | Example of use | Contrast colour |
|---|---|---|---|
| Red<br>04 E 53 | Stop<br>Prohibition | Stop signs<br>Emergency signs<br>Prohibition signs such as no smoking<br>Fire-fighting equipment and its location | White |
| Yellow<br>08 E 51 | Caution<br>Risk of danger | Indication of hazards (fire, explosion, radiation, chemicals)<br>Warning for thresholds, low passages, obstacles, projections | Black |
| Green<br>14 E 53 | Emergency signs<br>Safe conditions | Escape routes<br>Emergency exits<br>Emergency showers<br>First aid<br>Rescue stations | White |
| Blue<br>18 E 53 | Mandatory action<br>Information | Obligation to wear safety equipment<br>Location of telephone | White |

NCS 6 elementary colours

**Plate 19** Elementary colours

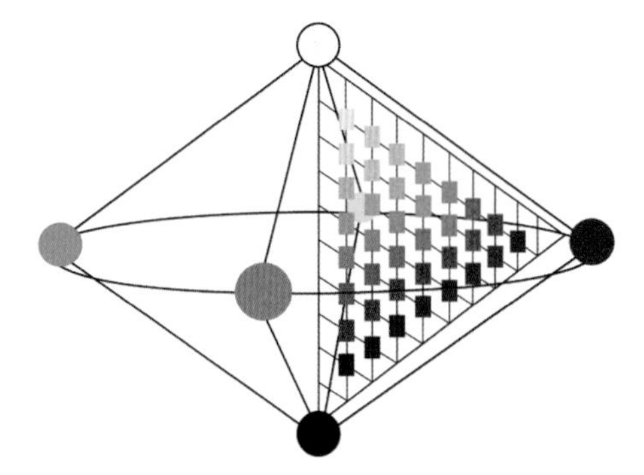

**Plate 20** NCS colour space

to black which is 10% and to the maximum chromaticness, which is 50%.

The hue Y90R describes the degree of resemblance between Yellow and Red (Y and R). Y90R describes Yellow with 90% redness and 10% yellowness.

Pure grey colours have no hue and are given nuance notations followed by -N. There is a scale from 0300-N, which is white, to 9000-N, which is black.

The letter S that precedes the NCS notation means that the NCS sample is from NCS Edition 2. The colour standards for this edition were created with non-toxic pigments.

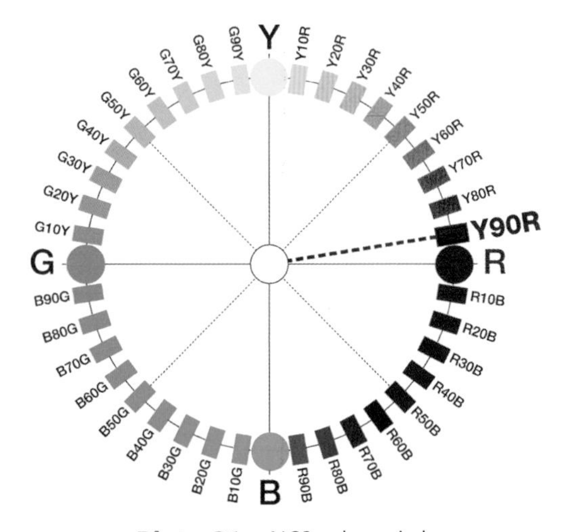


**Plate 21** NCS colour circle

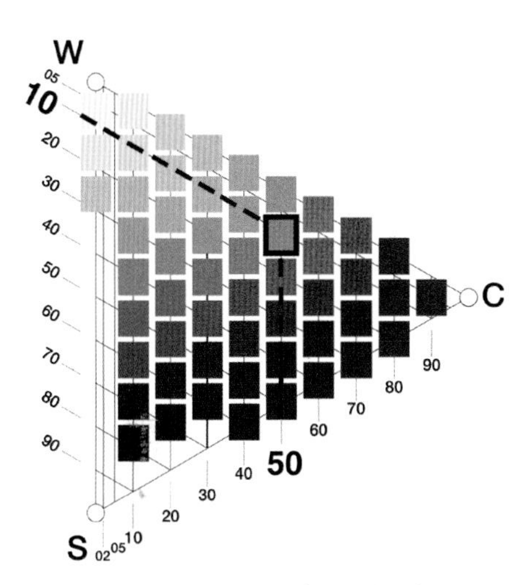


**Plate 22** NCS colour triangle

## The NCS colour space (Plate 20)

Within this three-dimensional model, all imaginable surface colours can be plotted and given an NCS notation.

## The NCS colour circle (Plate 21)

By taking a horizontal section through the centre of the model we see this circle where the four chromatic colours are placed like points of the compass with the space between divided into 100 equal steps.

The 10% steps here represent the pages of the NCS atlas and show the colour hues.

## The NCS colour triangle (Plate 22)

The NCS triangle is a vertical section through the NCS model at one of the 10% steps. Here it is Y90R.

The section shows all the colours on the NCS atlas page for the hue Y90R between White (W), Black (S) and the full chromatic colour (C). S 1050-Y90R has been highlighted.

# 8.4 Terms related to colour schemes

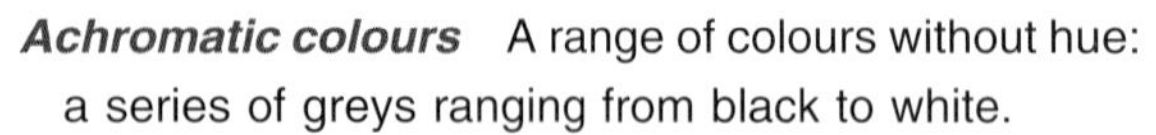

***Achromatic colours*** A range of colours without hue: a series of greys ranging from black to white.

***Advancing colours*** Warm colours taken from the red, yellow or orange part of the colour circle. They give the appearance of coming out or advancing towards the eye.

***Analogous colours*** (Fig. 8.8) Harmonious or pleasing colours closely related to each other on the colour circle, e.g. yellow, yellow-green and green.

***Complementary colours*** (Fig. 8.9) Any two colours which lie opposite each other on the colour circle and when mixed will produce grey or a neutral colour.

***Contrasting colours*** Colours different in hue, value or chroma when seen together, e.g. the contrast of red against green, light colours against dark colours, strong colours against greyed colours.

***Contrasting harmony*** Colours taken from various parts of the circle which appear pleasant when used together.

***Cool colour*** Colours from the green and blue part of the colour circle, which give the appearance of coolness.

***Discordant colours*** Colours used out of their natural tonal order (see Natural order of colour), e.g. pale blue with deep orange.

***Harmonious colours*** Colours which appear pleasant when used together in a scheme. Common combinations are:

(a) Monochrome
(b) Analogous
(c) Contrasting
(d) Split complementary

***Juxtaposition of colour*** The effect one colour has on another when viewed side by side. The complementary colour of each will intensify or distort the effect of its neighbour: e.g. when red and yellow are seen together, the red will take on a purplish tinge while the yellow will appear greenish.

***Monochrome or dominant hue*** (Fig. 8.10) A colour scheme based on tints and shades of one colour.

***Natural order of colour*** Yellow is the lightest colour in the colour circle and purple-blue the darkest. Used in this order, they are usually pleasing to the eye. Reverse this order and colours become discordant.

***Pastel colours*** Colours which have black and white added. Usually refers to pale colours or tints with a predominance of white added.

***Retiring colours*** Cool colours taken from the blue-green part of the colour circle. They give the appearance of receding or retiring away from the eye (opposite to advancing).

***Shades*** Colour with black added.

***Split complementary*** (Fig. 8.11) A colour scheme based on one hue contrasted with two hues either side of its complementary colour.

**Fig. 8.8** Analogous colours

**Fig. 8.9** Complementary colours

**Fig. 8.10** Monochrome

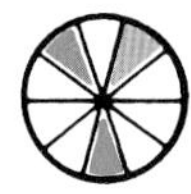

**Fig. 8.11** Split complementary colours

***Tints*** Colours with white added.

***Tones*** The lightness or darkness of a colour, i.e. the value rating on the Munsell scale.

***Warm colours*** Advancing colours from the red-yellow-orange part of the colour circle, which give the appearance of warmth.

# PART 9

# Glazing

# 9.1 Types of glass and glazing compound

Glass is available in many types, grades, qualities, colours and sizes. Two main groups are used in domestic and industrial buildings and it is from these groups that the decorator may be called on to replace small, broken panes of glass:

(a) Transparent glass that allows light to pass through it and that can also be seen through, i.e. plate glass or clear sheet glass.
(b) Translucent glass that allows light to pass through it but cannot be seen through, i.e. patterned glass.

## 9.1.1 Clear sheet glass

Drawn sheet glass used to be available in varying qualities but has largely been superseded by float glass.

***Horticultural*** 3 mm lower-grade glass for greenhouse work.

### Float glass

The most common flat sheet glass. The liquid glass is floated on a bath of molten tin to produce clear glass of regular thickness, with few imperfections.

## 9.1.2 Patterned glass

There are many types of patterned glass available in nominal thickness of 3 or 5 mm. They have two main uses:

(a) To provide a decorative effect in doors, screens and partitions.
(b) To provide privacy in rooms such as bathrooms and toilets and for partitions and screens. A clear image cannot be seen through them, but the degree of privacy depends on the type and depth of the pattern. They are classified in five groups graded from A to E, A being the clearest and E the most obscure.

## 9.1.3 Safety glazing materials

To ensure safety in the use of glass, all builders, architects and specifiers need to comply with the safety section of BS 6262: 1994 *Code of practice for glazing in buildings*, which isolates risk areas in buildings and sets out minimum requirements to ensure safety.

**Table 9.1** Grades of clear sheet glass

| ***Nominal*** | ***Approximate weight*** | ***Maximum area*** | |
|---|---|---|---|
| Thickness | | Glazed | |
| mm | kg/m$^2$ | m$^2$ | Use |
| 4 | 9.5 | 0.2 | Glazing small areas in domestic buildings |
| 6 | 15.0 | 1.8 | Glazing large areas, mirrors, shelves, etc. |
| Thicker glass available | | | |
| 10 | | 3.3 | |
| 12 | | 5.0 | |

*Note:* Not less than 6 mm annealed glass should be used in doors and side panels.

The code recommends that where many people, especially children, are moving about, glazing situated lower than 800 mm from floor level should be of a safety glazing material of which there are four main types:

***Wired glass*** Clear or patterned glass enveloping a fine wire mesh. Sometimes called fire-resistant glass as it will act as a shield against the spread of flames for up to 30 minutes, depending on the size and type of frame. If during a fire the glass breaks, the wire holds the glass in position.

*Note:* Wired glass is no longer considered to be a safety glass for panels below the height of 800 mm. Georgian wired polished plate is always used for fire doors.

***Laminated glass*** Two or more sheets of ordinary glass between which are inserted interlayers of a tough, resilient plastic material or resin. When broken, the fragments remain in place, adhering to the plastic.

***Toughened glass*** Tempered glass. Four to five times stronger than ordinary glass, requiring considerable impact to break. When broken, it disintegrates into small, relatively harmless pieces.

*Note:* Toughened glass cannot be cut and must be ordered in exact sizes required.

***Glazing plastics*** Usually acrylic, polycarbonate, or rigid PVC. They all resist considerable impact and are not dangerous when broken.

## 9.1.4 Glazing compounds

Four main types of glazing compound are used.

### Linseed oil putty

Made from linseed oil, whiting and a special filler. It hardens by oxidation. Used for bedding and fronting in primed timber frames on domestic and industrial work including roof work. Needs to be protected by painting within 2 weeks, after the surface has hardened sufficiently to take the paint. If required, it should be softened with linseed oil only.

### Metal casement putty

Made from linseed and other drying oils, and a special grade of whiting. It hardens by oxidation. Used for bedding and fronting glass in metal and concrete frames on domestic and industrial work including roof work. Needs to be painted within the period specified by the manufacturer – usually 7–28 days. If required, it should be softened with special thinner supplied by the manufacturer.

### Glazing compounds

Made from drying and non-drying oils, whiting and usually butyl rubber. They set but remain permanently flexible. Used on timber frames with face beads, concrete frames, metal beadwork, double-glazed units and laminated safety glass. Need to be painted within a period specified by the manufacturer. If required, they should be thinned with special thinner supplied by the manufacturer.

### Multi-purpose compound

Formulated on calcium carbonate fillers and selected oils. Intended for 'face glazing' of primed softwood/hardwood and steel window frames.

# 9.2 Handling and storage of glass

## 9.2.1 Handling glass

(i) It is safer to carry more than one pane at a time as this prevents the sheet whipping (Fig. 9.1a). However, caution should be applied in carrying any more than two panes because of the weight and also potential slippage of central panes (a common occurrence).

(ii) Always make sure that the path ahead is clear and free from scaffolding, low doors and other hazards (Fig. 9.1b).

(iii) Never stop suddenly or step backward: anyone following behind may be injured (Fig. 9.1c).

(iv) Small panes of glass should be carried balanced centrally under the arm (Fig. 9.1d).

(v) Larger panes should be carried with one hand underneath and the other in the front (Fig. 9.1e). *Note:* The left hand should be three-quarters of the way up the pane of glass to avoid losing control of the top of the pane.

(vi) Glass should be carried almost vertical and close to the body (Fig. 9.1f).

## 9.2.2 Storage (Fig. 9.2)

Glass can easily be broken if not stored correctly, and the face of glass can be damaged if it is stacked damp. The following rules should be observed:

(i) Never store lying flat: glass should be stood on one long edge, on timber or felt blocks, at an angle of approximately 80° or almost upright.

(ii) There should be no spaces between sheets; this will cause warping and the glass will eventually break.

(iii) Glass must be stored dry and protected from the wind – indoors if possible.

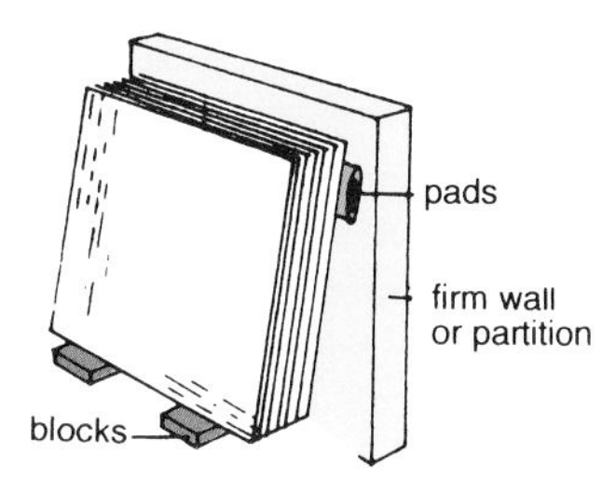


**Fig. 9.2** Storage of glass

**Fig. 9.1** Handling glass

# 9.3 Measuring and cutting glass

Figure 9.5 shows the names of various sizes referred to in glazing. *Tolerance* is the amount of variance allowed when measuring and cutting glass, usually ±1 mm on the glazing size: e.g. a height size of 360 mm should be not less than 359 mm and not more than 361 mm.

## 9.3.1 Measuring for glass

(i) When measuring for any type of glass, it is standard practice to give the width size first. This ensures that when decorative glass is being used, the pattern is always glazed the correct way, e.g. pattern vertical.

(ii) If the frame is unsymmetrical or oddly shaped, a small sketch should be made and the front marked *face* (Fig. 9.3). This ensures that the glass is cut correctly to fit it.

*Note:* All glazing sizes should be expressed in millimetres (mm).

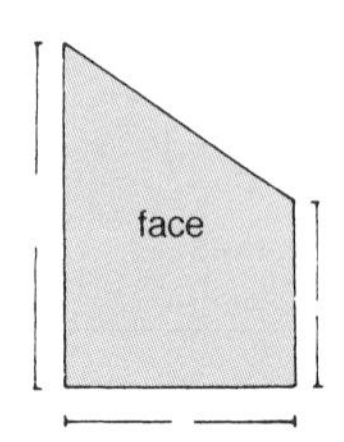


**Fig. 9.3** Measurement of glass

## 9.3.2 Planning for cutting glass (Fig. 9.4)

Before attempting any cutting, an economic plan must be decided upon to eliminate as much waste as possible. The required size should be considered in conjunction with the size of sheet available.

Well-planned cutting should produce strips of waste glass not exceeding 50 mm to 100 mm wide.

## 9.3.3 Relevant glazing sizes (Fig. 9.5)

***T/S (tight size)***

The actual full measurement between the glazing bars.

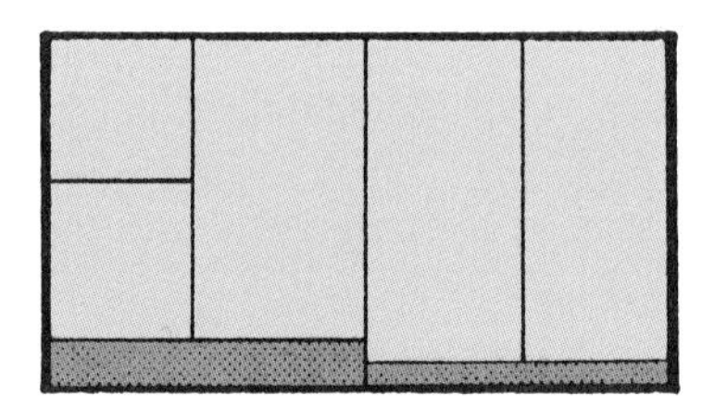

**Fig. 9.4** Planning for cutting glass

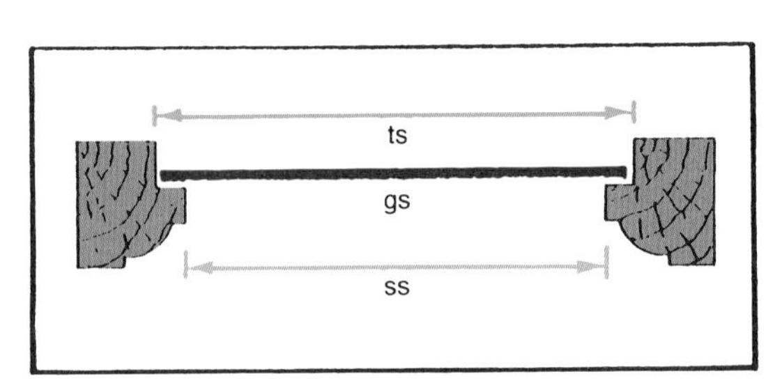


**Fig. 9.5** Glazing sizes

***S/S (sight size) or (daylight size)***
The measurement of space between the glazing bars which allows light through.
***G/S (glazing size)***
The actual measurement of the glass allowing for the edge clearance of 2 mm all round.

## 9.3.4 Cutting glass

Glass should always be cut on a flat, rigid, padded surface. If a glazier's felt-covered cutting bench is not available, a flat surface padded with newspaper will suffice.

# 9.4 Methods of glazing

## 9.4.1 Timber frames

The two main methods of glazing timber frames are (a) with putty or glazing compound and (b) with beads.

**Putty glazing** (Fig. 9.6)

The timber frame, which must be clean and primed, is bedded with a 'back putty'. The glass, cut with a 2 mm edge clearance, is fitted and small sprigs (small headless nails) are tapped in at intervals of approximately 460 mm around the frame. These hold the glass in until the putty sets. The front putty is then run and faced at an angle to provide a weather-proof fillet.

**Glazing with beads** (Fig. 9.7)

The timber frame must be clean and primed. The back putty is run, the glass fitted and the beads are bedded and fixed with either screws or panel pins.

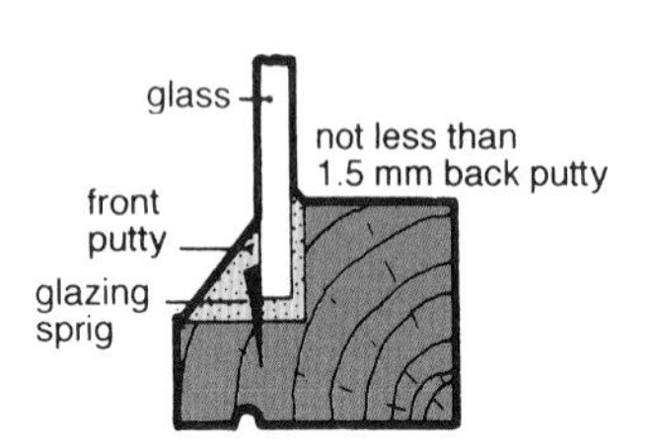


**Fig. 9.6** Putty glazing

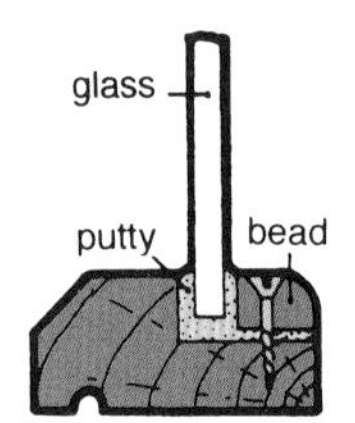


**Fig. 9.7** Glazing with beads

*Note:* Hardwood frames should be either primed or sealed with exterior-quality varnish, and non-setting compounds used for glazing.

## 9.4.2 Metal frames (Fig. 9.8)

This process is similar to glazing timber frames, but with the following differences:

(i) Metal casement putty is used.
(ii) A metal primer is used.
(iii) The glass must always be cut with an edge clearance of 2 mm all round to allow for expansion and contraction of the metal and to allow for the metal clips. Otherwise, the glass may be pinched and may crack.
(iv) In place of sprigs, spring clips are used to hold the glass in position. One end of the clip is fitted into small holes in the metal frame, the other end is sprung against the glass.

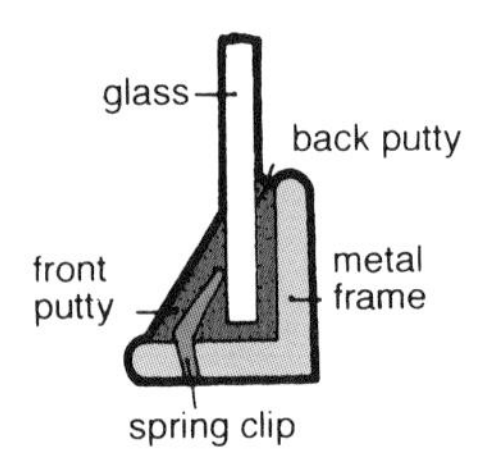


**Fig. 9.8** Glazing metal frames

## 9.4.3 Setting or location blocks (Fig. 9.9)

Panes of glass exceeding 0.2 $m^2$ should be set on blocks (often PVC) to ensure that the glass fits squarely into the frame (Fig. 9.9).

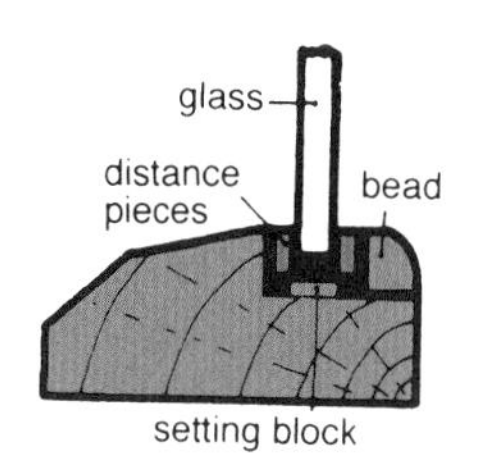


**Fig. 9.9** Setting blocks

# 9.5 Double glazing

Double glazing consists of two panes of glass with a space ranging between 6 mm and 20 mm. This provides better sound and heat insulation.

(i) sound insulation, to reduce the amount of noise entering or leaving a building.
(ii) heat insulation, to reduce the amount of heat energy escaping from a building.

## Double glazing

*Double glazing* is where a patent double-glazed unit is fitted as standard.

*Secondary double glazing* is where an additional layer is added to an existing single-glazed component.

## Types

***Double window*** Two single-glazed timber or metal frames, spaced apart in the same wall opening and independently mounted.

***Coupled windows*** A single-glazed frame fitted with an auxiliary frame coupled to it, allowing both to move together. Can be separated for cleaning purposes.

***Dual glazing*** A secondary frame of metal, plastic or wood fitted to an existing frame. Fitted on slides or hinges and can be separated for cleaning and maintenance. Can also be screwed down as a more permanent structure.

***Factory-made sealed units*** Aluminium or uPVC frames containing hermetically sealed double-glazed units. Prevents the infiltration of dust or water vapour into the cavity between the glass.

# PART 10

# Useful websites

# 10.1 Useful websites

**Brushes**

| | |
|---|---|
| Hamilton Acorn | www.hamilton-acorn.co.uk |
| A S Handover Ltd | www.handover.co.uk |
| Harris Brushes & Tools | www.t-class.net |
| Purdy Brushes | www.purdycorp.com |
| Wrights of Lymm | www.wrightsoflymm.co.uk |

**Colour**

| | |
|---|---|
| Munsell Colour | www.applepainter.com |
| Natural Colour System | www.ncscolour.co.uk |

**Cornices & Covings**

| | |
|---|---|
| Copley Décor | www.copleydecor.com |
| Hayles & Howe | www.haylesandhowe.com |
| NMC Coving | www.nmc-uk.com |
| Stevensons | www.stevensons-of-norwich.co.uk |

**Paints**

| | |
|---|---|
| Blackfriar Paints | www.blackfriar.co.uk |
| Craig & Rose Paints | www.craigandrose.com |
| Crown Paints | www.crownpaint.co.uk |
| Dulux Trade | www.duluxtrade.com |
| Earthborn Paints | www.earthbornpaints.co.uk |
| Farrow & Ball Paints | www.farrow-ball.com |
| ICI/AkzoNobel Paints | www.icipaints.com |
| International Paints | www.international-paints.co.uk |
| Johnstones Paints | www.johnstonestrade.com |
| Jotun Paints | www.jotun.com |
| Permoglaze Paints | www.permoglaze.co.uk |
| Ronseal Wood Finishes | www.ronseal.co.uk |
| Sadolin | www.sadolin.co.uk |
| Sandtex | www.sandtextrade.co.uk |
| Sigma Coatings | www.sigmacoatings.com |
| Sikkens Paints | www.sikkens.co.uk |
| Sto Paints | www.sto.co.uk |
| Tor Coatings | www.tor-coatings.com |
| Zinsser Primers | www.zinsser.com |

### Scaffolding

| | |
|---|---|
| Harsco (SGB) | www.harsco-i.co.uk |
| National Access & Scaffold Federation | www.nasc.org.uk |
| Youngman Scaffold | www.youngmangroup.com |

### Spray Painting

| | |
|---|---|
| Gray Campling | www.graycampling.co.uk |
| ITW European Finishing | www.itweuropeanfinishing.com |
| Lion Industries | www.lionindustries.co.uk |
| Spray Equipment Operator Training | www.spraytrain.com |
| Wagner | www.wagnerspraytech.co.uk |

### Trade Organisations

| | |
|---|---|
| Association of Painting Craft Teachers | www.apct.co.uk |
| British Coatings Federation Ltd | www.coatings.org.uk |
| Construction Skills | www.cskills.org |
| Health & Safety Executive | www.hse.gov.uk |
| Lead Paint Safety Association | www.lipsa.org.uk |
| Painting & Decorating Association (PDA) | www.paintingdecoratingassociation.co.uk |
| Scottish Association of Painting Craft Teachers | www.sapct.org |
| The Faculty of Decoration | www.facultyofdecoration.co.uk |
| The Paint Research Association | www.pra.org.uk |
| The Traditional Paint Forum | www.traditionalpaintforum.org.uk |
| Wallpaper History Society | www.wallpaperhistorysociety.org.uk |
| Worshipful Company of Painter Stainers | www.painters-hall.co.uk |

### Wallpaper

| | |
|---|---|
| Anaglypta | www.anaglypta.co.uk |
| Cole & Son | www.cole-and-son.com |
| IHDG Wallcoverings | www.ihdg.com |
| Lincrusta Walton | www.lincrusta.com |
| Mav Wallcoverings | www.erfurtmav.com |
| Muraspec | www.muraspec.com |
| Tektura | www.tektura.com |
| Wallcoverings Association | www.wallcoverings.org |

### Micellaneous

| | |
|---|---|
| Fiddes (Wood Finishes) | www.fiddes.co.uk |
| Geocel (Fillers) | www.geocel.co.uk |
| Mangers | www.mangers.co.uk |
| The Green Building Store | www.greenbuildingstore.co.uk |
| The Stencil Store | www.stencilstore.com |

# Index

***Note*** In addition to the Contents, each of the ten parts of this book is preceded with a detailed list. If the required reference is not found in the Index, refer first to the Contents, and then to the appropriate lists preceding the book part.